Edited by
Raju Francis

Recycling of Polymers

Edited by Raju Francis

Recycling of Polymers

Methods, Characterization and Applications

WILEY-VCH
Verlag GmbH & Co. KGaA

Editor

Dr. Raju Francis
Mahatma Gandhi University
School of Chemical Sciences
Priyadarsini Hills
Kottayam 686560
India

Cover
Blue bottles: choness@gettyimages (No 475893605)
Background: PhotoAlto/James Hardy@gettyimages (No 107906856)

Library of Congress Card No.: applied for

British Library Cataloguing-in-Publication Data
A catalogue record for this book is available from the British Library.

Bibliographic information published by the Deutsche Nationalbibliothek
The Deutsche Nationalbibliothek lists this publication in the Deutsche Nationalbibliografie; detailed bibliographic data are available on the Internet at <http://dnb.d-nb.de>.

Print ISBN: 978-3-527-33848-1
ePDF ISBN: 978-3-527-68903-3
ePub ISBN: 978-3-527-68909-5
Mobi ISBN: 978-3-527-68902-6
oBook ISBN: 978-3-527-68900-2

Typesetting SPi Global, Chennai, India
Printing and Binding Markono Print Media Pte Ltd., Singapore

Printed on acid-free paper

3.1.8.1 Introduction *67*
3.1.8.2 Pyrolysis *69*
3.1.8.3 Co-pyrolysis *70*
3.1.8.4 Catalytic Cracking *71*
3.1.9 Chemical Recycling of Polystyrene *75*
3.1.9.1 Introduction *75*
3.1.9.2 Recycling Methods for Polystyrene Products *76*
3.1.9.3 Future Prospects *83*
3.1.10 Chemical Recycling of Poly(vinyl chloride) *83*
3.1.10.1 Introduction *83*
3.1.10.2 Mixed Plastic Recycling Processes *86*
3.1.10.3 Mixed PVC Wastes World Initiatives *88*
3.1.10.4 The BASF Feedstock Recycling Process *88*
3.1.10.5 Veba Combi Cracking Process *90*
3.1.11 Chemical Recycling of Poly(methyl methacrylate) *90*
3.1.11.1 Introduction *90*
3.1.11.2 Dissolution/Reprecipitation *91*
3.1.11.3 Chemical/Feedstock Recycling *92*
References *93*
3.2 Methods of Recycling of Polymers: Condensation Polymers *101*
Beena Sethi
3.2.1 Introduction *101*
3.2.2 Chemical Recycling of Nylon *101*
3.2.2.1 Introduction *101*
3.2.2.2 Recycling Methods *101*
3.2.3 Chemical Recycling Involving Depolymerization of Nylons Which Can Be Carried Out by Hydrolysis or Ammonolysis of Nylon 6,6 and Nylon 6 *102*
3.2.3.1 Hydrolysis of Nylon 6 *102*
3.2.3.2 Hydrolysis of Nylon 6,6 and Nylon 4,6 *103*
3.2.3.3 Ammonolysis of Nylon 6,6 *103*
3.2.3.4 Recovery of Nylon 6,6 Monomers *104*
3.2.3.5 Catalytic Pyrolysis *105*
3.2.3.6 Applications of Depolymerized Nylon 6 *105*
3.2.4 Chemical Recycling of Polycarbonate *105*
3.2.4.1 Introduction *105*
3.2.4.2 Recycling Techniques *106*
3.2.5 Advantages of Recycling and Reuse of Polymers *110*
References *112*

4 Recycling of Plastics *115*
Preetha Balakrishnan and Meyyappallil Sadasivan Sreekala
4.1 Introduction *115*
4.2 Plastic Waste Management Scenario *117*
4.3 Ways of Recycling *119*

4.3.1 Reuse *120*
4.3.2 Mechanical Recycling *121*
4.3.3 Chemical Recycling *121*
4.4 Poly(Lactic Acid) *122*
4.5 Poly(Vinyl Chloride) *125*
4.6 Polyethylene *126*
4.7 Polypropylene *128*
4.8 Polystyrene *129*
4.9 Poly(Ethylene Terephthalate) (PET) *129*
4.10 Applications *134*
References *135*

5 Recycling of Rubber *141*
Valiya Parambath Swapna and Ranimol Stephen
5.1 Introduction *141*
5.2 Rubber *142*
5.3 Recycling of Rubber Products *143*
5.3.1 Chemical Process *143*
5.3.2 Physical Methods *145*
5.3.2.1 Mechanochemical Techniques *145*
5.3.2.2 Microwave Technique *146*
5.3.2.3 Ultrasonic Technique *146*
5.3.2.4 Twin-Screw Extruder *148*
5.3.3 Biological Process *148*
5.4 Applications of Recycled Rubber *152*
5.4.1 Sound-Insulation Materials *152*
5.4.2 Civil Engineering Applications *153*
5.4.3 Oil Absorbent *154*
5.4.4 Energy Production *154*
5.4.5 Zinc Fertilizer *155*
5.5 Concluding Remarks *155*
References *156*

6 Fibers *163*
Raju Francis, Nidhin Joy, Anjaly Sivadas, and Geethy P. Gopalan
6.1 Introduction *163*
6.2 Natural Fibers *164*
6.2.1 Kenaf *165*
6.2.2 Cotton *167*
6.2.3 Sisal *170*
6.2.4 Asbestos *174*
6.3 Synthetic Fibers *176*
6.3.1 Nylon *177*
6.3.2 Polyester *182*
6.3.3 Glass Fiber *187*

6.3.3.1 Glass Fiber-Reinforced Plastics *188*
6.3.3.2 Mechanical Process *188*
6.3.3.3 Thermal Process *188*
6.3.3.4 Chemical Recycling *190*
6.3.4 Carbon Fiber *192*
6.3.4.1 Mechanical Recycling *192*
6.3.4.2 Thermal Recycling *193*
6.3.4.3 Chemical Recycling *195*
6.4 Conclusion *198*
References *198*

7 Recycling of Polymer Blends and Composites (Epoxy Blends) *209*
Jyothi V. Sunny
7.1 Introduction *209*
7.2 Polymer Blends and Composites *209*
7.2.1 Methods of Recycling *213*
7.2.1.1 Mechanical Recycling *213*
7.2.1.2 Chemical Recycling *215*
7.2.1.3 Thermal Recycling *216*
7.3 Characterization and Application of Recyclates *218*
7.4 Conclusions *219*
References *219*

8 Recycling of Other Layered Mixed Plastics or Resins: Polyurethanes *223*
Jyothi V. Sunny
8.1 Introduction *223*
8.2 Mechanical Recycling *226*
8.3 Chemical Recycling *227*
8.3.1 Glycolysis *228*
8.3.2 Hydrolysis *229*
8.3.3 Aminolysis *229*
8.4 Thermochemical methods *230*
8.4.1 Pyrolysis *230*
8.4.2 Gasification *230*
8.4.3 Hydrogenation *230*
8.5 Energy Recovery by Incineration *231*
References *232*

9 Ecoprofiles of Recycled Polymers at a Glance *235*
9.1 Advantages of Recycled Polymers on the Environment *235*
Raju Francis and Anjaly Sivadas
9.1.1 Introduction *235*
9.1.2 Poly(ethylene terephthalate) (PET) *236*
9.1.3 High-Density Polyethylene (HDPE) *237*

9.1.4 Poly(vinyl chloride) (PVC) *239*
9.1.5 Polypropylene (PP) *240*
9.1.6 Polystyrene (PS) *241*
9.1.7 Other Polymers *242*
9.1.8 Conclusion *245*
References *245*
9.2 Toxic or Environmental Effects of Recycled Polymers *248*
Raju Francis, Nidhin Joy, and Anjaly Sivadas
9.2.1 Introduction *248*
9.2.2 Will Recycling Reduce the Amount of Waste? *249*
9.2.2.1 Recycling of Waste Electrical and Electronic Equipment (WEEE) *250*
9.2.2.2 Recycling of Tires *251*
9.2.2.3 Recycling of Plastics *251*
9.2.2.4 Recycling of Polymers *251*
9.2.2.5 Health Problems *252*
9.2.2.6 Recycling by Polymer Incineration *252*
9.2.3 Conclusion *253*
References *253*

Index *257*

Preface

Polymer products are indispensible to humans because of their several advantages, such as easy processability into various shapes, low cost, lightweight, and durability, over conventional products. But the irony is that some of these advantages make polymeric materials a threat to life on Earth through widespread and irreparable damage to environment. This comes mainly because some of us still believe that polymer products are of the "use and throw" type. Because of this, our soil, water, and air are catastrophically affected. Therefore, it is high time to think and work on alleviating the serious ill effects of polymers and attempt to regenerate our environment for future generations. One of the possible remedies that is being considered and debated by the general public, scientists, and academicians is polymer recycling. This is because all other alternatives are either extremely dangerous or economically unviable. One can see that the two common substitutes for polymer recycling are (i) the simple burning of used polymers in open air, which is more dangerous to the environment, and (ii) the use of biodegradable polymers, which is uneconomical. Therefore, "recycle and reuse" is considered the best option for a sustainable environment.

Secondly, recycling minimizes the need for raw materials so that the rainforests can be preserved. Great amounts of energy are used when making products from raw materials. Recycling requires much less energy and therefore helps us preserve natural resources.

This book *Recycling of Polymers* is a collection of recent research and academic studies on the methods of recycling, followed by applications and, finally, the merits and demerits of recycled polymer products. It is noteworthy that this book encompasses almost all categories of polymers, namely plastics, elastomers, and fibers, and, in addition, also blends, composites, and resins.

This book consists of nine chapters. The first chapter mainly presents the overall idea that recycling is one of the best options for making a positive impact on the world in which we live. It gives a general idea about its importance, why we should do recycling, what are the sources of recycling, various stages of recycling, and so on.

Chapter 2 (Parts 2.1 and 2.2) provides the different types of additives that are commonly used for recycling. Additives play a leading role in the success of commercial plastics, elastomers, rubbers, coatings, and adhesives. It also describes the common additives used in the recycling of polymers. This includes a study of the different classes of additives that are employed alone or in combination with other additives in the polymer recycling or manufacturing process. After describing the different additives that are not included in the first part of this chapter, a quick look into the recent trends, advancements, and the future of additives is included in the second part.

The third chapter includes the method of recycling of polymers. Part 3.1 of the chapter comprises a significant review of the chemical recycling of the generally used addition polymers such as polypropylene (PP), polystyrene (PS), low-density polyethylene (LDPE), high-density polyethylene (HDPE), poly(vinyl chloride) (PVC), and poly(methyl methacrylate) (PMMA), and Part 3.2 includes chemical recycling of condensation polymers such as poly(ethylene terephthalate) (PET), polycarbonate (PC), nylon, and so on.

Fourth chapter reviews the recycling of thermoplastic waste from some traditional polymers such as polyethylene (PE), PP, PS, PVC, PS, PET, and so on.

Chapter 5 includes the production and world consumption of rubber products and applications of recycled rubber. The recycling of rubber products is not a trivial process because their crosslinked structure restricts reprocessing. Efficient methods of devulcanization include chemical, mechanical, biological, thermal, microwave, and ultrasonic techniques.

Chapter 6 mainly focuses on the recycling of fibers. The most commonly recycled natural and synthetic fibers are included in this chapter. Natural polymers are biodegradable. They can be blended with plastics to produce materials that are more biodegradable while retaining the more desirable features of conventional plastics. Synthetic polymers are non-biodegradable. So this chapter mainly gives an idea about the recycling and use of recycled products of synthetic fibers.

Chapter 7 deals with the recycling of polymer blends and composites. Epoxies are thermoset polymers and are very difficult to degrade. Therefore, the different types of recycling techniques used for the epoxy thermosets are presented in this chapter. Examples of recycled epoxy thermosets that are converted into useful products are highlighted.

Chapter 8 deals with the recycling of polyurethanes. Mainly polyurethanes are used to obtain rigid and flexible foams. Nowadays, recycling of polyurethanes is drawing more and more attention worldwide because of the variety of products developed with them for various applications.

Chapter 9 gives an idea on the benefits of recycling and the impact of some significant recycled polymers on the environment. First part of this chapter discusses the advantages of recycling with the help of six major recycled polymers. Recycled polymers leads to the following positive impacts: (i) they save the Earth, (ii) they conserve energy, (iii) they help in mitigating global warming and in reducing pollution, (iv) they minimize waste products placed in landfills, (v) they help save money, (vi) they reduce the need for allied activities such as transportation and mining, and (vii) they spread awareness for the environment. The second part of this chapter evaluates the effects of recycled polymers from three angles – environmental, human health, and economic.

India
August 11, 2016

Raju Francis

List of Contributors

S. Anil Kumar
Mahatma Gandhi University
NSS Hindu College
Department of Chemistry
Changanacherry
Kottayam
Kerala

Preetha Balakrishnan
Mahatma Gandhi University
International and Inter
University Centre for
Nanoscience and
Nanotechnology
Kottayam
Kerala
India

Raju Francis
Mahatma Gandhi University
School of Chemical Sciences
Priyadarsini Hills
Kottayam
Kerala 686560
India

Geethy P. Gopalan
Mahatma Gandhi University
School of Chemical Sciences
Priyadarsini Hills
Kottayam
Kerala 686560
India

Nidhin Joy
Mahatma Gandhi University
School of Chemical Sciences
Priyadarsini Hills
Kottayam
Kerala 686560
India

Beena Sethi
Department of Chemistry
K. L. Mehta D. N. College for
Women
New Industrial Township
K. L. Mehta Marg, N.H-3
Faridabad
Haryana 121001
India

Anjaly Sivadas
Mahatma Gandhi University
School of Chemical Sciences
Priyadarsini Hills
Kottayam
Kerala 686560
India

M. S. Sreekala
Department of Chemistry
Sree Sankara College – Kalady
Sankar Nagar, Mattoor
Ernakulam
Kerala 683574
India

Ranimol Stephen
Department of Chemistry
St. Joseph's College
(Autonomous)
Devagiri, Calicut
Kerala 673008
India

Jyothi V. Sunny
BASF Corporation
889 Valley Park Drive
Shakopee, MN 55379
USA

V.P. Swapna
Department of Chemistry
St. Joseph's College
(Autonomous)
Devagiri, Calicut
Kerala 673008
India

S. Vishnu Sankar
Mahatma Gandhi University
NSS Hindu College
Department of Chemistry
Changanacherry
Kottayam
Kerala

Abbreviations

ABS	Acrylonitrile/butadiene/styrene
AC	Acidification
AIBN	Azobisisobutyronitrile
ADN	Adiponitrile
BLL	Blood lead level
BPA	Bisphenol A
CD	Circular disk
CPE	Chlorinated polyethylene
CSBR	Conical spouted bed reactor
DCP	Dicumylperoxide
DFE	Design for environment
DMC	Dimethyl carbonate
DMF	Dimethylformamide
DSC	Differential scanning calorimetry
DTST	Dynamic thermal stability time
EPDM	Ethylene–propylene–diene monomer rubber
EPR	Ethylene–propylene rubber
EPS	Expanded polystyrene
FCC	Fluid catalytic cracking
FRP	Fiber-reinforced plastic
FTIR	Fourier transform infrared
GC	Gas chromatography
GF	Glass fiber
GHG	Greenhouse gas
GOP	Vaccum gas oil product
GW	Global warming
HALS	Hindered amine light stabilizers
HCl	Hydrogen chloride
HDPE	High-density polyethylene
HDT	Heat distortion temperature
HIPS	High-impact polystyrene
HMDA	Hexamethylenediamine
HT	Human toxicity

IPP	Isopropenyl phenol
IR	Infrared
LCA	Life-cycle analysis
LDPE	Low-density polyethylene
LHV	Lower heating value
LLDPE	Linear low-density polyethylene
LPG	Liquefied petroleum gas
MBS	Methacrylate/butadiene/styrene
MFI	Melt flow index
MMA	Methyl methacrylate
MSW	Municipal solid waste
MW	Microwave
MWD	Molecular weight distribution
NE	Nutrient enrichment
OPS	Oriented polystyrene
PAH	Polyaromatic hydrocarbons
PAHs	Polycyclic aromatic hydrocarbons
PBCDDs	Polybrominated–chlorinated dibenzo-*p*-dioxins
PC	Polycarbonate
PCDFs	Polychlorinated dibenzofurans
PE	Polyethylene
PEHD	Polyethylene high density
PET	Poly(ethylene terephthalate)
PHAs	Polyhydroxyalkanoates
PLA	Poly(lactic acid)
PMMA	Poly(methyl methacrylate)
POF	Photochemical ozone formation
POP	Persistent organic pollutant
PP	Polypropylene
PS	Polystyrene
PSBD	Poly(styrene–butadiene)
PSW	Plastic solid waste
PT	Persistent toxicity
PVC	Poly(vinyl chloride)
R-PVC	Rigid poly(vinyl chloride)
SAPO	Silicoaluminophosphate
SEP	Styrene–ethylene–propylene block copolymer
TBE	Tetrabromoethane
TDF	Tire-derived fuel
TG	Thermogravimetry
TGA	Thermogravimetric analysis
TPH	1,3,5-Triphenylhexane
VCC	Veba Combi cracking
VOC	Volatile organic compound
WEEE	Waste electrical and electronic equipment

1
Introduction

Raju Francis, Geethy P. Gopalan, and Anjaly Sivadas

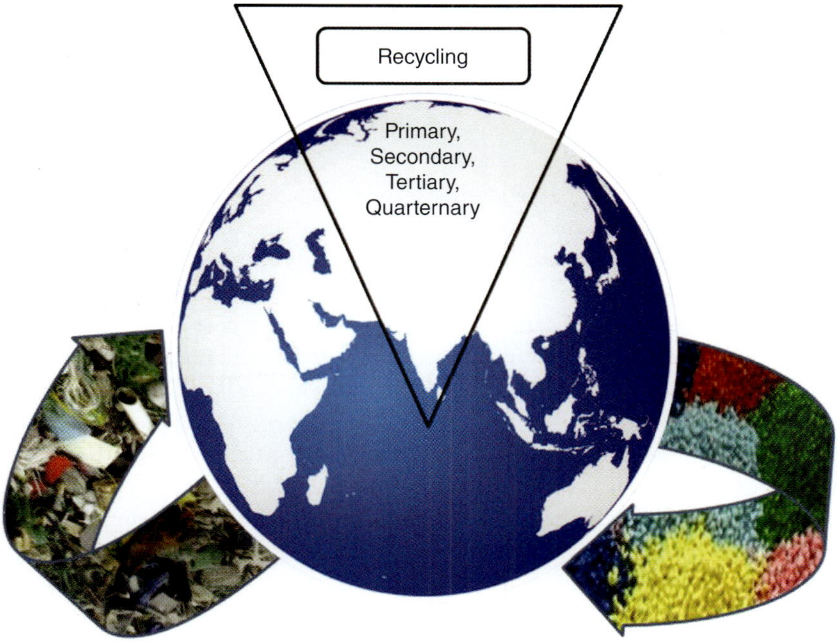

"Recycling saves energy, preserves natural resources, reduces greenhouse-gas emissions, and keeps toxins from leaking out of landfills."
–*Marc Gunther*

Recycling of Polymers: Methods, Characterization and Applications, First Edition. Edited by Raju Francis.

1.1 Introduction

1.1.1 Why Recycling?

During the past decades, the enormous population increase worldwide, together with the need for people to adopt improved conditions of living, has led to a dramatic increase of the consumption of polymers (mainly plastics). Materials appear interwoven with our consumer society, where it would be hard to imagine living without plastics, which have found a myriad of uses in fields as diverse as household appliances, packaging, construction, medicine, electronics, and automotive and aerospace components. The unabated increase in the use of plastics has led to an increase in the quantity of plastics ending up in the waste stream, which has stimulated intense interest in the recycling and reuse of plastics [1]. Worldwide, the production of plastics was 168 million tons in the year 1999 and approximately 210 million tons in 2010 .

Since the treatment of plastic wastes has become a serious problem, the development of effective recycling processes is urgently needed [2].

1.1.2 Sources of Waste

Plastics play an important role in almost every aspect of our lives. Plastics are used to manufacture products of daily use such as beverage containers, toys, and furniture. The widespread use of plastics demands proper end-life management [3]. A large number of items can be easily recycled in most curbside programs, including all kinds of paper and cardboard, glass of all colors and types, plastic bottles, aluminum cans, and yard trimmings. In addition, a number of localities offer drop-off programs for recycling other items, such as household hazardous wastes (paints, cleaners, oils, batteries, and pesticides), automobile items (tires, used engine oil, car batteries, antifreeze), wood construction materials, certain metals, appliances, and consumer electronics [4].

The largest amount of plastics is found in containers and packaging (e.g., soft drink bottles, lids, shampoo bottles), but they also are found in durables (e.g., appliances, furniture) and nondurables (e.g., diapers, trash bags, cups, utensils, and medical devices). Commercial waste is often produced by workshops, craftsmen, shops, supermarkets, and wholesalers. Agricultural waste can be obtained from farm and nursery gardens outside the urban areas. This is usually in the form of packaging (plastic containers or sheets) or construction materials (irrigation or hosepipes). Municipal waste can be collected from residential areas (domestic or household waste), streets, parks, collection depots, and waste dumps [5].

Around 50% of plastics are used for single-use disposable applications, such as packaging, agricultural films, and disposable consumer items; between 20% and 25% for long-term infrastructure such as pipes, cable coatings, and structural materials; and the remainder for durable consumer applications with intermediate lifespan, such as in electronic goods, furniture, and vehicles [6].

1.1.3 Plastics

Plastics are made up of polymers and other materials that are added to give the polymer increased functionality. The polymer content in a plastic can vary widely from less than 20% to nearly 100%. Those plastics consisting virtually entirely of polymers are termed *prime grades*. The level and type of the other additives used depend on the application for which the plastic is intended. Plastics are inexpensive, lightweight, and durable materials, which can readily be molded into a variety of products that find use in a wide range of applications. As a consequence, the production of plastics has increased markedly over the last 60 years [6]. Thermosets and thermoplastics are the two major classifications of plastics. This distinction is based on both the molecular structure and the processing routes that can be applied. It also relates to recycling routes, as each category needs a different approach to utilize its recovery potential. Thermoplastics and thermosets will now be discussed.

- **Thermoplastics**
  These materials melt and flow when heated and solidify when cooled. On subsequent reheating, they melt and regain the ability to flow. This means that they can be used again and hence recycled by remelting them. Thermoplastics are used to make consumer items such as drinks containers, carrier bags, and buckets.
- **Thermosets**
  These materials are processed by melting, often in a similar manner to thermoplastics. However, once formed and cooled, they cannot be reprocessed; they decompose before they can melt. This is because they are chemically crosslinked by a process termed *curing*. The material becomes stiff and brittle with a highly dense molecular network [7].

1.1.4 Recycling of Plastics

Recycling of plastics is one method for reducing environmental impact and resource depletion. Recycling can therefore decrease energy and material usage per unit of output, leading to improved eco-efficiency. The only way to decrease the environmental problems caused by polymeric waste accumulation produced from day-to-day applications of polymer materials such those used in packaging and construction is by recycling. This helps to conserve natural resources because most polymer materials are made from oil and gas [8].

Recycling is the final result of the intermediate stages of collection, sorting by type, and processing of polymers. It reduces the quantity of residues in landfills and those indiscriminately discarded in the environment. Thus, it also leads to a reduction of problems such as the spread of diseases as well as contamination of soil, air, and water bodies [9]. It is one of the most important options currently available to reduce these impacts and represents one of the most dynamic areas in the plastics industry today. It provides opportunities to reduce oil usage, carbon dioxide emission, and the quantities of waste requiring disposal.

Recycling plastics encompasses four phases of activity, namely collection, separation, processing, and manufacturing and marketing. Because only the use of clean, homogeneous resins can produce the highest quality recycled plastic products in the existing secondary process (material recycling) and high-value chemical products in the existing tertiary process (feedstock recycling) [10], an effective separation of mixed plastics waste is necessary.

1.1.5
Municipal Solid Waste

The growth of plastics waste has a great impact on the management of municipal solid waste (MSW) by landfilling and incineration, because the available capacity for landfill of MSW is declining and plastics incineration may cause emission and toxic fly and bottom ash containing lead and cadmium [10]. Plastics waste recycling is a method of reducing the quantity of net discards of MSW. Although the benefits have not been quantified, plastics recycling also offer the potential to generate demonstrable savings in fossil fuel consumption, both because the recycled plastics can supplement and even compete with "virgin" resins produced from refined fossil fuel and because the energy required to yield recycled plastics may be less than that consumed in the production of the same resins from virgin feedstock. Therefore, plastics waste recycling conserves both material and energy and provides a comparatively simple way to make a substantial reduction in the overall volume of MSW [11].

The major plastics recycled are polyolefins (high-density polyethylene (HDPE), low-density polyethylene (LDPE), and polypropylene (PP)) and poly(ethylene terephthalate) (PET), poly(vinyl chloride) (PVC), polystyrene (PS), and polycarbonate (PC). The recyclable polymers and the recycling codes are shown in Table 1.1.

There are several options for how this can be done: primary recycling, mechanical or secondary recycling, tertiary or chemical recycling, and energy recovery or quaternary recycling (Figure 1.1).

- Primary recycling involves the use of the same product without essential changes in a new use cycle (e.g., refillable packaging after cleaning).
- Mechanical recycling implies the application of the material used, without changing the chemical structure, for a new application.
- Chemical recycling implies the chemical structure of the material is changed, which means that the resulting chemicals can be used to produce the original material again [12].

Table 1.1 Various polymers with their characteristic recycling codes for particular applications.

Symbol	Acronym	Full name and uses
1	PET	Polyethylene terephthalate - Fizzy drink bottles and frozen ready meal packages.
2	HDPE	High-density polyethylene - Milk and washing-up liquid bottles
3	PVC	Polyvinyl chloride - Food trays, cling film, bottles for squash, mineral water and shampoo.
4	LDPE	Low density polyethylene - Carrier bags and bin liners.
5	PP	Polypropylene - Margarine tubs, microwave-able meal trays.
6	PS	Polystyrene - Yoghurt pots, foam meat or fish trays, hamburger boxes and egg cartons, vending cups, plastic cutlery, protective packaging for electronic goods and toys.
7	Other	Any other plastics that do not fall into any of the above categories. For example melamine, often used in plastic plates and cups.

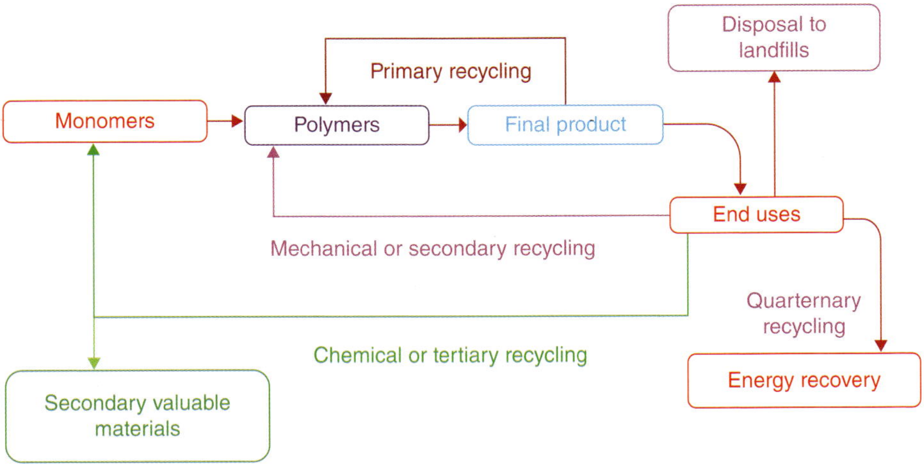

Figure 1.1 Four methods of recycling.

- Energy recovery refers to the recovery of plastic's energy content. Incineration aiming at the recovery of energy is currently the most effective way to reduce the volume of organic materials. Although polymers are actually high-yielding energy sources, this method has been widely accused as ecologically unacceptable owing to the health risk from airborne toxic substances, for example, dioxins (in the case of chlorine containing polymers).

1.1.6
Various Stages of Recycling Plastic Wastes

There are various stages of recycling:

- *Collection*: Plastic waste is collected from different locations. This can be achieved by keeping special containers at home, public places, farms, and so on. These wastes are then collected by professional waste collectors and transported to the recycling sites.
- *Cleaning*: The cleaning stage consists of washing and drying the plastic items. Cleaning is important since clean waste materials fetch better prices and they improve the quality of end products. Plastics can be washed at various stages of recycling process: before, after, or even during sorting.
- *Sorting*: This involves not only the separation of the polymers from recoverable foreign bodies but also the separation of these polymers themselves.
- *Size reduction*: It aims to reduce the size of the waste, which in turn facilitates not only in the separation of different polymers but also recovery of the micronized powder which is used to feed processing machines. The end products of shredding can be irregularly shaped pieces of plastics, which can be sold to reprocessing industries and workshops.

After processing, these materials are further subjected to various techniques such as extrusion, injection molding, blow molding, and film molding. Finally, the processed materials are converted into various products such as pipes, tubes, bags, sheets, and miscellaneous items.

1.1.7
Additives

Polymer industry cannot survive without additives. Additives in plastics provide the means whereby processing problems, property performance limitations, and restricted environmental stability are overcome. In order to get a technical effect additives used to incorporate into the plastics. So additives are expected to be the key part of the finished particle. A few examples of additives are antistatic agents, antioxidants, emulsifiers, antifogging agents, impact modifiers, fillers, plasticizers, lubricants, solvents stabilizers, UV absorbers, release agents and thickeners. It might be either inorganic (e.g., oxides, salts, fillers), organic (e.g., alkyl phenols, hydroxybenzophenones), or organometallic (e.g., Ni complexes, Zn accelerators, metallocarboxylates) [13].

Benefits of adding additives in plastics significantly shows varying properties with one or more directions such as stiffness, and strength, general durability, thermal resistance, impact resistance, resistance to flexure and wear, acoustic isolation and so on. In the broadest sense, these are essential ingredients of a manufactured polymeric material. An additive can be a primary ingredient that forms an integral part of the end product's basic characteristics or a secondary ingredient that functions to improve performance and/or durability.

The other recyclable materials are fibers, rubbers, mixed plastics, blends and composites, and so on. The recycling techniques, use of additives, and reusing applications are discussed in the following chapters.

Rubber recycling is growing in importance worldwide because of increasing raw material costs, diminishing resources, and the growing awareness of environmental issues and sustainability [14]. The rubber industry faces a major challenge in finding a satisfactory way to deal with the increasing quantities of rubber goods that reach the end of their useful life and are rejected from factories as scrap. The main source of waste rubber is discarded rubber products, such as tires, rubber hoses, belts, shoes, flash, and so on [15].

Reclaimed rubber is the product resulting when waste vulcanized scrap rubber is treated to produce a plastic material that can be easily processed, compounded, and vulcanized with or without the addition of either natural or synthetic rubbers. Regeneration can occur either by breaking the existing crosslinks in the vulcanized polymer, or by promoting scission of the main chain of the polymer, or a combination of both processes. Reclaiming of scrap rubber is, therefore, the most desirable approach to solve the disposal problem. Reclamation is done from vulcanized rubber granules by breaking down the vulcanized structure using heat, chemicals, and mechanical techniques. Reclaimed rubber has the plasticity of new unvulcanized rubber compound, but the molecular weight is reduced so reclaimed compounds have poorer physical properties when compared to new rubber [16].

Natural fibers are obtained from plants and animals whereas synthetic fibers are obtained by chemical processing of petrochemicals. Natural fibers have recently attracted the attention of scientists and technologists because of the advantages that these fibers provide over conventional reinforcement materials; the development of natural fiber composites has been a subject of interest during the past few years [17–19]. These natural fibers are of low cost, low density, and high specific properties. These are biodegradable and nonabrasive unlike other reinforcing fibers. Also, they are readily available and their specific properties are comparable to those of other fibers used.

Fiber-reinforced plastics (FRPs) are inherently difficult to separate into the base materials, that is, fiber and matrix, and the FRP matrix into separate usable plastic, polymers, and monomers. These are all concerns for environmentally informed design today, but plastics often offer savings in energy and cost in comparison to other materials. Also, with the advent of new and more environmentally friendly matrices such as bioplastics and UV-degradable plastics, FRP will similarly gain environmental sensitivity [20].

One of the biggest challenges posed by FRPs is their recycling. Many different recycling techniques have been studied during the last two decades, such as mechanical processes (mainly grinding) [21–24], pyrolysis and other thermal processes [25, 26], and solvolysis [27–29].

1.1.8 Mixed Plastics

Another type of plastics used for recycling is mixed plastics. Mixed plastics contain different types of plastics with different processing behavior and stability. Usually, these plastics are not compatible (or thermodynamically miscible) with each other, and the resulting properties are very often inferior to those of the parent polymers. In its broadest sense, mixed plastics constitute a a mixture of plastic resins or a mixture of package/product types which may or may not be the same plastic type or color category, and may not have been fabricated using the same manufacturing techniques.

1.1.9 Composites

Composites are generally considered high-value, high-performance materials that are employed in producing end products of high net worth. The term *composite* can be used to describe a large number of multiphase materials, consisting of a wide variety of matrix materials along with a correspondingly large array of different fillers and reinforcements. Composites can be easily recycled. Additionally, composites have been demonstrated to often have a better ecological track than traditional materials such as steel, aluminum, and concrete [30].

1.2 Conclusion

Recycling or reuse is one approach for end-of-life waste management of plastic products. It makes increasing sense economically as well as environmentally, and recent trends demonstrate a substantial increase in the rate of recovery and recycling of plastic wastes. This process has advantages and disadvantages. The foremost advantage of recycling is that it helps in protecting the environment in the most balanced manner. It helps in conserving important raw materials and protecting natural habitats for the future. Protecting natural resources such as wood, water, and minerals ensure their optimum use. Governments and various environmental organizations regularly emphasize the many benefits of recycling. First and foremost, recycling reduces the amount of waste that must be placed into landfills or incinerated. The recycling of metals, glass, and other materials reduces the pollution that would be caused by the manufacturing of products from virgin materials. Using recycled materials also saves energy because it takes less energy to use recyclables than to make a product from raw materials.

References

1. Hamad, K., Kaseem, M., and Deri, F. (2013) Recycling of waste from polymer materials: an overview of the recent, works. *Polym. Degrad. Stab.*, **98**, 2801–2812.
2. Goto, M. (2009) Chemical recycling of plastics using sub- and supercritical fluids. *J. Supercrit. Fluids*, **47**, 500–507.
3. Thaddaeus, J. (2015) The potentials of plastic wastes recycling in Maiduguri metropolis. *Int. J. Adv. Res.*, **3**, 624–634.
4. Miller, D.A. (2010) *Garbage and Recycling*.
5. Lardinois, I. and van de Klundert, A. (1995) *Plastic Waste, Option for Small Scale Resource Recovery*, TOOL.
6. Hopewell, J., Dvorak, R., and Kosior, E. (2009) Plastics recycling: challenges and opportunities. *Philos. Trans. R. Soc. London, Ser. B*, **364**, 2115–2126.
7. Goodship, V. (2007) *Introduction to Plastics Recycling*, 2nd edn, Smithers Rapra Technology Limited, Shawbury.
8. Cui, J. and Forrsberg, E. (2003) Mechanical recycling of waste electric and electronic equipment a review. *J. Hazard. Mater. B*, **99**, 243–263.
9. Lavee, D. (2007) Is municipal solid waste recycling economically efficient? *Environ. Manage.*, **40** (6), 926–943.
10. Curlee, T.R. and Das, S. (1991) *Plastic Wastes (Management, Control, Recycling and Disposal)*, Noyes Data Corporation, New Jersey.
11. Alter, S.H. (1986) Disposal and reuse of plastics, in *Encyclopedia of Polymer Science and Engineering*, vol. 5 (eds H.F. Mark, N.M. Bikales, C.G. Overberger, and G. Menges), John Wiley & Sons, Inc, pp. 103–128.
12. Tukker, A. (2002) Plastics waste – Feedstock recycling, chemical recycling and incineration. *Rapra Rev. Rep.*, Report 148, **13** (4), 3–4.
13. Bart, J.C.J. (2005) *Additives in Polymers: Industrial Analysis and Applications*, John Wiley & Sons, Ltd, Chichester, pp. 836.
14. Myhre, M., Saiwari, S., Dierkes, W., and Noordermeer, J. (2012) Rubber recycling: chemistry, processing, and applications. *Rubber Chem. Technol.*, **85**, 408–449.
15. Scheirs, J. (1998) in *Polymer Recycling: Science, Technology and Applications* (ed. J. Scheirs), Chichester: John Wiley & Sons Ltd, London, pp. 125–137.
16. Abraham, E., Cherian, B.M., Elbi, P.A., Pothen, L.A., and Thomas, S. (2011) *Recent Developments in Polymer Recycling*, (eds. A. Fainleib and O. Grigoryeva) Transworld Research Network, pp. 47–100.
17. Schneider, J.P., Myers, G.E., Clemons, C.M., and English, B.W. (1995) Biofibres as reinforcing fillers in thermoplastic composites. *Eng. Plast.*, **8** (3), 207–222.
18. Colberg, M. and Sauerbier, M. (1997) Injection moulding of natural fibre reinforced plastics. *Kunstst-Plast Europe*, **87** (12), 9.
19. Schloesser, T. and Knothe, J. (1997) Vehicle parts reinforced with natural fibres. *Kunstst-Plast Europe*, **87** (9), 25–27.
20. Smallman, R.E. and Bishop, R.J. (1999) *Modern Physical Metallurgy and Materials Engineering*, 6th edn, Butterworth-Heinemann, Oxford.
21. Molnar, A. (1995) Recycling Advanced Composites. Final Report for the Clean Washington Center (CWC), Report No. IBP–95–3, December 1995.
22. Steenkamer, D.A. and Sullivan, J.L. (1998) On the recyclability of a cyclic thermoplastic composite material. *Composites Part B*, **29**, 745–752.
23. Palmer, J., Ghita, O.R., Savage, L., and Evans, K.E. (2009) Successful closed-loop recycling of thermoset composites. *Composites Part A*, **40**, 490–498.
24. Pickering, S.J. (2006) Recycling technologies for thermoset composite materials – current status. *Composites Part A*, **37**, 1206–1215.
25. Markovic, V. and Marinkovic, S. (1980) A study of pyrolysis of phenolic resin reinforced with carbon fibres and oxidized PAN fibres. *Carbon*, **18**, 329–335.
26. Torres, A., de Marco, I., Caballero, B.M., Laresgoiti, M.F., Legarreta, J.A., Cabrero, M.A. *et al.* (2000) Recycling by pyrolysis of thermoset composites: characteristics of the liquid and gases fuels obtained. *Fuel*, **79**, 897–902.

27. Yoon, K.H., Di Benedetto, A.T., and Huang, S.J. (1997) Recycling of unsaturated polyester resin using propylene glycol. *Polymer*, **38**, 2281–2285.
28. Yuyan, L., Guohua, S., and Linghui, M. (2009) Recycling of carbon fibre reinforced composites using water in subcritical conditions. *Mater. Sci. Eng., A*, **520**, 179–183.
29. Bai, Y., Wang, Z., and Feng, L. (2010) Chemical recycling of carbon fibers reinforced epoxy resin composites in oxygen in supercritical water. *Mater. Des.*, **31**, 999–1002.
30. Sp, S., Kant, T., and Desa, Y. (2008) Application of polymer composites in civil construction: a general review. *Compos. Struct.*, **84**, 114–124.

2
Common Additives used in Recycling of Polymers

2.1
Review on Different Additives Used in Polymer Recycling
Sivasankarapillai Vishnu Sankar and Sivasankarapillai Anil Kumar

2.1.1
Introduction

Polymer recycling is a process involving both environmental and economic issues. The choice of the recycling technique is influenced by factors such as the purity of the polymer stream, chemical composition, and the nature of additives present in the polymeric material. During the past two decades, polymer recycling methods have attracted considerable interest because they open a way to reduce environmental issues caused by the accumulation of polymer waste generated from day-to-day use of polymer materials. The various methods include reuse, mechanical recycling, and chemical recycling. Even though various strategies are available for the recycling of polymer materials, still there remain certain problems during the post-consumer recycling stage. These problems are due to the degradation of polymers during the processing steps and to the incompatibility of different polymer types. Additives can be used for reducing these problems in the recycling stage and to increase the properties of the recycled products. The primary goal of this chapter is to give an overview of the different types of additives used in polymer (plastic) recycling.

2.1.1.1
Challenges in Recycling – Need for Additives

An increase in the use of plastics has led to a corresponding increase in the amount of plastic waste, which plays a major role in environmental pollution. This has resulted in great interest in the recycling of polymer waste generated from our daily lives. With advances in polymer technology, the processing of plastic mixtures for recycling has been attempted with some success, but poor mechanical properties of the secondary materials and uncertain economic value limit their versatile reuse [1–14].

Recycling of Polymers: Methods, Characterization and Applications, First Edition. Edited by Raju Francis.

For homogeneous polymeric materials, two main methods can be adopted to reduce the negative effects during the recycling process [15]:

1. *Restabilization* avoids or retards the degradation process undergone by polymers.
2. *Addition of fillers and modifiers* improves the performance of thermoplastic polymers without increasing the final cost of secondary material.

The major challenge in the recycling of homogeneous polymers is only related to the degradation phenomena that take place during recycling [16]. These phenomena are generally much more dramatic than in virgin polymers since the oxygenated groups, formed during the processing or during the lifecycle, remarkably accelerate the degradation of plastic materials. This causes serious deterioration of the end properties of the secondary materials [15].

For heterogeneous polymers, the major problem associated with the recycling process is its heterogeneous composition due to the presence of polymers having different chemical structures; so compatibilization is a necessary step in the recycling process. Incompatibility gives rise to blends with poor properties, in some cases worse than those of the virgin polymers [16]. Thus, additives such as compatibilizers can be used to obtain secondary materials with acceptable properties. Therefore, it is clear that additives play a crucial role in the recycling and processing of polymer materials.

2.1.1.2 Equipment for Additive Processing

There are a few vital pieces of equipment and machinery that are essential for additive processing. The most effective of these for the incorporation and handling of additives into polymer networks are volumetric or gravimetric feeders and blenders. Volumetric devices measure the volume that passes through a metering disk, while gravimetric devices measure and control the weight of additives dispensed over a given time interval. Gravimetric type equipments are better when two or more additives are being dispensed at either the machine throat or in a blending system. The blending system can be a dosing or mixing unit, which functions by dividing the main component under consideration into different streams of material such that the streams can combine with the additives at the feed inlet, which helps produce a homogeneous mixture. Another important class of equipment is “Level sensors,” which can measure insufficient additive material in order to prevent unnecessary downtime.

2.1.2 Different Types of Additives

Polymer additives are materials that are added to polymers in the processing stage in order to improve or change the visual, process, environmental resistance, or degradation properties, thereby enhancing their overall performance and application potential. The discovery and research in the field of natural polymer materials has created a need for additives to modify their intrinsic properties so that the polymers have improved characteristics to allow for a wider variety of uses. Today, both natural and synthetic polymers depend on additives during processing. To create new blends, additives and polymer resins are mixed to produce improved materials. Standard mixtures are produced that can be further modified by adding various additives, which can develop a variety of chosen materials with different application potentials to meet individual processing needs.

The term *additive* means all substances, inorganic and organic, that can alter one or more properties of polymers [15]. Additives are essential components of plastic materials, providing maintenance and modification of the polymer's properties, performance, and long-term use. The main purpose of adding these compounds in the recycling stage is to prevent unwanted degradation phenomena during the recycling process and to improve the properties of the secondary materials generated in the recycling stage. The additives in plastics recycling play the same role as in their manufacturing process [15].

Additives can be categorized based on their primary objectives, which are to

- Add bulk or volume while controlling properties and costs of production
- Modify the intrinsic chemical or physical properties of the polymer material
- Reinforce the mechanical properties of the polymer, thereby increasing its impact strength.

Additive materials differ from fillers and reinforcements in following aspects:

- Fillers are added mainly to reduce cost and may or may not serve any other objectives
- Reinforcements are added to increase the structural and mechanical properties of polymers (e.g., fibers and flakes).

The role of additives on the recycling process can be summarized as follows:

- They prevent the thermomechanical degradation of materials occurring during recycling operations.
- They improve the properties of the secondary materials.
- They make heterogeneous polymers compatible.

The most important classes of polymer additives and their function can be listed as follows [15]:

- *Antioxidants and stabilizers* delay the degradative processes of polymers.
- *Mineral fillers* can decrease the cost and enhance the properties of the polymer.
- *Impact modifiers* increase, in particular, the impact strength.
- *Compatibilizers* improve the compatibility between incompatible polymeric materials.

In addition to these, there are various other additives such as lubricants, coloring agents, and so on.

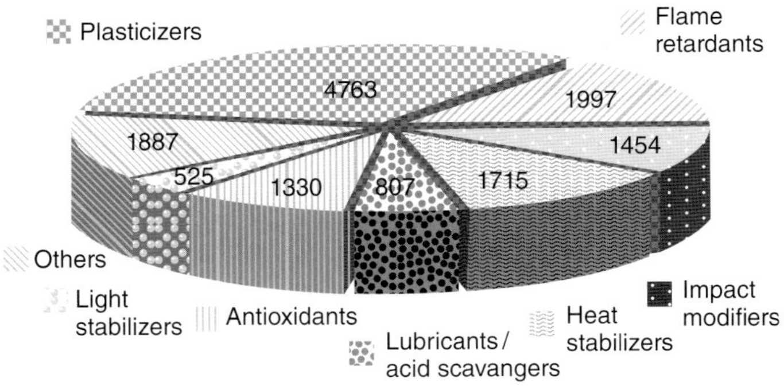

Figure 2.1.1 Turnover of additives in 2004. (Pfaendner [17] Reproduced with permission of Elsevier.)

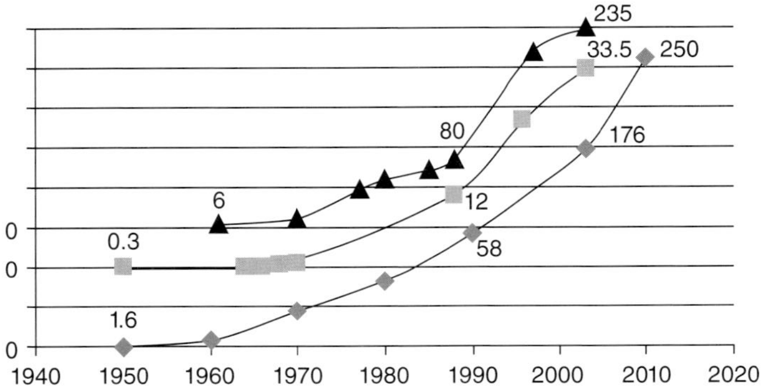

Figure 2.1.2 Consumption of plastics, light stabilizers, and antioxidants since 1950. (Pfaendner [17] Reproduced with permission of Elsevier.)

Increase in the use of plastics leads to an increase in the growth of additives and their consumption (Figures 2.1.1 and 2.1.2) [17].

The most commonly used additives in polymer recycling and processing methods can be classified as follows:

2.1.2.1 Stabilizing Agents

The role of stabilizing agents during polymer processing or recycling is to prevent radiation-induced and thermal degradation affecting the polymer product quality over time. The efficacy of stabilizers appears to decay with lifetime. Restabilization of the material is one of the key steps in polymer recycling by compounding recycled plastics with a fresh reagent. In addition to extending the polymer product lifetime, stabilizing agents also help in protecting virgin plastics and recycled materials from the inevitable degradation that occurs during the relatively short-lived and severe conditions that are associated with polymer processing operations [18].

For example, thermal stabilizers are employed to maintain the stability of the polymer up to the processing temperature, and ultraviolet (UV) absorbers and radical traps are incorporated to withstand photodegradation of the polymer. The nature of the stabilizing agent will depend on the nature of the polymer system into which it has to be incorporated. For example, polymeric phenolic phosphites represent a new family of polypropylene (PP) stabilizers acting through their phenolic group as free-radical scavengers and through their phosphite group as hydroperoxide decomposers.

2.1.2.1.1 Thermal Stabilization

A combination of heat and oxygen will cause oxidation of the polymer, resulting in degradation. The mechanism is the formation of free radicals, which are highly reactive chemical species. This reaction can be visually observed, as the products tend to show discoloration to yellow or brown. Additives called *antioxidants* can be used to arrest this mechanism. The chemicals most commonly employed for this purpose are hindered phenols, which act as peroxide radical decomposers.

Other additives are also employed apart from hindered phenols for these purposes. Additives called *phosphites* combined with hindered phenols have a synergetic effect. This combination is especially effective for polyolefins. The well-known example of the use of additives to prevent thermal degradation is the thermal stabilization of polyvinylchloride (PVC). The free radicals produced in this case are chlorine, leading to the formation of hydrochloric acid. The stabilizers must be able to stop these reactions, which can lead to acid corrosion of the processing equipment.

2.1.2.1.2 Photostabilization

Light, especially in the UV range, can induce photoxidation, which results in the degradation and cleavage of the polymer chains. To prevent this effect, three classes of additives are employed. They are usually called *UV absorbers*, *quenchers* (*scavengers*), and *radical traps*.

UV Absorbers These are among the oldest light stabilizers and work by absorbing the harmful UV radiation and converting it to heat energy. By absorbing these rays, they protect the vulnerable polymer chains. Examples of additives of this class are benzophenones and benzotriazoles.

For example, hydroxybenzophenone and hydroxyphenylbenzotriazole are effective UV stabilizers that can be conveniently used for applications requiring neutrality or transparency. Hydroxyphenylbenzotriazole is not very useful when employed in thin parts, below about 100 μm. Other UV absorbers include oxanilides for polyamides, benzophenones for PVC, and benzotriazoles and hydroxyphenyltriazines for polycarbonates.

Quenchers Quenchers contain chromophore groups (light-absorbing species) that can absorb energy and convert it into less harmful forms. Nickel compounds are the most commercially available quenchers and these are used in applications such as agricultural film production.

Radical Traps One of the major effects of light on a polymer material is the formation of radicals in the polymer system. Radicals are highly unstable reactive species and can cause rapid degradation of the polymer material. To prevent further damage to the plastic, these can be "scavenged" using radical traps. The most important radical traps are known as *hindered amine light stabilizers* (HALSs). It should be noted, however, that once the additive has been consumed, the degradation process will commence. For this reason, it is important that the correct levels of additives are employed. Interestingly, the mechanism of the radical trap can also be applied to flame-retardant technology in plastics. Thus, instead of the chemicals trapping damaging UV radiations, the additives also help in limiting the combustion process. Different chemicals are employed for this purpose, but the mechanisms of action are very similar.

The category of additives that act as radical traps usually contain the 2,2,6,6-tetramethylpiperidine ring system in their structure as seen in commercially available HALS products. Even though there are wide structural differences in these products, all of them act by trapping free radicals during the photoxidation of the system, thus preventing further degradation of the polymer system.

Polymers that have poor thermal stability, such as PP and PVC, have to be stabilized against thermomechanical degradation [15]. Klemchuk and Thompson [19] have reported that stabilization is necessary during reprocessing to prevent, or at least to retard, the degradation of these polymers and the subsequent deterioration of their rheological and mechanical properties. Ma and La Mantia [20] have investigated the variation in the molecular structure of PP during repeated reprocessing. They reported the change in the melt flow index (MFI) of a molding-grade PP as function of the number of injection molding steps and in presence of two stabilizers, namely B900 (CIBA), which is a mixture of Irganox 1076 and Irgafox 168, and P-EPQ (Sandoz), which is a phosphite stabilizer (Figure 2.1.3).

The dramatic increase in the MFI of the unstabilized sample suggests a drastic degradation of PP due to the thermomechanical stress during the processing. The MFI value is about twice that of the virgin sample, after the first recycling. The degradation is considerably reduced by adding the stabilizer before each injection molding step. In this case, the MFI slightly increases, indicating a small change in the molecular weight. It can be seen that the two stabilizers probably act in a similar way. The study also revealed that the stabilizer also helps in preventing the dramatic deterioration of some mechanical properties, as shown in Table 2.1.1 [15]. They found that without any stabilization, the PP sample shows a brittle fracture after five extrusion steps, while the elongation at break is very close to that of the virgin polymer if the stabilizer is added before each extrusion.

Thermal stability studies of PVC have revealed that the poor thermal stability of PVC requires the addition of heat stabilizers to prevent degradation to a large extent. The stabilizer in the PVC products is consumed both during processing

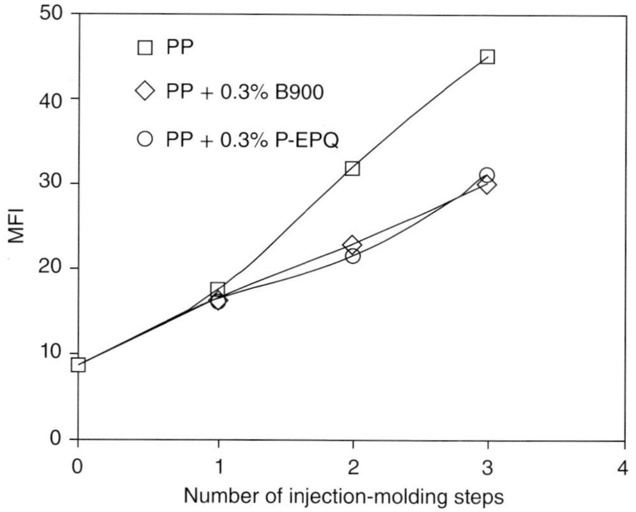

Figure 2.1.3 MFI vs the number of injection molding steps for stabilized and unstabilized PP. (La Mantia [15] Reproduced with permission of John Wiley and Sons.)

Table 2.1.1 Elongation at break of stabilized and unstabilized PP.

Sample	Elongation at break (%)
PP virgin	680
PP after five extrusions	20
PP + 0.3% B900 after five extrusions	520
PP + 0.3% P-EPQ after five extrusions	560

Source: La Mantia [15]. Reproduced with permission of John Wiley and sons.

and, sometimes, during their lifetime. The thermal stability is remarkably reduced if PVC is subjected to the reprocessing steps [15].

Thermal stability of the polymer can be best evaluated using the dynamic thermal stability time (DTST) index, which is the time at which the torque in a mixing test at constant temperature starts to rise [20]. Thus to enhance processability, it is essential to increase the DTST value by adding suitable stabilizing agents that have been consumed during both processing and the lifetime of the PVC products. The effect of a lead stabilizer (which is employed as a thermal stabilizing agent for PVC) on the processability of recycled PVC is shown in Figure 2.1.4, where the DTST values under some processing conditions are reported as a function of the stabilizer content [15].

It can be seen that the thermal stability of PVC is greatly enhanced by adding 1 phr of the stabilizer. Then the curve tends to flatten, and the processability does not remarkably improve by increasing the concentration of the stabilizer. The stabilizer is particularly effective at high temperature and rotational speeds. Indeed,

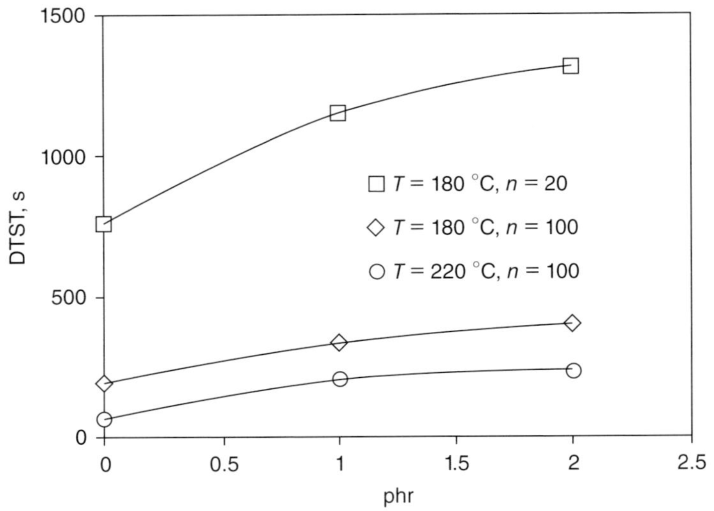

Figure 2.1.4 DTST of recycled PVC as a function of the stabilizer content. (La Mantia [15] Reproduced with permission of John Wiley and Sons.)

under these processing conditions, DTST increases nearly 3 times by adding 1 phr of the lead compound. On the contrary, the improvement is limited when the processing is carried out under softer conditions. From Figure 2.1.4, it is evident that the improvement in DTST at 180 °C and 20 rpm is only about 50% [15].

2.1.2.1.3 **Testing the Effects of Stabilizers**

In order to measure the stabilization of a plastic by a stabilizing additive, whether it is virgin or a recyclate, information is required on the effects of processing, heat, and light.

2.1.2.1.4 **Processing Stability**

A common technique to investigate processing stability is through repeated cycling of a sample, such as by extrusion or injection molding. Mechanical tests that rely on parameters such as tensile and impact strength as well as rheological tests such as MFI can then be used to monitor the changes occurring in the properties of the plastic.

2.1.2.1.5 **Heat Stability**

When discussing heat stability, there are a number of different monitoring methods that can be employed.

1. The temperature at which the plastic decomposes, which can be measured by techniques such as differential scanning calorimetry (DSC). DSC measures the heat output from the polymer as it is heated up or cooled down. It can be used across a wide range of temperatures, from −180 °C to above 600 °C. This technique also permits the measurement of changes occurring in the plastic,

which gives information on the polymer properties, such as the melting point and the temperature at which thermal degradation occurs.

2. The maximum temperature at which the polymer material can be effectively processed without affecting its color or other problems of the polymer formulation. These problems include decomposition of the polymer or other additives employed (e.g., colorants).
3. In many applications in daily life, plastics are employed where they may be subjected to prolonged use at high temperatures. For these types of applications, data is needed on how the material will behave in service. This is usually tested by aging a sample in an oven for an appropriate time interval to reflect the behaviour that is to be expected when the plastic is used in a component.

2.1.2.1.6 Light Stability

There are two techniques commonly employed to assess the light stability of a polymer. One is to utilize artificial weathering equipment as described in ISO 4892. These can simulate weathering cycles by controlling the light (including UV) intensity, heat, and humidity. In this method, polymers are exposed to detailed and programmed cycles, which are similar to those that the components would be exposed to in real applications.

The second method is to simply leave the samples exposed to natural environmental conditions. The effects occurring in the polymeric system can be measured in a number of ways. For example, the surface of the sample can be evaluated in terms of chalking, gloss, and surface texture as a function of the weathering time.

2.1.2.1.7 Stabilizer Combinations for Specific Applications of Polymers

The additives used in a particular plastic will depend very much upon the intended application. Outdoor applications, for example, will require both light and heat stability. Bottle crates, made from a single plastic such as high-density polyethylene (HDPE), have been extensively studied. These studies have revealed the interesting result that the use of 100% recyclate without further stabilization of the material results in the deterioration of mechanical properties, cracking, and color fading after 6 months. However, when HALS and UV absorbers were incorporated into these systems, they retained their properties for more than 4 years. Further studies on heat aging have found that restabilization using a combination of processing stabilizers, light stabilizers, and HALS effectively stabilized the recyclate for further reuse. From the above results, it can be concluded that for successful application of the recycled polymers, care must be taken to ensure that the recyclates are adequately protected by incorporating suitable additives, depending upon their future application potential.

2.1.2.2 Compatibilizers

As mentioned previously, compatibilizers are used in the recycling of heterogeneous plastic mixtures that are incompatible because of their chemical makeup.

Thus, they help in obtaining secondary materials with acceptable properties by processing their blends. Unfortunately, only a few pairs of polymer components can be at present compatibilized, while plastic wastes are frequently composed of several polymers. Interesting results have been achieved by using modified copolymers [15].

For example, literature shows the use of chlorinated polyethylene (CPE) as a compatibilizer for the HDPE/PVC pair [9, 13]. Styrene–ethylene–propylene block copolymer (SEP) and ethylene–propylene rubber (EPR) have been used for HDPE/PS [10] and HDPE/PP blends [11, 12], respectively. Thus it can be seen that the ability of such compatibilizers is due to the enhanced interfacial adhesion between two component polymers, because the compatibilizers usually possess units of the parent polymers. Abundant literature is available investigating the effect of compatibilizers for binary blends [21–25], but only a few studies have reported the effect of compatibilizers on the properties of ternary blends consisting of any of the parent polymers and the third polymer whose compatibility is not affected by the compatibilizer [26, 27].

Studies have been reported that the addition of a compatibilizer to a binary blend enhanced such mechanical properties as the impact strength, because of the additives locating at the interface between the phases and enhancing the stress transfer and the particle size reduction – emulsion effect [28, 29]. Ha *et al.* [30] reported the effect of a compatibilizer for binary blends on the properties of ternary blends composed of HDPE, PP (or PS, polystyrene), and PVC virgin polymers with a simulated waste plastics fraction. CPE, SEP, EPR, and their mixtures were tested as compatibilizers in their study. Figure 2.1.5 [30] shows the tensile strength of the ternary HDPE/PP/PVC blends with or without compatibilizers. By the addition of CPE or EPR, it was expected that interfacial

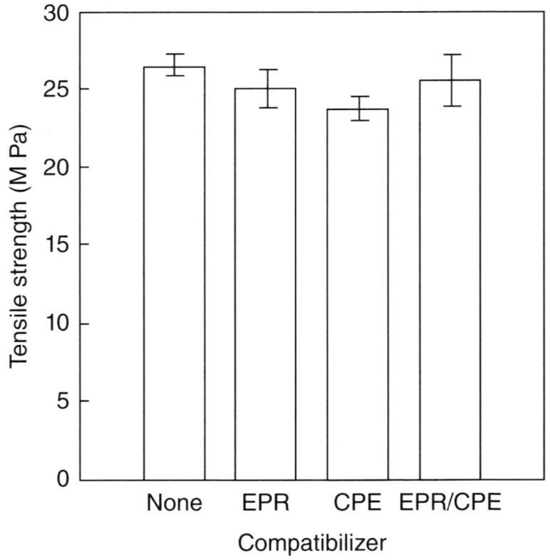

Figure 2.1.5 Effect of different compatibilizers on the tensile strength of HDPE/PP/PVC (8/1/1) blend. (Ha [30] Reproduced with permission of John Wiley and Sons.)

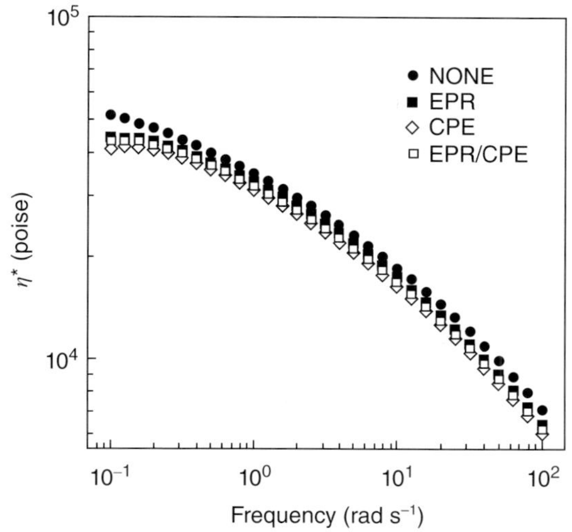

Figure 2.1.6 Complex viscosity of HDPE/PP/PVC (8/1/1) blend with different compatibilizers. (Ha [30] Reproduced with permission of John Wiley and Sons.)

adhesion would be improved, resulting in an increase in tensile strength. This can be attributed to the compatibilization effect of the CPE or EPR at the interface.

Figure 2.1.6 [30] shows a logarithmic plot of the complex viscosity of the HDPE/PP/PVC ternary blends with compatibilizers. All complex viscosities of the ternary blends with either of the compatibilizers were lower than that of the ternary blend without a compatibilizer. The result can be explained by the fact that the tensile strength as well as impact strength of the ternary blends containing EPR are higher than those of CPE-containing ternary blends.

Another example is polyethylene terephthalate (PET) and PP, which are thermoplastics having different chemical properties and are strongly incompatible. Mechanical properties of the blends are very poor, but a suitable compatibilizer can drastically change this picture. The thermoplastic rubber Kraton FG 1901X, a triblock copolymer consisting of PS end blocks and a partly hydrogenated polybutadiene midblock grafted with maleic anhydride, can be considered a good candidate for compatibilizer for this system [15].

2.1.2.3 Antioxidants

Antioxidants can stop or slow down the deterioration of mechanical properties due to thermomechanical degradation during the reprocessing steps [31]. In other words, the presence of antioxidants can strongly improve the resistance to oxidation of the polymer during reprocessing. Thus the function of antioxidants is to inhibit atmospheric oxidation during processing and usage.

Antioxidants used in polymers are further classified into primary and secondary, as described in the following.

2.1.2.3.1 **Primary Antioxidants**

They inhibit oxidation of polymers by scavenging free radicals (molecules cleaved from the polymer chain). Examples are phenols and arylamines.

2.1.2.3.2 **Secondary Antioxidants**

They stop oxygenated radical molecule propagation by decomposing them into stable products. Examples are phosphites and sulfur compounds.

Dintcheva *et al.* [31] studied the influence of antioxidants on the recycling of a waste polymer coming from a separate collection of plastic packagings in Germany. They found that the elastic modulus and tensile strength slightly improved in the presence of the antioxidant, while the elongation at break was significantly enhanced. Figure 2.1.7 [31] shows the flow curves of the stabilized materials reprocessed in the extruders. It can be seen that the flow curves of the stabilized materials are only slightly higher than those of the corresponding unstabilized samples. In contrast, the curve of the sample reprocessed in the mixer in the presence of the antioxidant is significantly higher than that of the unstabilized material, which confirms the drastic degradation phenomena taking place during mixing and the stabilizing action of the antioxidant.

Boersma [32] studied the behavior of phenolic antioxidants (di-*t*-butyl-hydroxyphenyl head and tails varying from methyl to octadecyl chains) in polycarbonate, a thermoplastic polymer, and ethylene propylene diene monomer (EPDM) rubber, an elastomer. The study showed that the mobility and solubility of antioxidants were important parameters in the degradation behavior of a polymer under thermal stress and that the prediction of these parameters will be very useful in screening the most suitable antioxidant for a given polymer product. Thus reliable predictions of these physical parameters will open up the

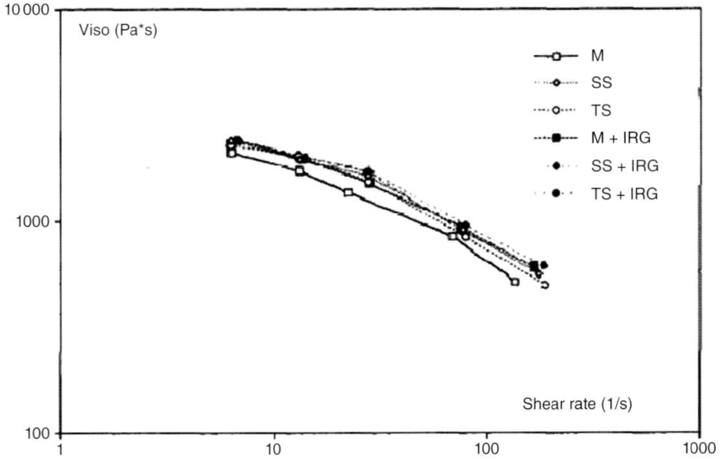

Figure 2.1.7 Flow curves of stabilized and unstabilized polymers. (Dintcheva [31] Reproduced with permission of Elsevier.)

possibilities of even designing the most appropriate antioxidant for a certain polymer.

2.1.2.4
Impact Modifiers

It is reported that some of the additives used in plastics formulations, such as flame retardants, fillers, heat distortion temperature (HDT) modifiers, and pigments, can have a detrimental effect on the impact resistance of the polymeric material. Thus, impact modifiers are additives used to "correct" the detrimental effects of these additives by boosting the impact performance of the material [33].

Some of the additives commonly used at concentrations sufficient to decrease the impact resistance to inadmissibly low levels include flame retardants, fillers, pigments, and HDT modifiers. It is possible to find substitutes for these types of additives, which have either increased efficiencies or decreased effects on the impact performance, resulting in obtaining an acceptable balance of properties in the polymer material under consideration. But in many cases, this is not possible. A remedial measure in some of these cases is to use an impact modifier to boost the impact resistance of the material to the desired level. Effects of an HDT modifier and an impact modifier on the impact resistance of a PVC bottle formulation at 23 °C are shown in Figure 2.1.8 [33].

There are several reasons why additives, particularly at high usage levels, have detrimental effects on the impact performance. The main reason is that a significant volume fraction of the polymer material, which can dissipate stress through mechanisms called *shear yielding* or *crazing*, gets substituted by the additive, which generally cannot deform and dissipate the stress easily. Thus the total ability of the material to dissipate stress decreases [33]. A second reason is that certain additives, such as polymeric additives that are miscible in the polymer matrix, may hinder the local chain motions of the polymer molecules

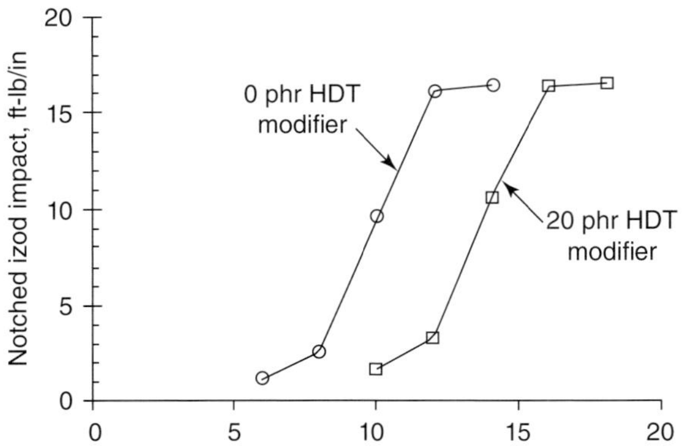

Figure 2.1.8 Effect of HDT modifier and impact modifier on PVC. (Stevenson [33] Reproduced with permission of Wiley.)

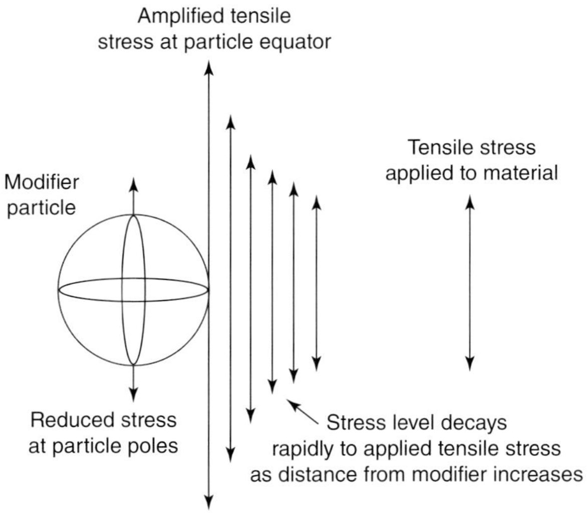

Figure 2.1.9 Stress distribution in polymer matrix surrounding a rubbery impact modifier particle. (Stevenson [33]. Reproduced with permission of Wiley.)

that enable them to shear yield, thereby sharply decreasing the impact resistance of the material [34].

The stress concentration that occurs in the polymer matrix around an impact modifier particle upon application of a tensile stress to the material is schematically represented in Figure 2.1.9 [33]. The tensile stress is amplified in the region surrounding the equator of the modifier particle. As the distance from the modifier particle increases, the amplified stress decays relatively rapidly to the level of the applied stress in the system. Also, the stress changes quickly on moving from the equator toward the poles of the modifier particle and is greatly reduced from the applied stress near the poles. This results in the formation of localized concentrations of stress, and this region when dispersed throughout the polymer matrix results in providing multiple sites in the polymer system at which shear yield and/or crazing of the polymer can be initiated simultaneously upon impact. This results in a structure with a large number of small crazes and/or shear bands instead of a structure with a small number of large crazes or shear bands, which is more prone to failure. An impact modifier can frequently compensate for the decrease in shear yielding or crazing caused by other additives by facilitating shear yielding and/or crazing of the polymer matrix.

There are a variety of different core/shell impact modifiers available on the market today. For example, for PVC compounds, three major types of core/shell impact modifiers are available: all-acrylic impact modifiers, methacrylate/butadiene/styrene (MBS) modifiers, and acrylonitrile/butadiene/styrene (ABS) modifiers. Each of these types of modifiers features a different combination of characteristics. It is important to recognize that an impact modifier can modify a number of other properties of the material such as its optical properties, weatherability, processability, tensile properties, flammability, heat distortion, and cost, in addition to the impact performance of the compound. Thus it is essential to select an impact modifier that provides an appropriate balance of these properties depending on the nature of the application [33].

2.1.2.5 Fillers and Modifiers

The purpose of adding these compounds in the processing stage is to improve the mechanical behavior of the polymer materials. Fillers improve some mechanical properties, modulus, and tensile strength, but worsen the processability and elongation at break. On the contrary, modifiers improve the elongation at break and impact strength. As for the fillers, in many cases a beneficial effect on the cost of the material is also obtained.

Studies are available indicating that the addition of glass fibers into incompatible PET/HDPE blends will cause remarkable improvement in the modulus, tensile strength, and impact strength of the material without much variation in the elongation at break, as shown in Table 2.1.2 [15].

From the table, it is evident that the properties of the PET/HDPE blends reinforced with 20% of glass fibers improve the modulus by about 50%, the impact strength by about 70%, and the tensile strength by about 110%, while the elongation at break remains almost unchanged. Increasing the glass fiber content enhances the mechanical properties of the blends but not the elongation at break. Studies have revealed that by adding glass fibers, significant improvements of some mechanical properties result, whereas the processability of the formulation deteriorates as a result of the increased viscosity [20].

The elongation at break and impact strength are mechanical properties that are more sensitive to degradation and incompatibility; moreover, the elongation at break cannot be improved by adding inert fillers. It can be achieved by the use of small amounts of modifiers such as elastomers in the formulations, which can dramatically enhance both the elongation at break and impact strength [15].

The effect of some elastomers on the mechanical properties has been well evidenced on an incompatible polymer mixture made of PET (~45%), PE

Table 2.1.2 Mechanical properties of glass fiber (GF)-filled PET/HDPE blends.

Blend	Modulus (GPa)	Tensile strength (MPa)	Elongation-at-break (%)	Impact strength ($J\,m^{-1}$)	Heat distortion temperature (°C)
PET/HDPE	1.6	12	1.3	19	120
PET/HDPE + 10% GF	1.95	21	1.4	28	158
PET/HDPE + 20% GF	2.35	26	1.5	33	233
PET/HDPE + 40% GF	3.1	30	1.4	39	239

Source: La Mantia [15]. Reproduced with permission of John Wiley and Sons.

Table 2.1.3 Mechanical properties of PET/PE/PVC blend and effect of the addition of 10% of an elastomer.

Blend	Modulus (GPa)	Tensile strength (MPa)	Elongation-at-break (%)	Impact strength ($J\,m^{-1}$)
PET/PE/PVC	5.6	4.3	1.5	21
PET/PE/PVC + SEBS	4.2	7.3	3.4	69
PET/PE/PVC + MBS	3.2	4.2	3.1	32
PET/PE/PVC + EPDM	3.7	5.8	5.3	91

Source: La Mantia [15] Reproduced with permission of John Wiley and Sons.

(polyethylene) (~35%), and PVC (~20%) from a collection of plastic containers. Table 2.1.3 [15] gives the values of the modulus, tensile strength, elongation at break, and impact strength reported for this formulation and for the same material with 10% of different elastomers.

The above data show that the elastic modulus decreases, which is expected considering the low values of the modulus of the elastomers. The opposite reason can explain the improvement of the other mechanical properties. These modifiers improve the elongation at break and impact strength because of the inherent properties of the rubbery compounds. Also, EPDM rubber is very effective, and, indeed, the elongation at break is more than 3 times and the impact strength more than 4 times higher than the values of the nonmodified blend.

2.1.2.6
Antistatic Agents

The choice of antistatic agent is crucial, and is determined by a wide range of factors such as the polymer type, processing conditions, and end application. They function by allowing polymeric materials to disperse static electricity. Antistatic agents are used where charge build-up on surfaces may cause problems. Since antistatic agents are hygroscopic, a conductive layer is formed on the polymer material by a thin film of water adsorbed from the atmosphere. Antistatic agents are commonly used in combination with anti-block and anti-slip products to maximize cost effectiveness and high performance in the film industry.

The commonly used antistatic agents in various polymer systems and their addition levels are listed in Table 2.1.4.

2.1.2.7
Coloring Agents

Coloring agents are important because they can greatly influence consumer appeal and may eliminate the need for an external color treatment (painting).The two types of these additives most often used are pigments and dyes.

Table 2.1.4 Commonly used antistatic agents.

Polymer	Commonly used antistatic agents	Typical addition levels (%)
Polyethylene (HDPE, LDPE, LLDPE)	Ethoxylated alkyl amines	0.1–0.3
	Ethoxylated lauramide	0.1–0.5
	Glycerol monostearate	1–2
Polypropelyne	Ethoxylated alkyl amines	0.1–0.3
	Ethoxylated lauramide	0.4–0.8
	Glycerol monostearate	1–2
PET	Sodium alkyl sulfonates	1–2
Polycarbonate	Sodium alkyl sulfonates	2.5–3.5

LDPE, low-density polyethylene; LLDPE, linear low-density polyethylene.

2.1.2.7.1 Pigments

They are insoluble in the polymer and dispersed as discrete particles in the plastic, and may be organic (carbon black) or inorganic (TiO_2). They usually give a dull color and render the polymer opaque.

2.1.2.7.2 Dyes

They are organic colorants that are soluble in the polymer and disperse easily. They are stronger than pigments, give brighter colors, and are more transparent.

The common coloring agents that are used in polymer recycling process to obtain different shades are the following:

- *White pigments*: Titanium dioxide (TiO_2) and zinc sulfide (ZnS)
- *Black pigment*: Carbon black

Coloring agents are often selected on the basis of their processing temperature:

- Cadmium pigments come in red-yellow, orange, and maroons and have excellent heat stability.
- Iron oxides come in red, yellow, brown, and black, and their heat stability depends on the color.
- Cobalt aluminates come in green and blue and have excellent heat resistance.

2.1.2.8 Flame Retardants

The purpose of flame retardants is to increase the ignition temperature of polymers and slow down the rate of burning. Flame retardants are a class of additives that give thermal stability to polymers. They can be classified into two types:

2.1.2.8.1 Additive Types

They are incorporated into the polymer by dispersion and remain separate from the polymer. Examples are as follows:

- Aluminum trihydrate (ATH), Magnesium hydroxide
- Halogenated compounds such as polybrominated diphenyloxides
- Phosphorous additives.

2.1.2.8.2 Reactive Types

They function as co-monomers and are part of the polymer system. Reactive types often replace some of the side groups in polymer chains. Either class of flame retardants is often used with a secondary additive, which increases its effectiveness, such as antimony and zinc.

Their mode of action is different, which depends on the nature of flame retardants:

- *Protective charring systems* form a barrier between the unburned polymer and the flame (phosphorous based systems).
- *Halogen systems* prevent burning by producing halogen gases, which can bind free radicals from the burning polymer. They also provide some protective barrier from oxygen at the burning surface (brominated and chlorinated systems).
- *Hydrate systems* act by "quenching" the burning surface by producing water that is trapped in their moieties (aluminum trihydrate, magnesium hydroxide).

2.1.2.9 Lubricants

Lubricants are used to facilitate the processing of polymers and may also serve in the application of polymers.

Lubricants can be widely classified into two types:

- *Internal lubricants* serve to reduce the melt viscosity of the polymer and thus lower the energy required for processing the polymer.
- *External lubricants* serve to reduce the friction between the polymer and the processing equipment.

The selection of an appropriate lubricant type for a polymer system depends on many factors such as the polarity, melt temperature, compatibility, interaction with the other additives, and so on.

Examples of the commonly used lubricant additives are alcohols, waxes, fatty acids, amides such as bis(stearoyl)ethylenediamine, and so on.

2.1.2.10 Plasticizers

Plasticizers are additives that increase the plasticity or flexibility of a plastic material. They soften and make inherently rigid polymers flexible but do not evaporate during the processing stage. They are generally used in polymers that contain polar moieties. In addition to providing flexibility to polymer applications, plasticizers lower the processing temperature and glass transition temperature and serve as internal lubricants, thus increasing movement and durability of the polymer system. The main feature of low molecular weight plasticizers is that they can be pumped, sprayed, and poured over a wide range of temperatures. But medium to high molecular weight plasticizers need to be heated under certain conditions. Since plasticizers are too rigid to be processed without assistance, they are commonly used in combination with PVC. Plasticizers can be helpful for production, but they should not be used for materials in food packaging applications because of their toxicity.

Most plasticizers are monomeric nonvolatile organic liquids or low-melting-point solids, such as dioctyl phthalate or stearic acid; polymeric plasticizers (e.g., phthalic and adipic polyester) are usually very viscous liquids. They are mostly used in PVC (80% market share), mainly as phthalate esters of C8, C9, and C10 alcohols. Waskitoaji *et al.* [35] have studied the application of ester compounds based on palm oil as secondary plasticizers in PVC. The modulus of elasticity and elongation of the plastics can be used to analyze the effects of addition of secondary plasticizers on PVC-based plastics.

2.1.2.11 Antibacterial or Antimicrobial Additives

Most polymers are resistant to microbial attack in their virgin form. On the other hand, additives such as plasticizers, lubricants, and even some heat stabilizers can induce and enhance the growth of microorganisms. Antibacterial additives are used to prevent the growth of microorganisms so that polymeric materials are protected from biological degradation. These classes of additives operate by interfering with the metabolism of microorganisms by blocking their enzyme systems. To be effective, the additive must be able to migrate to the surface of the material so that it can interact with the microorganisms that cause degradation of the polymer material [36].

Examples of initially used antimicrobial additives include arsenic, sulfur, and copper compounds. Other antibacterial or antimicrobial additives such as zinc pyrithione, chlorhexidine, and 1-hexadecylpyridinium chloride are also used in appropriate polymer formulations.

2.1.2.12 Coupling Agents

The primary purpose of coupling agents is to increase the interaction between polymer and filler materials. They create chemical links between molecules to improve bonding. When the coupling agents bind to a polymer, they can enhance the adhesion between the two materials. Since coupling agents promote bonding between the phases, they can be used to bond materials that are normally incompatible. This can be very useful when trying to create new polymer blends and for recycling old polymer material [37].

Maleated polyolefins are possibly the largest group of coupling agents, and they consist of mostly PE or PP with maleic anhydride functional groups grafted onto the polymer backbones. Peroxide reagents are often used for grafting, which takes place at tertiary carbons in the polymer chain or at terminal olefinic groups. Examples of other coupling agents employed in wood–plastic composites include organosilanes, fatty acid derivatives, long-chain chlorinated paraffins, and polyolefin copolymers with acid anhydrides incorporated into the polymer backbones.

The effect of a novel coupling agent, an alkyl ketene dimer, on the mechanical properties of wood–plastic composites, has been reviewed recently [38].

2.1.3 Conclusion

Additives are essential components of polymer processing, providing maintenance and/or modification of polymer properties, performance, and long-term use. The extension of polymer properties by the use of additives has played a significant role in the growth of plastics. At the beginning of the plastics age, the role of additives was mainly to retain polymer properties and to make them resistant to the heat treatment during the transforming process. But next-generation of additives such as antioxidants, heat and light stabilizers, and so on, help in the extension of service life and modification of the mechanical and physical properties. Recent developments in the field of high-performance additives address more stringent or new requirements, more severe processing, and the use conditions and environmental concerns, but still focus mainly on maintaining the properties of polymers. The driving force for new additive development is the increasing awareness of the environment, resulting in environmentally friendly additive solutions. This area covers the introduction of new members of the additive family, such as oligomeric/polymeric stabilizers, graftable stabilizers, halogen-free flame retardants, as well as additives for the mechanical recycling of polymers. In short, plastic additives are and will be essential components in polymer formulations and will provide innovative solutions for the challenges in polymer recycling methods in the present era of environmental concern.

References

1. Ehrig, R.J. (ed.) (1992) *Plastics Recycling*, Hanser, New York, pp. 1–18.
2. Nein, C., Salafranca, J., Rubio, J., and Cacho, J. (1998) *Food Addit. Contam.*, **15**, 842.
3. Marklund, S., Rappe, C., and Tsklind, M. (1987) *Chemosphere*, **16**, 29.
4. Leaversuch, R. (1989) *Mod. Plast.*, **66**, 19.
5. Kreisher, K.R. (1990) *Mod. Plast.*, **67**, 60.
6. Ajji, A. (1995) *Polym. Eng. Sci.*, **35**, 64.
7. Akkapeddi, M.K., Mason, C.D., and van Buskirk, B. (1993) *Polym. Prepr.*, **34**, 848.
8. Paul, D.R., Vinson, C.E., and Locke, C.E. (1972) *Polym. Eng. Sci.*, **12**, 157.
9. D.R, P., Locke, C.E., and Vinson, C.E. (1973) *Polym. Eng. Sci.*, **13**, 202.
10. Lindsey, C.R., Paul, D.R., and Barlow, J.W. (1981) *J. Appl. Polym. Sci.*, **26**, 1.

11. Bartlett, D.W., Barlow, J.W., and Paul, D.R. (1982) *J. Appl. Polym. Sci.*, **27**, 2351.
12. Chiuad, W.E. and Fang, S.J. (1985) *J. Appl. Polym. Sci.*, **30**, 1573.
13. Park, H.D., Park, K.P., Cho, W.J., Ha, C.S., and Kwon, S.K. (1996) *Polym. Recycl.*, **2**, 277.
14. Hajian, M., Sadrmohaghegh, C., and Scott, G. (1984) *Eur. Polym. J.*, **20**, 135.
15. La Mantia, F.P. (1998) *Macromol. Symp.*, **135**, 157–165.
16. La Mantia, F.P. (1996) Basic problems in plastic recycling, in *Recycling of PVC and Mixed Plastics* (ed. F.P. La Mantia), ChernTec Publishing, Toronto.
17. Pfaendner, R. (2006) *Polym. Degrad. Stab.*, **91**, 2249–2256.
18. Tsenoglou, C., Kartalis, C.N., Papaspyrides, C.D., and Pfaendner, R. (2001) *J. Appl. Polym. Sci.*, **80**, 2207–2217.
19. Klemchuk, P.P. and Thompson, T. (1992) Stabilization of recycled plastics, in *Emerging Technologies in Plastic Recycling*, ACS Symposium Series, vol. **513** (eds G.D. Andrews and P.M. Subrarnanian).
20. Ma, W.G. and La Mantia, F.P. (1996) *J. Appl. Polym. Sci.*, **59**, 759.
21. Radonjic, G., Musil, V., and Smit, I. (1998) *J. Appl. Polym. Sci.*, **69**, 2625.
22. Jurkowski, B., Kelar, K., and Ciesielska, D. (1998) *J. Appl. Polym. Sci.*, **69**, 719.
23. Samios, C.K. and Kalfoglou, N.K. (1998) *Polymer*, **39**, 3863.
24. Kang, T.K., Kim, Y., Lee, W.K., Park, H.D., Cho, W.J., and Ha, C.S. (1999) *J. Appl. Polym. Sci.*, **72**, 989.
25. Rana, D., Lee, C.H., Choe, S., Cho, K., and Lee, B.H. (1998) *Korean Polym. J.*, **6**, 158.
26. Zhong, Z.K., Zheng, S.X., Yang, K.J., and Guo, Q.P. (1998) *J. Appl. Polym. Sci.*, **69**, 995.
27. Blom, H.P., The, J.W., and Rudin, A. (1998) *J. Appl. Polym. Sci.*, **70**, 2081.
28. Galli, P., Danesi, S., and Simonazzi, T. (1984) *Polym. Eng. Sci.*, **248**, 544.
29. Yang, D., Zhang, B., Yang, Y., Fang, Z., and Sun, G. (1984) *Polym. Eng. Sci.*, **248**, 612.
30. Ha, C.S., Park, H.D., and Cho, W.J. (2000) *J. Appl. Polym. Sci.*, **76**, 1048–1053.
31. Dintcheva, N.T., Jilov, N., and La Mantia, F.P. (1997) *Polym. Degrad. Stab.*, **57**, 191–203.
32. Boersma, A. (2006) *Polym. Degrad. Stab.*, **91**, 472–478.
33. Stevenson, J.C. (1995) *J. Vinyl Add. Tech.*, **1**, 41.
34. Havriliak, S. Jr., and Williams, D.R. (1990) *Polymer*, **31**, 115.
35. Waskitoaji, W., Triwulandari, E., and Haryono, A. (2012) *Procedia Chem.*, **4**, 313–321.
36. Murphy, J. (2001) *Additives for Plastics*, 2nd edn, Elsevier Science Inc., New York.
37. Fox, J. (2008) *Analysis of Polymer Additives in the Packaging Industry*, Packaging Science, p. 9.
38. Zhang, H. (2014) *Mater. Des.*, **59**, 130–134.

2.2
Recent Trends and Future of Polymer Additives in Macromolecular Recycling Technology: A Brief Overview

Sivasankarapillai Vishnu Sankar and Sivasankarapillai Anil Kumar

2.2.1
Introduction

Additives form the backbone of commercial polymer manufacturing industry, but also play a vital role in the enhancement of properties of a variety of polymer systems like elastomers, rubbers, coatings, and adhesives. Without the presence of additives, many polymeric materials would be of limited use and cannot be used in the critical areas that require improved performance of polymer systems. This chapter deals with the advancement of additives, mainly miscellaneous classes of additives that are not described in the first part of this chapter, and also gives an overview in the advancement of additive technology, such as the analytical methods for additives. Finally, the chapter aims to discuss the challenges and future of recycling.

An overview of developments across all areas of polymer recycling additives is lacking, and a unified approach should therefore be of considerable importance to overcome the challenges arising in the polymer recycling process. With a rapidly emerging field such as this one, this chapter can only be considered as a work in progress. Thus the purpose of this chapter is to give a brief overview of the advancements in additives, as well as the scope and future of the area.

2.2.2 Miscellaneous Additives

Even though the major recycling additives satisfy functions such as stabilization and enhancement of the lifetime of the recycling materials, there are certain other classes of additives added to polymers that are meant to increase the overall performance of recycled and virgin polymer systems in their wide applications.

2.2.2.1 Nucleating Agents

A nucleating agent is an additive that forms nuclei in a polymer melt environment, which enhance or stimulate the growth of crystals in the polymer system [1]. A nucleating agent can also be considered as a clarifying agent or clarity-improving agent in the polymer system. The issues of nucleating agents have been reviewed in the literature [2].

Nucleating agents can be classified as inorganic nucleating agents, sorbitol compounds, phosphates, and coupled nucleating agents.

2.2.2.1.1 Inorganic Nucleating Agents

They are inorganic materials used in polymer processing for nucleating action. For example, inorganic substances such as alumina, aluminum hydroxide, aluminum powder, titanium dioxide, and calcium fluoride are used as nucleating agents in polycyanoaryl ethers. Similarly, kaolin, talc, mica, silica, calcium carbonate, a metal salt of an aliphatic acid, and so on, are frequently used for polyesters.

Examples of commonly used nucleating agents are talcum, titanium dioxide, magnesium oxide, 4-*tert*-butylbenzoic acid, adipic acid, diphenylacetic acid, sodium succinate, sodium benzoate, and ionomers [1, 3, 56–63]. Pyrogenic silicic acid (Aerosil®) has been experimented as a nucleating agent for polyaniline polymer system [11]. Aerosil® acts in multiple ways as a reinforcement agent, an antiblocking agent, and also as a rheology improver. Other examples of nucleating agents used for polyester systems are talc, calcium fluoride, sodium phenyl phosphonate, alumina, and also finely divided poly(tetrafluoroethylene).

2.2.2.1.2 Sorbitol Compounds

Dibenzylidene sorbitol has a fibril-like structure and acts as an effective nucleating agent, which facilitates the crystallization of polyolefinic system during the processing stage [12]. The mode of action of this compound is that when the concentration of dibenzylidene sorbitol reaches a critical value, the fibrils will self-assemble into a three-dimensional network structure when there is a decrease of temperature in the system but before the stage where crystallization takes place. The network of the fibrils may facilitate the subsequent complementary process of nucleation and crystallization growth in the polymer. An oriented deformation of the dibenzylidene sorbitol network will act as a template for the anisotropic crystallization of polypropylene (PP), which can result in providing a high lamellar orientation level. An example of this additive is 2,4-bis-(3,4-dimethylbenzylidene)sorbitol.

2.2.2.1.3 Phosphates

Organic phosphates have been observed to have a very high nucleating efficiency even at very low concentration. This can cause a substantial increase in the flexural modulus of the material [13]. Organic phosphorous nucleating agents are better than sorbitol compounds in improving the mechanical properties of i-PP. But sorbitol compounds can significantly enhance the transparency of the system. However, organic phosphorous nucleating agents and sorbitol nucleating agents have similar effects on the peak temperature of crystallization [14].

2.2.2.1.4 Coupled Nucleating Agents

This class of nucleating agents includes coupled semicrystalline homopolymers or copolymers of ethylene, propylene, or other α-olefins. These polymers are usually coupled together by one or more coupling agents, such as carbenes [15], nitrenes [16], or azide coupling agents [17]. Coupled nucleating agents can be used in the same manner and under the same processing conditions as conventional nucleating agents.

2.2.2.2 Reinforcing Agents or Fillers

These additives are meant for increasing the strength of polymers by acting as reinforcement for the polymer matrix. They can be inorganic or organic, depending on the polymer environment in which they are added. By the addition of this material, the polymers become composites. When the reinforcement is a nano object, they are called *nanocomposites*. It is expected that nanocomposites can provide superior physical properties because of the high surface-to-volume

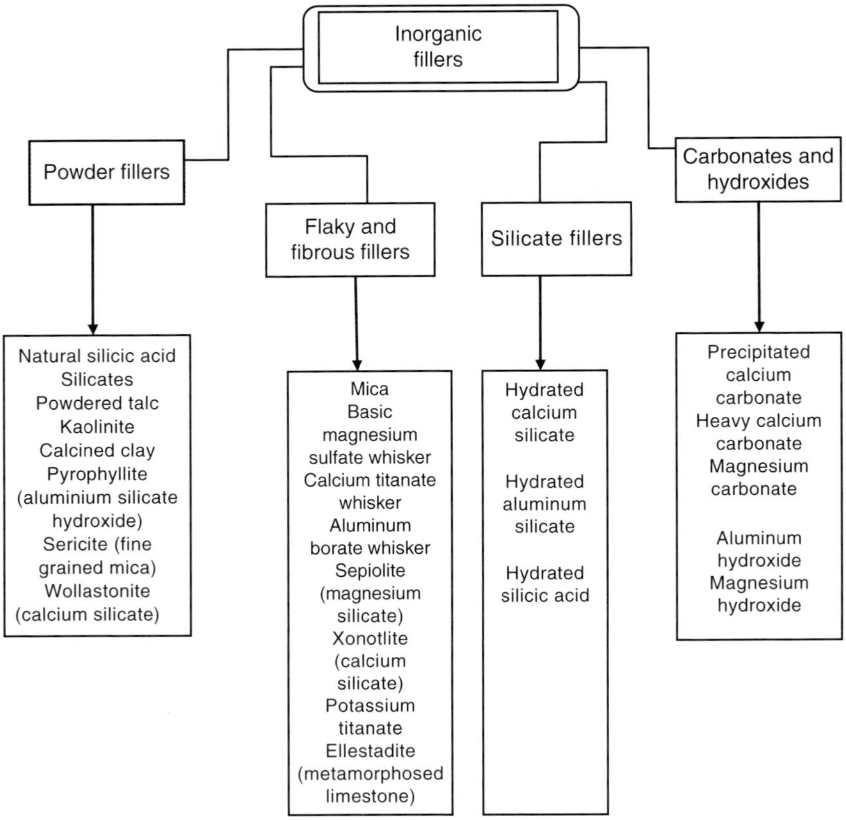

Figure 2.2.1 Commonly used inorganic fillers in polymer formulations.

ratio of nanometer-scale reinforcing fillers embedded in the matrix, compared to the conventional fiber- or particle-reinforced composites. In many cases, these additives are surface-modified to increase the interaction between the dispersed phases. An overview of the inorganic substances that are used as fillers is schematically shown in (Figure 2.2.1) [18].

Reinforcements and fillers differ slightly in their materialistic aspect. They can be distinguished by their functionalities on the polymer systems to which they are added. Fillers are tiny particles or objects that can contribute only slightly to the mechanical strength. But reinforcements are materials that can improve properties such as impact resistance, stiffness, and mechanical strength of the polymer material in which they are dispersed. Some materials such as glass are used both as reinforcement and filler.

The different types of reinforcements are selected on the basis of the value they add to the polymer material, like increased conductivity, flame retardation, and mechanical strength. Some common materials that are used for these purposes are discussed below.

2.2.2.2.1 Carbon Nanotubes

A carbon nanotube is a graphene sheet rolled into a cylindrical structure, typically several nanometers in diameter, and whose the ends are capped with half fullerene balls. They are unique nanostructures with interesting electrical and mechanical properties, which are due to the close relation between the carbon nanotube and graphite and to their one-dimensional aspect. The strength of the sp^2 carbon–carbon bonds in the nanotubes gives them amazing mechanical and conductive properties. In many cases, the nanotubes are surface-modified by using surfactants, and these functionalized carbon nanotubes can be well dispersed in the polymer matrix.

2.2.2.2.2 Nanoclays

Most of the clay minerals fall under the category of layered silicates or phyllosilicates because their silica layers and alumina sheets are joined together in varying proportions and in a specific manner [19]. Two-dimensional arrays of silicon–oxygen tetrahedra and aluminum– or magnesium–oxygen/hydroxyl octahedra are the major building blocks of the clay minerals, which are superimposed in different ways. Nanoclays are used as fillers in thermoplastics and thermosets. The nanoclay may be usually applied in combination with another chemical ingredient, such as a crosslinking agent, which helps to provide a distinctive and overall synergistic effect on mechanical properties of the polymer [20]. Since nanoclays are inorganic substances, they are organically modified by incorporating an organic moiety in their structure to give compatibility with the interacting polymer system.

2.2.2.2.3 Carbon Blacks and Carbon Fibers

Carbon blacks and carbon fibers are conducting additives but have only limited use due to their color. Commonly used carbon blacks have average particle sizes of less than 50 nm. Vapor-grown carbon fibers have diameters of 5–50 nm. Conductive carbon fibers can be used for modifying the electrostatic discharge properties of polymeric resins. Usually, carbon fibers have a diameter of 3–15 μm and may contain graphene ribbons parallel to the fiber axis. They are produced commercially by pyrolysis of organic precursors such as phenolics, polyacrylonitrile, and so on. In general, carbon fibers are sized, with an initial length of about 0.05–5 cm. These modified fibers are conventionally coated on at least a portion of their surfaces with a sizing composition.

Although nano reinforcement enhances overall material performance, the main challenge facing the manufacture of nanocomposites is the aggregation of nanoparticles during the manufacturing process. Severe particle aggregation is detrimental to the thermal and mechanical properties of the nanocomposites. The homogeneous dispersion of the nanoparticles in the dispersed matrix is a very critical in obtaining high-performance nanocomposites [21].

Apart from these, thermally conductive fillers, which include graphite, aluminum nitride, silicon carbide, boron nitride, and alumina, are suitably selected and employed as additives depending on the polymeric system and the nature of application of the obtained polymer material.

2.2.2.3 Optical Brighteners

They are also called *fluorescent brightening agents*. Brightening agents are used for reducing the yellow color developing in polymers as a result of aging and usage. Some examples of polymers that need brightening agents are PE (polyethylene), PVC (polyvinyl chloride), PU, PMMA (polymethylmethacrylate), PS (polystyrene), and copolymers.

The selection of appropriate optical brighteners depends on several factors of the substance such as the following:

- Maximum whitening effect – which depends on the polymer system
- Hue – generally neutral to blue is preferred
- Compatibility of the additive with the matrix resin to be incorporated
- Light fastness.

The essential requirement required for a fluorescent compound to be used as an optical brightener is the presence of an aromatic or aromatic heterocyclic condensed ring system in their structure. An important speciality of these compounds is the existence of an uninterrupted chain of conjugated double bonds associated with the aromatic ring. The number of such conjugated double bonds depends on the nature of substituent in the structure as well as the planarity of the fluorescent system of the molecule [22]. Most brightener compounds belong to derivatives of stilbene or 4,4′-diamino stilbene, biphenyl, five-membered heterocycles (triazoles, oxazoles, imidazoles, etc.), or six-membered heterocyclic systems (coumarins, naphthalamides, triazines, etc.) [23].

To be effective, the optical brightener must dissolve in the polymer. Optical brighteners can be inorganic or organic.

2.2.2.3.1 Inorganic Brighteners

These compounds absorb in the visible range of electromagnetic spectrum, for example, titanium pigments that absorb in the near-visible to UV range. Their efficiency is low compared to anatase pigments, which absorb nearly 40% of the UV light in the wavelength range of ~380 nm.

2.2.2.3.2 Organic Brighteners

Only a few classes of organic compounds are known to be suitable for use as brightening agents for polymers. Examples of these classes include [22]

- Bis(benzoxaziol)s
- Phenylcoumarins
- Bis-styryl biphenyls.

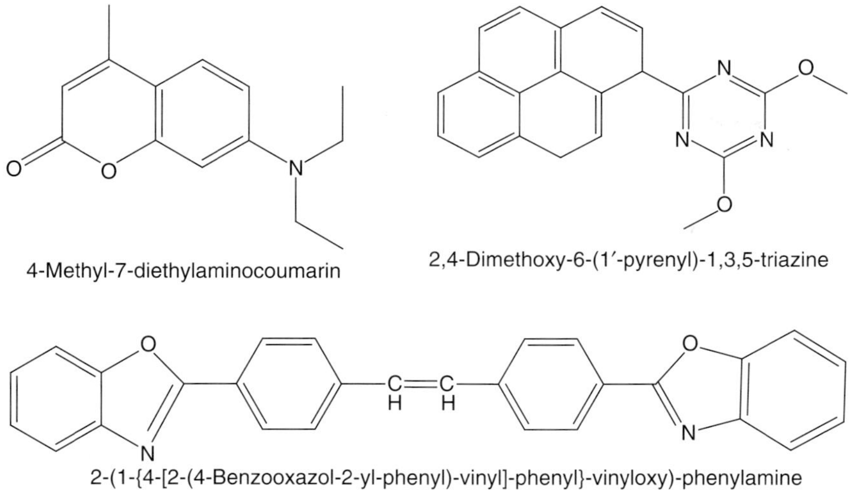

Figure 2.2.2 Chemical structure of organic brighteners.

Some of the compounds and their structures that come under these classes and are used as organic brighteners are given in Figure 2.2.2.

2.2.2.3.3 Enhancing Stability of Optical Brighteners – Melt Extrusion Technique

The thermal stability of optical brighteners can be improved by melt extrusion. When melt blending or extrusion is applied, the optical brighteners should have stability up to temperatures as high as 310–330 °C [18]. In polyesters, such as polyethylene terephthalate (PET), the tendency for acetaldehyde formation can be reduced by blending the polymer with a low molecular weight polyamide (PA). This is desirable for the application of PET as drinking bottles. However, the addition of PA causes yellowing of the products. Thus optical brighteners are added to solve this problem [23].

2.2.2.4 Surface Improvers

This class of additives can also be considered as polymer processing aids. A number of surface defects arise during the manufacture of extruded polymers, referred to as *sharkskin, snakeskin,* and *orange peel,* which are all related to the rheology of the polymer melt and in particular the melt fracture of the polymer [24].

Melt fracture is a flow phenomenon that occurs when the molten polymer flows through the die, and starts at the point of die entry, as evidenced by gross irregularities in the shape or surface of the extrudate. Thus melt fracture is considered to be the result of nonuniform or irregular elastic strains in the material at the die entrance. The shear rate on the surface of the polymer is so high that the surface of the polymer begins to fracture. That is, there is a slippage of the surface of the extruded polymer relative to the body of the polymer melt. The surface generally cannot flow fast enough to keep up with the body of the extrudate and a fracture in the melt occurs. These irregularities in the shape or surface of the extrudate are undesirable. For example, the irregularities in the polymer system will produce an unattractive pattern on blown films [22]. Processing aids are typically added to the polymer so that during the stage of melt processing, they will migrate to the surface of the polymer and cause a lubricating effect between the polymer and die surfaces, which results in high throughput with reduced melt fracture [25].

However, in some cases melt fracture is a desirable property since attractive patterns on the surface of extruded polymer products can be developed by forcing the molten polymer through a die which is coated with a controlled pattern of low-surface-energy material. Melt fracture of the polyethylene as it emerges from the die is produced at gaps in the coating material, which may be silicon, an inorganic compound, or a fluorine-containing polymer [26].

Different types of additives used for reducing the melt fracture during the polymer processing and recycling stages are described in the following:

2.2.2.4.1 **Fluorocarbon Compounds**

Fluorocarbon processing aids are known to partially alleviate melt defects in extrudable thermoplastic hydrocarbon-based polymer formulations and have the ability to permit faster and more efficient extrusion. Blatz developed the use of fluorocarbon polymer processing aids with melt-extrudable hydrocarbon polymers, wherein the fluorinated polymers are homopolymers and copolymers of fluorinated olefins with atomic fluorine to carbon ratio of at least 1 : 2, and also fluorocarbon polymers having melt flow characteristics similar to those of hydrocarbon polymers [27]. Studies show that vinylidene fluoride homopolymers can also be used to improve the extrusion of low-density polyethylene (LDPE) [28] and that multimodal fluoropolymers are superior in performance to unimodal fluoroplastics when functioning as polymer processing additives [29].

2.2.2.4.2 **Acrylics**

Acrylic-based processing aids can be employed to improve the extrusion of high-density polyethylene (HDPE). However, this processing aid shows only a marginal improvement in the extrusion of linear low-density polyethylene (LLDPE) with a narrow molecular weight distribution [30]. Also,acrylic polymers are generally much less expensive than fluoropolymers. Accordingly, there is an economic advantage in employing acrylics rather than fluoropolymers as processing additives. However, studies indicate that the simple addition of an acrylic polymer and a fluoropolymer to a polyethylene extrusion process produces antagonistic results, which implies that the performance produced by one is adversely affected by the simple addition of the other [24].

2.2.2.4.3 **Organic Salts**

Zinc stearate has been demonstrated to reduce the melt fracture in polyesters [31]. In this work, the authors mixed zinc salts with the polyester in a wide range of melt-processing equipment, such as Banbury mixers or extruders.

Either single-screw or twin-screw extruders may be used in the technique. A commonly employed process involves adding the zinc salt to the polyester to make a masterbatch of the zinc salt of an organic acid within the polymer base material such as a polyester or a polyolefin. Suitable polyolefins for this purpose include polypropylene, LLDPE, or HDPE. The masterbatch is mixed during the stage of extrusion into the desired polyester. The concentration of the zinc salt in the masterbatch was found to range from 5% to 30%.

2.2.2.5 Antiblocking Additives

This class of additives is employed to prevent the blocking effect in polymer systems. Blocking is the unwanted adhesion between layers of a plastic film, which may occur under pressure, usually during storage or use. Because of blocking, unwinding the material becomes difficult when it is a film roll, or opening it when it is a bag. Blocking can be prevented by the use of antiblocking agents, which are added to the composition that develops the surface layer of the film [32]. Commonly used antiblocking additives are silica, clay, limestone, talc, zeolites, glass, and synthetic waxes [33].

Organic antiblocking agents are also employed. For example, fatty acid amides are well-known additives that prevent unwanted adhesion in polymer networks. Unfortunately, the addition of such amides as antiblocking additives for polymer sheets, for example in glass laminates, adversely affects the optical characteristics of the polymer sheet, such as haze, transparency, and film clarity, as well as adhesion of the polymer sheet to glass. Consequently, the inclusion of fatty acid amide moieties between polymer sheet interlayers of glass laminates is essential for better adhesion.

Certain fatty acid amides, however, can be successfully used as antiblocking agents in polymer sheets without affecting or interfering with the optical properties of the polymer sheet or the adhesive properties of the polymer sheet to glass. These amides include erucamide, behenamide (docosanamide), *N*-oleyl palmitamide, stearyl erucamide, erucyl stearamide, hydroxystearamide, oleic acid diethanolamide, stearic acid diethanolamide, poly(ethylene glycol) oleic amide, and mixtures of these amides. Secondary mono-amides are particularly preferred, such as *N*-oleyl palmitamide, which is an amide with a double-bond geometry [34].

Some of the organic antiblocking agents that fall under this category are shown in Figure 2.2.3.

2.2.2.6 Blowing Agents (Foaming Agents)

Blowing agents, which are also called *foaming agents*, are a class of additives that decompose to form a gas that will expand a polymer's cellular structure during

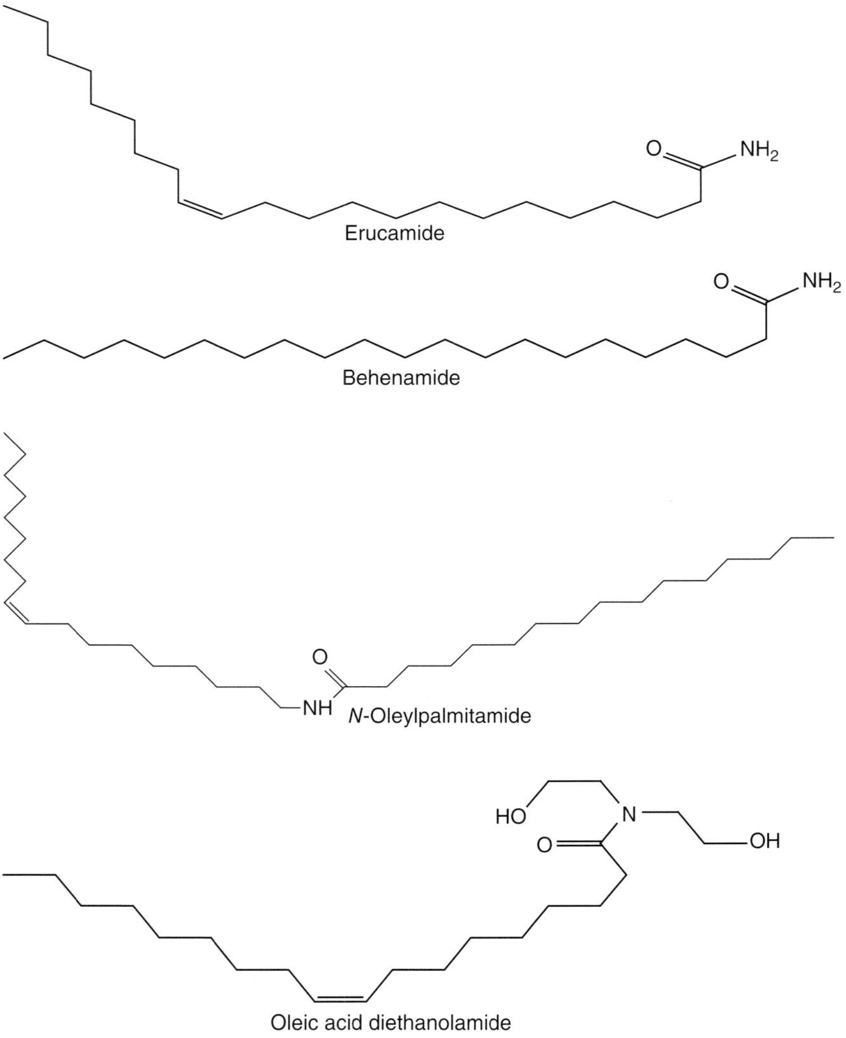

Figure 2.2.3 Structure of antiblocking agents.

the recycling or processing stage. In other words, these additives cause the formation of cells or voids by releasing gas. Thus they can be used to create foams and expandable materials that are light weight and can effectively provide thermal and/or shock protection. A common example is expanded polystyrene (EPS), which is widely used in packaging applications.

In this process, polystyrene is combined with a blowing agent in a mold and then heated so that the blowing agent is activated by temperature and causes the expansion of polystyrene [35]. EPS has a limitation in that some foams use chlorofluorocarbons (CFCs), which are harmful greenhouse gases and can cause depletion of the ozone layer. Murphy demonstrated an EPS that does not contain CFCs, which is made by using pentane gas as the blowing agent [36].

2.2.2.6.1 Mechanism of Blowing Action

Bates showed that the foaming action of these additives can be described as a four-step process described as follows:

- In the first step, the foaming agent must be homogeneously dispersed in the polymer during the liquid or melt state (it may also be incorporated into the chemical structure of the polymer).
- In second stage, the foaming agent releases the gas, leading to the formation of many individual small cells in the polymer system.
- In third step, the individual cells are allowed to expand into larger cells.
- Finally, the polymer is stabilized by cooling, crosslinking, or by increasing the viscosity to stop the cell growth.

Blowing agents are classified into physical blowing agents, (which include aliphatic hydrocarbons like *n*-pentane cyclopentane, *n*-hexane, cyclohexane, *n*-heptane, toluene, dichloromethane, etc.), which can be incorporated already at the stage of polymerization, thereby causing the blowing agent to be entrapped in the polymerized bead, and chemical blowing agents (e.g., azodicarbonamide(1,l′-azobisformamide),*N*,*N*′-dinitrosopentamethylenetetramine, *p*-toluenesulfonyl hydrazide, 4,4′-oxybis(benzenesulfonyl hydrazide)), which act by a chemical decomposition reaction producing volatile gases that cause blowing of the polymer matrix in which it is entrapped [22].

2.2.2.7 Antifogging Agents

These additives are employed to reduce or prevent the fogging associated with the polymer resins, particularly used as agricultural films for greenhouse culture or tunnel culture. Most of the polymer formulations used for this purpose include soft ethylene resin films, which are about 20–250 μm thick and comprise, as a base resin, PVC, branched LDPE, EVA copolymers, LLDPE, and so on.

The most essential properties required for a polymer material to be used as agricultural films are weather resistance, antifogging properties, heat-retaining properties, and transparency. The atmosphere in greenhouses or tunnels enveloped in agricultural film is saturated with water vapor that evaporates from the soil or plants, and this water vapor condenses on the inner surface of a cold film, causing fogging of the polymer material. Water droplets on the film can largely reduce the incident sunlight due to irregular reflection, and they also adversely affects plant health since the droplets fall on the plants, resulting in frequent occurrence of plant diseases [22].

Antifogging additives are incorporated into polymer films to overcome these problems. The modified plastic films do not prevent condensation but they work in an interesting manner. When water vapor condenses on such films, the antifogging additives migrate to the surface of the film and cause the condensation to spread evenly over the surface of the film, resulting in run-off instead of forming droplets on the surface, by creating a large surface area.

Antifogging additives can be incorporated into the polymer matrix in their pure form or as masterbatches or concentrates. Typical antifogging additive concentration ranges between 1% and 3%. They have the property of migrating over the surface of the film. In a monolayer film, the antifogging additive migrates in both directions – toward the inside of the agricultural film where the antifogging effect is desirable, but also to the outside of the film where it is unnecessary. On the outside of the polymer film, there is a possibility of "leaching" of the substance by rain water [3]. Common antifogging additives are glycerol monooleate, poly(glycerol ester)s, sorbitan esters, ethoxylated sorbitan esters, nonyl phenol ethoxylate, and ethoxylated alcohols [37]. Antifogging additives can be applied to the surface by coating. Coating of surfactant molecules may form a weak attachment to polymeric films or foils, particularly with polyethylene films, which is an undesirable property since they can be washed away by the action of heat and humidity. The typical structure of an ethoxylated sorbitan ester is shown in Figure 2.2.4 [22].

Antifogging agents are usually incorporated into the films with substances like non-ionic, anionic, and cationic surface active agents to enhance and stabilize their activity. Studies show that other methods for providing antifogging properties to agricultural films, apart from the conventional coating and incorporation method, include chemical modification of the ethylene-based resin or the ethylene resin film surface by introducing a polar group, such as a hydrophilic

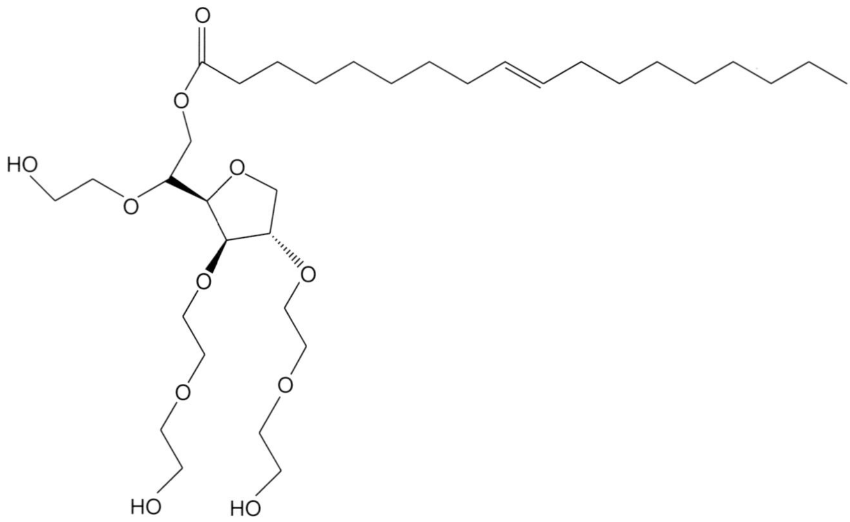

Figure 2.2.4 Structure of ethoxylated sorbitan ester (antifogging additive).

group. However, this technique is not of high interest because of its high economic demand at the present time and because of technical difficulties [22].

2.2.3 New Trends in Additives Technology

Additives can be considered as a tool for opening new markets and enhancing the value of existing polymers and their applications. By convention, chemical additives rely upon plastics for their existence. Thus the different factors that affect polymer consumption are important considerations for the additive industry. Recently, new stages in the life cycles of both plastics and additives have moved in a new direction, resulting in important changes. Even though rapid growth is over, polymer additives still present an “above-average” growth potential, despite intensified global competition. Investments in this field are shifting from new products to process development. Only a few new products are being introduced, and they are generally more of an evolutionary than a revolutionary character. The major factor in this field is that the use of additives can provide a cheap and effective solution for new material development by overcoming drawbacks. Although the rate of increase of the number of additives is slow, the number of possible formulations of existing materials is enormous, which thus can lead to new material discovery.

Other possible driving force for additive development is problems or drawbacks with existing additives and the need to find a complementary use for a side product or expand the applicability of a product. Decreased volatility, enhanced heat stability, and rising environmental considerations during processing and use are the other targets of new product formulations. These combined efforts will result in providing many new products. Only large producers can afford to invest in research and development associated with the testing and processing of new additives.

The competition between materials will considerably affect the additives industry. For example, if metallocene polyolefins displace the PVC market, then the demand for heat stabilizers and plasticizers might decrease, while the need for products such as fluoropolymer processing aids, antiblocks, and slip additives might increase. At present, PVC as the main user of additives is under threat as never before. However, “green” PVC, containing more environmentally friendly flame retardants (FRs) and plasticizers as well as non-heavy-metal stabilizers, is now becoming acceptable. Thus it is clear that the additives industry exists in an atmosphere of rapid change, competitive pressures, and continuous developments in technology.

The various opportunities and threats associated with technical factors regarding additives can be summarized as follows [38]:

Technical opportunities

- Better dosage of traditional grades (pellets vs powder)
- One-pack systems

- Better specifications
- Reduction in concentration variability
- New product developments
- Additives with new functions (e.g., UVAs as FRs or FRs as thermal stabilizers)
- Additives operating at an earlier stage during polymer degradation
- Multifunctional formulations
- Increased interaction between the additive and polymer
- Functionalization of polymer backbone (anchoring)
- Polymeric additives, polymer-bound (grafted) additives
- Environmentally friendly additives
- Plastic waste management
- Nanocomposites (for polar thermoplastics)
- Reactive additives
- Intelligent additives (go where action is needed)
- Surface HALS (hindered amine light stabilizer)
- Avoidance of antagonism between additives
- Patent expirations
- Lower analytical costs.

Technical threats

- Pressure on brominated FRs (Europe)
- Miniaturization
- New polymer developments
- e-Beam processing.

Recently, a significant development has occurred involving the addition of fragrance to masterbatch technology. Fragrance encapsulated in a polyurethane microcapsule system can be applied to a fabric by standard textile finishing processes. Target markets for these formulations include automotive interiors and packaging for health and beauty aids and cosmetics. Hyperbranched polymers, as multifunctional nanocarriers, may be used for performing boosting applications in paper-coating formulations and textile dyes [39, 40]. In the scientific concepts of new materials, the role of additives is high (e.g., in thermoplastic elastomers).

The trend of introducing new classes of additives continues toward making use of synthetic or vegetable-oil-based additive materials rather than animal-fat-derived additives, because of the factors related to environmental concerns. There is a rising trend in additives technology to develop combinations particularly designed for a specific application, in order to solve narrow focused problems. Additives for plastics are being used more widely. There are greater selections among existing alternatives, and also there are more ways to extend the functionalities of additives beyond their traditional uses. Additives are increasingly expected to expand their applicability; for example, they should not only provide a protective barrier for the packaged product but must also be involved in the total design of both the package and its contents [38].

As described previously, it is not reasonable to expect that new polymer innovations can easily avoid the use of additives. In this respect, e-beam processing of plastics constitutes an alternative to chemical additives, but studies indicate that this approach is very limited in scope [41].

The current status of various classes of additives can be summarized as follows [38]:

Impact Modifiers
- *Products.* Acrylic, EPDM/EPR (EPDM, ethylene propylene diene monomer rubber and EPR, ethylene-propylene rubber), MBS/ABS/MABS (MBS, methacrylate/butadiene/styrene and ABS, acrylonitrile/butadiene/styrene)
- *Driving forces.* Growth in PVC consumption (construction applications)
- *Threats.* Negative sentiments of PVC.

Organic peroxides
- *Products.* Wide variety
- *Driving forces.* Growth of resin production (PE, PVC, thermosets)
- *Threats.* New technology (metallocenes) displacing (reducing) needs.

Plasticizers
- *Products.* Mainly phthalates and trimellitates
- *Driving forces.* Growth in PVC consumption
- *Threats.* Inter-polymer competition (PVC replacement).

Antioxidants (AOs)
- *Products.* Phenolic primary AOs, organophosphite secondary AOs
- *Driving forces.* Global production of polyolefins
- *Threats.* None.

Light stabilizers
- *Products.* Hindered amine technology, benzotriazoles, benzophenones
- *Driving forces.* Consumer demand; most resin systems
- *Threats.* Reducing volume, surface coating, higher performing resins.

Heat stabilizers
- *Products.* Organotin, mixed metal, lead-based
- *Driving forces.* Growth in PVC consumption (construction applications)
- *Threats.* PVC's image.

Flame retardants
- *Products.* Brominated and phosphorous-based products, specialty inorganics
- *Driving forces.* Demands for safer end use
- *Threats.* Governmental regulations; shift to alternative chemistries or resins.

Let us now take a look at the advances in various classes of recycling additives.

2.2.3.1 Advances in Stabilizers

Advance in stabilizer technology has been quite significant in recent decades. New innovations aims at improvements and modifications such as better maintenance of melt-flow rates, lower initial color and better color maintenance, higher performance at the processing temperature, enhanced long-term thermal and light stability, inhibition of gas fade discoloration, enhanced additive compatibility, resistance to interactions with other additives, reduced taste and odor, suppression of visual imperfections (gels), more product forms, and better price/performance ratio [42].

Most of the development in the additive field in the last 20 years came from the evolution of the so-called secondary structure of additives, that is, development of stabilizers with higherMW (molecular weight), better compatibility, lower volatility, and so on, that is, all the properties directly connected with the physical aspects of polymer stabilization [38].

2.2.3.2 Advances in Flame Retardants (FRs)

Flame retardance is another active field of polymer product innovation. Some 150–200 FRs cover most of the requirements of the market (1997 value: US$ 2.2 billion or 14% of the plastics additives market). The markets for FRs are growing relative to average industrial markets, with mineral FRs growing twice as fast as other FRs. Because the wide range of applications, the types of FRs are also large; end-product specifications are the major driving force behind these innovations. The requirements of fiber manufacturers are particularly demanding [43], now being met by Reoflam® FG-372 (FMC). Other industry drivers here are the gradual phasing out of brominated materials and increased demand for halogen-free systems. In addition, stricter fire regulations require systems that offer slower heat release rate in fires, along with decreased smoke generation, and low FR additive toxicity, smoke toxicity, and corrosion. In some countries, brominated materials are still being widely used in TV manufacturing. Literature shows that new brominated FRs continue to be introduced, for example, tris(tribromophenyl)cyanurate (FR-245) for styrene copolymers and tris(tribromoneopentyl)phosphate (Dead Sea Bromine) aimed at use with PP, and the brominated polystyrene Saytex HP-7010 (Albemarle) for PBT, PET, and PAs [44].

The threatened ban on halogen FRs has not been effected in North America. Still, non-halogens are making inroads into selected areas [45]. In Europe, non-halogen FRs are thriving, and halogen-based additives are under increasing restrictions because of environmental perceptions, corrosiveness, and toxicity of the smoke produced. The technological challenge is to achieve acceptable fire retardance with non-halogen materials without negatively affecting the mechanical and physical properties of the compound. Studies indicate that the search for non-halogen FRs has been successful in several polymer systems, for example,

Clariant's APP-based Exolit AP and Bayer's DMPP-based Levagard VP SP 51009 FRs. The serendipitously discovered novel non-heavy-metal, non-halogenated FR synergists are based on *N*-alkoxyamine (NOR) chemistry [46, 47].

Other recently reported systems are the non-halogen phosphorous FR FP-500 (Amfine) for PC/ABS (polycarbonate/acrylonitrile/butadiene/styrene) and PPO blends, the reactive phosphite Struktol PD3710 (Schill and Seilacher) for epoxies [48], and non-halogen, non-phosphorous FRs for the cable industry. Mechanistic studies on polymer combustion and fire retardance are necessary to develop new nontoxic and environmentally friendly FRs (e.g., intumescent systems) with high effectiveness. On the other hand, high-throughput approaches have been proposed [49]. Halogenated material formulations, usually in combination with antimony oxide, work by quenching radical reactions in the gaseous phase. Hydrated fillers, like $Al(OH)_3$ and $Mg(OH)_2$, function by simply releasing water to the system on heating. Intumescent FR systems (usually phosphate-based), which are particularly suited to olefinic polymers, function by creating an expanded char layer on the surface of the polymeric material exposed to heat, thereby retarding the flame. Examples of these classes include halogen-free FR additives based on phosphorous, which function by developing a protective char.

Recently there have been new developments in the class of FR additives that create environmentally safe fire-proofing agents, which are focused on (in)organic Si compounds, nanocomposites, borates, hydroxides, and melamine derivatives. Development of FRs is moving toward synergistic systems in which several means of retardance are used, for example, combinations of phosphorous with halogens. There are one-pack FR systems, which are now available on the additives market (e.g., Fyrebloc, GLCC). Also in this field, inter-additive competition (e.g., between phosphorous, nitrogen, silicon, and metal oxide/hydroxide FR systems) is both an opportunity and a threat to the FR additive classes.

2.2.3.3 Advances in Plasticizers

For additives functioning as plasticizers, migration is a main point of focus, with the result that certain phthalate types have been gradually phased out and replaced by high-MW polymeric types, such as ethylene copolymers. Mesamoll (Bayer) is a phthalate-free plasticizer. PVC is being reformulated from DOP, Pb, and Sb_2O_3 to no DOP, no Pb, ATH/MDH, and low Sb_2O_3 [50]. Although the literature indicates the evaluation of novel materials as external plasticizers for PVC, only few of these are likely to be commercialized. The major drawbacks of these are the costs of the obligatory tests for HSE effects, which now constitute a major development cost.

2.2.3.4 Advances in Coloring Agents

Coloring agents are a class of additives that go through great advancements in their functionalities. Studies indicate that replacement of heavy-metal-based pigments is really possible via organic pigments for the coloration of polyolefins [51]. Some recent advancements are based on high-dispersibility quinacridone, diketopyrrolopyrrole (DPP), anthraquinone, isoindolinone, benzimidazolone, and so on, and liquid color concentrates. Recently introduced new pigments are Ciba Irgacolor Yellow 2 GTF and Ciba Cromophtal Yellow 3 RLR. Typical inorganic powder pigments are, for example, bismuth vanadate, Cr/Sb/Ti oxide, Mn/Sb/Ti oxide, Cr/Fe/Zn oxide, Co/Al oxide, and others [38].

2.2.3.5
Advances in Fillers

In the design of fillers, the control of particle size and particle-size distribution (from micro to meso and nano dimension) are the key factors that are increasingly exploited for the preparation of novel polymer composites. The driving force for this approach includes increase in interface area, and hence the volume fraction of the polymer–filler interfacial zone, and better homogeneity of the particle distribution. Other advancements consist of *in situ* generation of filler particles by the hydrolysis of inorganic/organic hybrid precursors [52, 53], surface modification, encapsulation, multifunctional fillers, and improved compounding technology. Current novel material classes of research interest include engineered mineral fibers, graphite nanotubes/nanoplatelets/nanofibers, and conductive fillers.

2.2.3.6
Advances in Other Additive Classes

Technological advances are also being reported for other additive materials. These are concerned with peroxide initiators for unsaturated polyesters [54]. Also, chlorinated fluorocarbons have been virtually removed from mold release agents and rigid polyurethane insulation foams. Fluoropolymer-based processing aids (such as vinylidene/hexafluoropropylene copolymer with/without PE oxide) have been introduced [55]. Some recent developments also include a permanent (nonmigratory) antistatic agent (Irgastat P22) and very efficient antifog additives (Atmer 651, Atmer 691) [38]. Polymeric antistatic agents and their utilization have been reviewed recently [56]. Other additives are also undergoing significant development. Millad 3988 (Milliken Chemical) is a new nucleating and clarifying agent for PP. Recently, new and very effective biocides (e.g., Irgosan PA) were also introduced. New dispersion aids for pigment concentrates (Ceridust waxes, Clariant) have been commercialized [57].

The use of co-polymeric system is essentially a new concept that is free from low-MW additive materials. However, a random copolymer, which includes additive functions in the chain of their polymeric system, usually results in a relatively costly solution; yet industrial examples have been reported (Borealis, Union Carbide) [38]. Locking an FR functionality into the polymer backbone will result in preventing migration. Organophosphorus functionalities have been incorporated in PA backbones to modify their thermal behavior [58].

2.2.3.7
Multifunctional Additives

Multifunctional additives have been introduced recently and have attained significant interest, like Nylostab S-EED (melt, light, and long-term heat stabilizer, fiber dyeability), which is used in PA fiber manufacture, and Lowinox MD 24 (metal deactivator and antioxidant). Liu *et al.* [59] have reported a series of bifunctional stabilizers (UV-HALS, UV-AO, and HALS-AO) containing *ortho*-hydroxy benzophenone, hindered amine, or hindered phenol groups, which are effective in protecting PP against photoxidation and thermal oxidation. Hindered amine thermal stabilizers (HATSs), such as Hostavin N-30, are other examples of additives with multifunctional properties which function as both as antioxidants and UV stabilizers. Various other polyfunctional stabilizers for polymers have been mentioned in the literature [60, 61].

Multifunctional additives in plastic recycling have been reviewed [62]. The different functionalities can be listed as follows:

Flame retardant – smoke suppressor
Flame retardant – plasticizer
Flame retardant – lubricant
Flame retardant – thermal stabilizer
Flame retardant – thermal stabilizer – light stabilizer
Plasticizer – thermal stabilizer
UV absorber – thermal stabilizer
UV absorber – antioxidant
UV stabilizer – antioxidant – processing aid
UV stabilizer – flame retardant – smoke suppressor
Antioxidant – thermal stabilizer
Antioxidant – thermal stabilizer – thickening agent
Antioxidant – plasticizer
Melt stabilizer – light stabilizer
Nucleating agent – optical whitener
Slip agent – antiblocking agent
Pigment – filler
Filler – nucleating agent
Metal deactivator – antioxidant flame retardant
Antiplasticizer – processing aid
Compatibilizer – carrier resin.

2.2.4
Conclusion

As this is an era of rapidly changing social attitudes toward the sustainability of the environment and natural resources and also a time period of health and safety concerns, there is an increased demand for "safer" plastics and "more friendly" additives in ecological and environmental aspects. Polymer recycling itself is an innovation, which is a result of environmental concern and health factors. The characteristics of additive materials used in polymer recycling processes must have to satisfy these features of hygienic and eco-toxicological concerns to avoid risk factors. The analysis of additives is thus an important field that helps to assess the nature of additive materials in polymeric systems and thus to monitor the impact of additives in the environment. Processing of additives mainly aims at the application potential of polymer–additive formulations, which includes the performance of polymer materials incorporated with additives. Thus the innovations and developments in the field of eco-friendly additives in twenty-first century helps in balancing the environmental interactions of the polymeric system with enhancements in the performance of the system, thereby fulfilling the ultimate target of polymer recycling.

References

1. Mehta, A.K., Datta, S., Li, W., and Iyer, S.S. (2009) Heterogeneous polymer blends and molded articles therefrom. US Patent 7 476 710, assigned to ExxonMobil Chemical Patents Inc. (Houston, TX), Jan. 13, 2009.
2. Gahleitner, M. and Wolfschwenger, J. (2001) in *Encyclopedia of Materials: Science and Technology* (eds K.H.J. Buschow, R. Cahn, M.C. Flemings, B. Ilschner, E.J. Kramer, S. Mahajan, and P. Veyssiere), Elsevier, Oxford, pp. 7239–7244.
3. Piccinelli, P., Vitali, M., Landuzzi, A., Da Roit, G., Carrozza, P., Grob, M., and Lelli, N. (2008) Permanent surface modifiers. US Patent 7 408 077, assigned to Ciba Specialty Chemicals Corp. (Tarrytown, NY).
4. Fanegas, N., Gbmez, M., Marco, C., Jimhez, I., and Ellis, G. (2007) Influence of a nucleating agent on the crystallization behaviour of isotactic polypropylene and elastomer blends. *Polymer*, **48** (18), 5324–5331.
5. Burmaster, D., Hodges, O., Lumus, J.L., Kelly, L.A., and Murphy, M. (2007) Polypropylene having improved clarity and articles prepared there from. US Patent 7 241 850, assigned to Fina Technology, Inc. (Houston, TX).
6. ASTM (2009) Standard test method for haze and luminous transmittance of transparent plastics, in *ASTM Standard, Book of Standards*, Vol. 08.01 ASTM D1003-07e1, ASTM International, West Conshohocken, PA.
7. Tobita, E., Nomura, K., and Kawamoto, N. (2008) Nucleating agent composition and crystalline polymer compositions containing the same. US Patent 7 442 735, assigned to Asahi Denka Co., Ltd. (Tokyo, JP).
8. Yi, Q.F., Wen, X.J., Dong, J.Y., and Han, C.C. (2008) A novel effective way of comprising a [beta]-nucleating agent in isotactic polypropylene (i-pp): polymerized dispersion and polymer characterization. *Polymer*, **49** (23), 5053–5063.
9. Zhao, S., Cai, Z., and Xin, Z. (2008) A highly active novel p-nucleating agent for isotactic polypropylene. *Polymer*, **49** (11), 2745–2754.
10. SpecialChem S.A. (2009) Nucleation Solutions to Improve Polyolefins Performance and Processing, http://www.specialchem4polymers.com/tc/Polyolefin-Nucleators/2009 (accessed 11 May 2016).
11. Presenz, U., Schmid, S., Hartmann, R., and Luck, H.R. (2005) Method for producing a polyamide compound. US Patent 6 881 477, assigned to EMS-Chemie AG (Domat/Ems, CH).

12. Wang, K., Zhou, C., Tang, C., Zhang, Q., Du, R., Fu, Q., and Li, L. (2009) Rheologically determined negative influence of increasing nucleating agent content on the crystallization of isotactic polypropylene. *Polymer*, **50** (2), 696–706.
13. Marco, C., Gomez, M.A., Ellis, G., and Arribas, J.M. (2002) Highly efficient nucleating additive for isotactic polypropylene studied by differential scanning calorimetry. *J. Appl. Polym. Sci.*, **84** (9), 1669–1679.
14. Zhang, Y.F. (2008) Comparison of nucleation effects of organic phosphorous and sorbitol derivative nucleating agents in isotactic polypropylene. *J. Macromol. Sci. Part B Phys.*, **47** (6), 1188–1196.
15. Mathur, N.C., Snow, M.S., Young, K.M., and Pincock, J.A. (1985) Substituent effects on the rate of carbene formation by the pyrolysis of rigid aryl substituted diazomethanes. *Tetrahedron*, **41** (8), 1509–1516.
16. Abramovitch, R.A., Chellathurai, T., Holcomb, W.D., McMaster, I.T., and Vanderpool, D.P. (1977) Intramolecular insertion of arylsulfonylnitrenes into aliphatic side chains. *J. Org. Chem.*, **42** (17), 2920–2926.
17. Stevens, J.C., Vanderlende, D.D., and Ansems, P. (2007) Crystallization of polypropylene using a semi-crystalline, branched or coupled nucleating agent. US Patent 7 250 470, assigned to Dow Global Technologies Inc. (Midland, MI).
18. Tanizaki, T., Nakahara, T., and Kubo, M. (2001) Poly (4-methyl-l-pentene) resin laminates and uses thereof. US Patent 6 265 083, assigned to Mitsui Chemicals, Inc. (Tokyo, JP).
19. Wong, E.W., Sheehan, P.E., and Lieber, C.M. (1997) Nanobeam mechanics: elasticity, strength, and toughness of nanorods and nanotubes. *Science*, **277** (5334), 1971–1975.
20. Acquarulo, L.A. Jr., O'Neil, C., and Xu, J. (2006) Optimizing nano-filler performance in polymers. US Patent 7 034 071, assigned to Foster Corporation (Putnam, CT).
21. Thomas, S. and Stephen, R. (eds) (2010) *Rubber Nanocomposites: Preparation, Properties and Applications*, Wiley-VCH Verlag GmbH.
22. Karl Fink, J. (ed.) (2010) *A Concise Introduction to Additives for Thermoplastic Polymers*, Wiley-VCH Verlag GmbH.
23. Dalton, J.S.N., Stafford, S.L., and Hewa, J.D. (1999) Polyester and optical brightener blend having improved properties. US Patent 5 985 389, assigned to Eastman Chemical Company (Kingsport, TN).
24. Chisholm, P.S., Tikuisis, T., Goyal, S.K., Checknita, D., and Bohnet, N.K.K. (1998) Melt fracture reduction. US Patent 5 854 352, assigned to Nova Chemical Ltd. (Calgary, CA).
25. May, D.D. and Zimmermann, B. (2008) Titanium dioxide-containing polymers and films with reduced melt fracture. US Patent 7 338 995, assigned to E.I. du Pont de Nemours and Company (Wilmington, DE).
26. Su, T.K. (1986) Method and apparatus producing decorative melt fracture patterns on polymeric products. US Patent 4 615 858, assigned to Mobil Oil Corporation (New York, NY), Oct. 7, 1986.
27. Blatz, P.S. (1964) Extrudable composition consisting of a polyolefin and a fluorocarbon polymer. US Patent 3 125 547, assigned to E.I. DuPont de Nemours and Co.
28. Chu, S.C. and Shaw, R.G. (1988) Compositions of linear polyethylene and polyvinylidene fluoride for film extrusion, and films thereof. US Patent 4 753 995, assigned to Mobil Oil Corporation (New York).
29. Dillon, M.P., Woods, S.S., Fronek, K.J., Lavallee, C., Amos, S.E., Weilandt, K.D., Kaspar, H., Hirsch, B., Hintzer, K., and Scott, P.J. (2001) Polymer processing additive containing a multimodal fluoropolymer and melt processable thermoplastic polymer composition employing the same. US Patent 6 277 919, assigned to Dyneon LLC (Oakdale, MN).
30. Heitz, W.D. (1990) Paraloid extrusion aids for high molecular weight hdpe film resins. US Patent 4 963 622, assigned to Union Carbide

Chemicals and Plastics Company Inc. (Danbury, CT).
31. Cobb, D.C. and Mercer, J.W. Jr., (2000) Process for making thermoplastic profiles having reduced melt fracture. US Patent 6 100 320, assigned to Eastman Chemical Company (Kingsport, TN).
32. Cretekos, G.F. and Wagner, J.R. Jr., (2002) Block-resistant film. US Patent 6 472 077, assigned to Exxon Mobil Oil Corporation (Fairfax, VA).
33. Keung, J.K., Donovan, K.M., and Balloni, R. (1987) Heat sealable film and method for its preparation. US Patent 4 692 379, assigned to Mobil Oil Corporation (New York).
34. Chen, W., Smith, A.N., and Karagiannis, A. (2009) Poly(vinyl butyral) pellets. US Patent 7 491 761, assigned to Solutia Incorporated (St. Louis, MO).
35. Soroka, W. (2002) *Fundamentals of Packaging Technology*, 3rd edn, Institute of Packaging Professionals, Naperville, IL.
36. Murphy, J. (2001) *Additives for Plastics*, 2nd edn, Elsevier Science Inc., New York.
37. Sudo, K., Mori, S., Okada, T., Ikeno, H., Shichijo, K., and Ohnishi, S. (1993) Agricultural film. US Patent 5 262 233, assigned to Mitsubishi Petrochemical Co., Ltd. (Tokyo, JP).
38. Bart, J.C.J. (2005) *Additives in Polymers: Industrial Analysis and Applications*, John Wiley & Sons, Ltd.
39. Bruens, C., Nieland, R., and Stanssens, D. (1999) New performance additives from DSM. *Polym. Polym. Compos.*, **7** (8), 581–587.
40. DSM Venturing & Business Development (2002) Profile Sheet 7, Hybrane ®, DSM, Heerlen.
41. Stern, M. (2000) Proceedings SPE ANTEC 2000, Orlando, FL, pp. 1772–1776.
42. Voigt, W. and Todesco, R. (2002) New approaches to the melt stabilization of polyolefins. *Polym. Degrad. Stab.*, **77**, 397–402.
43. Buszard, D.L. (1998) Proceedings Polypropylene '98, Zurich.
44. De Schrijver, D. (2003) in *Proceedings 3rd European Additives and Colors Conference* (ed. J.M. Líegois), SPE Benelux, Antwerp, pp. 46–51.
45. Wigotsky, V. (2001) Flame retardants. *Plast. Eng.*, **57** (2), 22–33.
46. Srinivasan, R., Gupta, A., and Horsey, D. (1998) Proceedings International Conference on Additives in Polyolefins, Houston, TX, pp. 69–83.
47. Troutman, M.V. (2002) Proceedings 11th International Conference on Additives, Clearwater Beach, FL.
48. Sprenger, S. and Utz, R. (1998) Proceedings AddCon World '98, London, Paper 18.
49. Nyden, R.M. and Gilman, J.W. (2000) *Recent Adv. Flame Retard. Polym. Mater.*, **11**, 19–24.
50. Schmidt, R. (1998) Proceedings AddCon World '98, London, Paper 17.
51. Ansias, M.P. (1999) Sostituzione di pigmenti a base dimetallo pesante nella colorazione delle poliolefine. *Pitturee Vernici, Eur. Coat.*, **75** (12–13), 39–41.
52. van Damme, H. (2001) Preprints Euro-Fillers'01, Łodź, Poland, Paper L-01.
53. Soma, I. (1998) *Purasuchikkusu*, **49** (10), 20–27.
54. Maltha, A. (1998) Proceedings Organic Peroxides Symposium (AKZO Nobel), Apeldoorn, Paper 5.
55. Amos, S.E. (1997) Proceedings Polyolefins X, Houston, TX, pp. 133–143.
56. Sumi, H. (1998) *Purasuchikkusu*, **49** (10), 34–40.
57. Sale, N. (2003) in *Proceedings 3rd European Additives and Colors Conference* (ed. J.M. Líegois), SPE Benelux, Antwerp, pp. 117–123.
58. Wheeler, J.W., Zhang, Y., and Tebby, J.C. (1998) *Spec. Publ. R. Soc. Chem.*, **224**, 252–265.
59. Liu, N., Pan, J.-Q., and Lau, W.W.Y. (1998) *Polym. Degrad. Stab.*, **62**, 307–314.
60. Minsker, K.S., Biglova, R.Z., Davydov, E.Y., and Zaikov, G.E. (1998) *Plast. Massy*, **4**, 25–27.
61. Habicher, W.D., Pawelke, B., Bauer, I., Yamaguchi, K., Kosa, C., Chmela, S., and Pospíšil, J. (2001) Synthesis and antioxidative properties of novel

multifunctional stabilizers. *J. Vinyl Add. Tech.*, 7 (1), 4–18.

62. Herbst, H. and Pfaendner, R. (1998) Proceedings ARC '98 (Annual Recycling Conference), SPE, Brookfield, CT, pp. 205–214.

63. Matsuo, S., Murakami, T., Bando, T., and Nagatoshi, K. (1989) Reinforced resinous composition comprising polycyano arylene ether. US Patent 4 812 507, assigned to Idemitsu Kosan Company Limited (Tokyo, JP).

3
Methods of Recycling

3.1
Methods of Recycling of Polymers: Addition Polymers

Beena Sethi

3.1.1
Introduction

It has been estimated that 100 million tons of plastics are produced worldwide each year [1]. The primary reasons for the popularity and widespread application of many polymers are their exceptionally high stability and durability [2, 3]. Widely used plastics do not naturally degrade to a large degree when released into the environment [2, 4, 5].

The dramatic increase in production and lack of biodegradability of commercial polymers, particularly commodity plastics used in packaging (e.g., fast food), industry, and agriculture, has focused public attention on a potentially huge environmental accumulation and pollution problem that could persist for centuries [6]. Several communities are now more sensitive to the impact of discarded plastic on the environment, including their deleterious effects on wildlife and on the aesthetic qualities of cities and forests because improperly disposed plastic materials persist in our environment and harm life. On the other side, the burning of polymers and plastics produces persistent organic pollutants (POPs) such as furans and dioxins [7].

There are four mechanisms by which plastics degrade in the environment: photodegradation, thermo-oxidative degradation, hydrolytic degradation, and biodegradation by microorganisms [8]. It is known that natural degradation of plastic begins with photodegradation because ultraviolet light from the sun provides the activation energy required to initiate the incorporation of oxygen atoms into the polymer, which leads to thermo-oxidative degradation [8, 9]. This causes the plastic to become brittle and to break into smaller and smaller pieces, until the polymer chains reach sufficiently low molecular weight to be metabolized by microorganisms [8]. These microbes either convert the carbon of the polymer chains to carbon dioxide, or incorporate it into biomolecules [2, 8].

Recycling of Polymers: Methods, Characterization and Applications, First Edition. Edited by Raju Francis.

However, this entire process is very slow, and it can take 50 or more years to fully degrade plastics [10].

A polymer is described as a large molecule built up from numerous smaller molecules known as *monomers*. A polymer may be linear, slightly branched, or highly interconnected. There are two main systems used to classify polymers: one based on their response to thermal treatment (thermoplastics and thermosets), and the other based on the nature of the chemical reactions employed in the polymerization (condensation and addition polymers) [11]. Addition polymerization involves the combination of simple molecules without the formation of any new products. In contrast, condensation polymerization is the combination of simple or dissimilar molecules, with the formation of byproducts such as water or ammonia [11].

The most useful decomposition products can be obtained by understanding the structures of the different types of polymers found in municipal solid waste (MSW), and their mechanisms of degradation can be altered by the presence of catalysts [11].

The current strategies to deal with solid plastic waste (around 62% of total available solid waste is collectable) are still based on landfilling, incineration without energy recovery, and recycling. Because of the production of a huge amount of plastic solid waste (PSW), many municipalities are facing disposal problems such as emission of toxic substances (dioxins and furanes) on incineration and shortage of landfill sites [12].

Although polymers are actually high-yielding energy sources, incineration has been widely accused as ecologically unacceptable owing to the health risk from airborne toxic substances, for example, dioxins (in the case of chlorine-containing polymers). In the last decade, many environmental regulations have been implemented for a more sustainable recycling-oriented society [13]. The objective of a plastic management policy, in accordance with the principles of sustainable development, should be not only the reuse of polymeric materials but also the production of raw materials (monomers), from which they could be reproduced, or other secondary valuable products (Figure 3.1.1) [14].

Nowadays, in the movement toward achieving zero discharge and sustainability, already commonplace terms such as *carbon-neutral*, *renewable resources*, and *circulative utilization* are being joined by other keywords from the field of polymer chemistry, such as *recycling*, *reversible*, *re-workable*, *de-crosslinking*, and *biomass-based*, all of which are indicative of a gradual but steady paradigm shift toward sustainable technology. In fact, to carry out the remit of the basic law, all plastics will ultimately be required to be recycled. As regards recyclability, most commodity plastics are relatively stable, making monomer recovery poor. There are, however, some plastics, the so-called recyclable plastics, possessing the important and superior attribute of being able to be converted into valuable raw chemicals. One special category of plastics, termed *depolymerizable plastics*, is capable of being converted back to the corresponding monomers. For example, poly(methyl methacrylate) (PMMA), poly(ethylene terephthalate) (PET), Nylon-6, and poly(lactic acid) (PLA) are well-known examples of depolymerizable plastics [12].

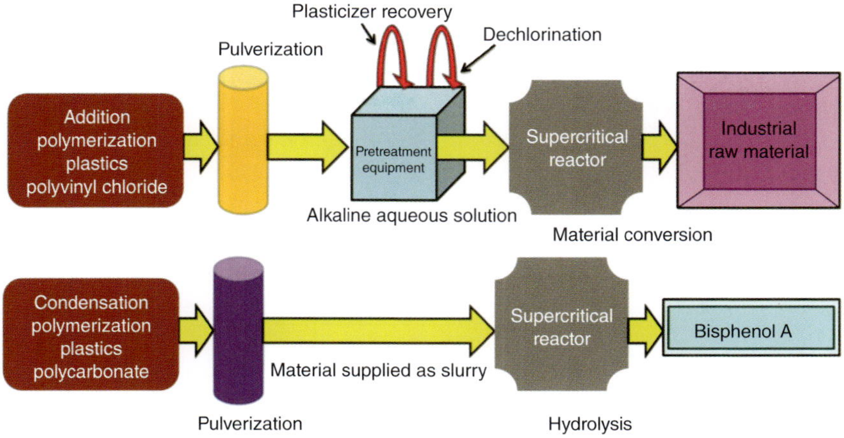

Figure 3.1.1 Recycling of polymers.

In this sense, the recycling of waste polymers through various technologies has been successfully demonstrated and continues to be developed [14]. Four main approaches have been proposed, as presented in Scheme 3.1.1 [15, 16].

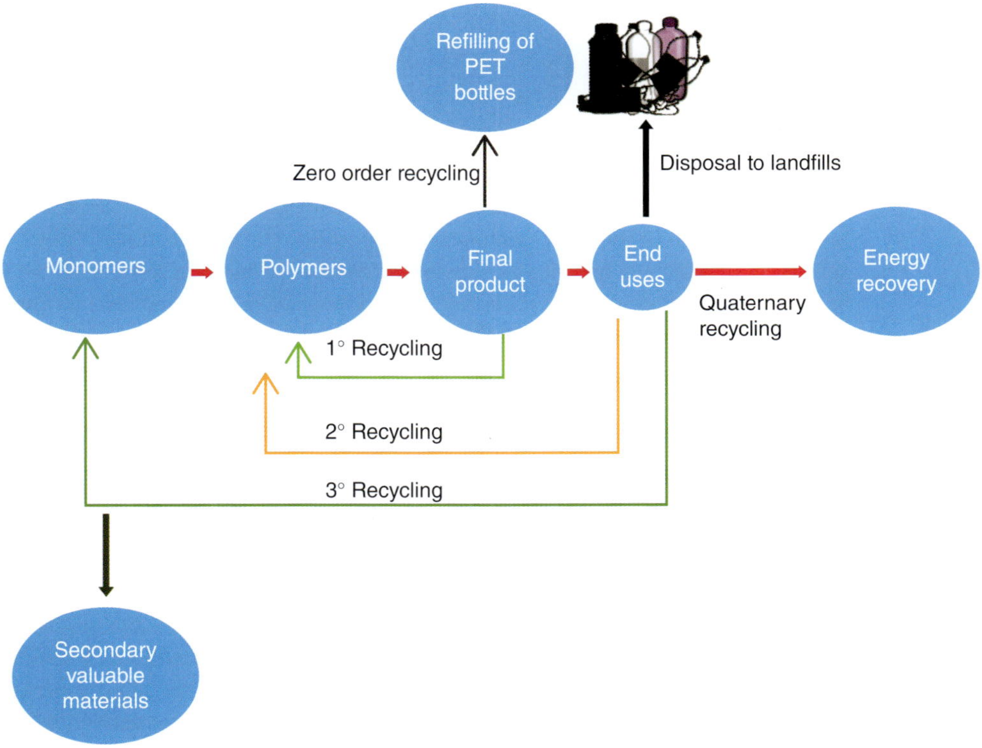

Scheme 3.1.1 Types of plastic waste recycling.

3.1.2
Primary Recycling

Primary recycling refers to the *in-plant* recycling of the scrap material of controlled history, that is, reuse of products in their original structure [17]. This process remains the most popular, as it ensures simplicity and low cost, dealing, however, only with the recycling of clean, uncontaminated, single-type waste [14]. However, this is possible for some wastes, and there is also an obvious limit on the number of cycles for each material.

3.1.3
Mechanical Recycling (or Secondary Recycling)

Mechanical recycling (or secondary recycling) is the reprocessing of materials of waste plastics by physical means, like cutting, shredding, washing, and so on, into plastic products. In this approach, the polymer is separated from its associated contaminants, and it can be readily reprocessed into granules by conventional melt filtration extrusion. Size of the waste plastic is reduced after it is sorted, cleaned, and dried, and then directly processed into end products or flakes of consistent quality, which can be further used for manufacturing other goods. The succeeding steps for recycling can vary from operation to operation and the end use [18]. The basic polymer is not altered during the process. Among the main issues of secondary recycling are the heterogeneity of the solid waste and the deterioration of product's properties in each cycle. This occurs because the molecular weight of the recycled resin is reduced because of chain-scission reactions caused by the presence of water and trace acidic impurities. Strategies for maintaining the polymer's average molecular weight during reprocessing include intensive drying, reprocessing with vacuum degassing, the use of chain extender compounds, and so on [14]. In mechanical recycling, only thermoplastics can be used because they can be re-melted and reprocessed into end products [18].

Thermoset plastic cannot be melted and reprocessed, so mechanical recycling is not suitable for recycling such plastics. Thermoset plastic can be processed through their use in cement kilns or tarring roads.

The steps involved in mechanical recycling are the following:

1. Collection/segregation
2. Cleaning and drying
3. Chipping/sizing
4. Coloring/agglomeration
5. Pelletization/extrusion
6. Manufacturing of the end product [18].

During the grinding or melting phases, the reprocessed material can be blended with virgin material to obtain superior results. Mechanical recycling requires homogenous plastics and relatively cleanmaterial. It is one of the traditional and most used methods in many countries for plastic recycling. Its cost is relatively low, but needs substantial initial investment [18].

3.1.4
Chemical or Feedstock Recycling (Tertiary Recycling)

Chemical or feedstock recycling (tertiary recycling) is defined as the process leading to the total depolymerization to the monomers, or partial depolymerization to oligomers and other chemical substances. The monomers could subsequently be re-polymerized to regenerate the original polymer [14]. In other words, chemical recycling is an advanced technology that transforms plastic materials into smaller molecules, mostly liquids or gases, suitable for use as feedstock material starting with monomers, oligomers, or mixtures of other hydrocarbon compounds, or as input for the production of new plastics and petrochemicals by means of heat or chemical agents [18]. The term *chemical* is used because a change occurs to the chemical structure of the polymer. The final output in chemical recycling is very useful as a fuel, which has been proved scientifically. It uses a technique called *depolymerization*, which can be very profitable and beneficial from the sustainability point of view [18].

Tertiary recycling is attracting much attention because of its possible use as an energy resource. Thermolysis is considered as one of the most promising processes of tertiary recycling, conducted at high temperature and in a non-oxidative environment [17]. Also, it upgrades waste plastics into valuable chemicals and into fuels of various states and purposes. Thermolysis is conducted with or without a catalyst. The advantage of the catalytic process is lower temperatures, faster degradation of the feedstock, and a narrower fractional composition of the products. The main issue with this is economic efficiency, because commercial catalysts generally are expensive, have a relatively short life cycle, and typically cannot be regenerated [19–28]. If the material used has energetic value, that is, high content of carbon and hydrogen and low content of heteroatoms, it becomes a valuable feedstock for the production of fuels. In this sense, petroleum is considered the most attractive and easily refined feedstock, as it gives conventional fuels such as automotive gasoline, jet fuel, and diesel. The products, however, have to meet a number of strict standards specified in EN 228, EN 590, and so on, [29–36]. Many methods for chemical recycling are available, such as direct chemical treatment involving gasification, melting in a blast furnace or coke oven, and degradation by liquefaction. For additional polymers, subcategories of chemical recycling are gasification, pyrolysis, liquid–gas hydrogenation, steam or catalytic cracking, and the use of PSW as reducing agent in blast furnaces. Addition polymers like polyolefins give a mixture containing numerous components to make use as fuels. Subcategories of chemical recycling for condensation polymers are hydrolysis (acidic, alkaline, and neutral), glycolysis, methanolysis, aminolysis, ammonolysis, and so on. Condensation polymers such as PET and Nylon undergo degradation to prepare monomer units, that is, feedstock or monomer recycling [18].

Tertiary recycling involves the following methods:

1. Hydrogenation
2. Gasification
3. Chemical de-polymerization
4. Thermal cracking
5. Catalytic cracking and reforming
6. Photodegradation
7. Ultrasound degradation
8. Degradation in microwave reactor.

Numerous methods of degradation to obtain petrochemicals are under investigation presently, and suitable conditions for gasification and pyrolysis are under extensive research.

3.1.5 Energy Recovery (Quaternary Recycling)

Energy recovery (quaternary recycling) refers to the recovery of plastics' energy content. This is an effective way to reduce the volume of organic materials. Incineration aiming at the recovery of energy is currently the most effective way to reduce the volume of organic materials. Though this method yields considerable energy from polymers, it is ecologically unacceptable owing to the health risk from airborne toxic substances, for example, dioxins (in the case of chlorine-containing polymers) [14]. Apart from the aforementioned methods, direct reuse of a plastic material (i.e., PET) could be considered as a *zero-order* recycling technique [37]. In many countries it is a common practice to refill and reuse PET bottles. However, this should be done with a great care since plastic bottles are more likely than glass to absorb contaminants, which could be released into the contents (especially food) when the bottle is refilled. Moreover, refilling a PET bottle with a drink of high alcoholic content may lead to the degradation of the macromolecular chains with unpredictable consequences.

A number of researchers have tested all the above-mentioned methods for recycling feedstock. All these methods have many disadvantages associated with them; for example, mechanical recycling is suitable only for thermoplastics, and small impurities change the properties and value of the polymer. Poorer quality polymer is obtained than the virgin polymer (already degraded polymer in waste). Mostly, recycled polymer is gray in color. Hydrogenation is associated with high cost of hydrogen gas, and high pressure of hydrogen is required during operation. Gasification involves the formation of highly viscous melted plastic along with corrosive compounds. Chemical de-polymerization also has disadvantages: it is difficult to apply to addition polymers, and a coupling of processes is required. In thermal cracking, a mixture of hydrocarbons is produced, and a low concentration of the original monomer is formed. During catalytic cracking and reforming, the catalyst gets deactivated by decomposed carbonaceous residue after some time because of poisoning of catalyst by chlorine and nitro compounds present in the waste plastic. The highly viscous melted plastic formed hinders heat flow [12].

Despite the great strides made in the effectiveness, efficiency, and precision of these polymers and technologies, further improvements are required to meet the practical requirements of a responsible and sustainable system for the recycling of containers, packaging materials, household electric appliances, and end-of-life vehicles, all of which are operated in compliance with the recycling laws of developed and developing countries [12]. So it can be concluded that, though there are a number of technologies available, there is a need to develop more economical, safe, eco-friendly, and sustainable methods [12].

In accordance with the principles of sustainable development (development that meets the needs of present generation without compromising the ability of future generations to meet their needs), the objective of a plastic management policy should not be only the reuse of polymeric materials but also the production of raw materials (monomers) from which they could be reproduced, or other secondary valuable products, which could be useful as feedstock for a variety of downstream industrial processes or as transportation fuel. In this sense, among the techniques proposed above for recycling of waste polymers, the most challenging method is chemical or feedstock recycling, and various technologies have been successfully demonstrated and continue to be developed for these purposes [38].

So far, there are three main methods for plastic waste handling:

1. Burying in landfills
2. Incineration
3. Recycling to a limited extent [39].

Each method has its own inherent limitations: for example, landfills take up space that can be utilized for more productive means such as agriculture [39]. This is compounded by the slow degradability of most plastics, which means that the occupied land is unavailable for long periods. Plastic components of landfill waste have been shown to persist for more than 20 years [40]. This is due to the limited availability of oxygen in landfills; the surrounding environment is essentially anaerobic [41, 42]. The limited degradation that is experienced by many plastics is largely due to thermo-oxidative degradation [8] and the anaerobic conditions in landfills only serve to further limit the degradation rates. Plastic debris in landfill also acts as a source of a number of secondary environmental pollutants [39]. Pollutants of note include volatile organics, such as benzene, toluene, xylene, ethyl benzenes, and trimethyl benzenes, which are released both as gases and contained in the leachate [43], and endocrine disrupting compounds, in particular bisphenol A (BPA) [44–46]. In addition to its endocrine disruption properties, BPA released from plastics in landfill has also been shown to lead to an increase in production of hydrogen sulfide by sulfate-reducing bacteria in the soil. High concentrations of hydrogen sulfide are potentially lethal [46].

Incineration of used plastic waste also has drawbacks [39]. Though plastic incineration overcomes some of the limitations placed on landfill, as it does not require any significant space and there is even the possibility of energy recovery in the form of heat [47], there is a significant trade-off in that incineration of plastics leads to the formation of numerous harmful compounds, most of which are released into the atmosphere [39]. Polyaromatic hydrocarbons (PAHs), PCBs, heavy metals, toxic carbon, and oxygen-based free radicals, not to mention significant quantities of greenhouse gases (GHGs), especially carbon dioxide, are all produced and released when plastics are incinerated [47–51]. The significant environmental drawbacks of plastic disposal via both landfill and incineration were the driving force behind the development of plastic recycling processes [52].

This chapter includes a critical review of chemical recycling of widely used polymers such as polypropylene (PP), polystyrene (PS), low-density polyethylene (LDPE), high-density polyethylene (HDPE), polyvinylchloride (PVC), and PMMA. Chapter 3.2 includes chemical recycling of condensational polymers such as PET, polycarbonate (PC), and nylon.

3.1.6 Chemical Recycling of Polyethylene (LDPE and HDPE)

3.1.6.1 Introduction

Advanced processes of chemical recycling of polyethylenes are similar to those employed in the petrochemical industry, for example, pyrolysis, gasification, liquid–gas hydrogenation, viscosity breaking, and steam or catalytic cracking [53]. Catalytic cracking and reforming facilitate the selective degradation of waste plastics. Solid catalysts such as silica alumina, ZSM-5 zeolites, and mesoporous materials effectively convert polyolefins into liquid fuel, giving lighter fractions as compared to thermal cracking [53].

PE is used as a potential feedstock for fuel (gasoline)-producing technologies. PE can be thermally cracked into gases, liquids, waxes, aromatics, and char. The relative amounts of gas and liquid fraction are very much dependent on the type of polymer used. Less crystalline or more branched polymers are less stable in thermal degradation, so higher decomposition is observed in PP, followed by LDPE and HDPE [54]. PE (as well as other vinyl polymers) degrade via a four-step free-radical mechanism: radical initiation, de-propagation (as opposed to propagation in the case of polymerization), intermolecular and intramolecular hydrogen transfer followed by β-scission (initial step in the chemistry of thermal cracking of hydrocarbons and the formation of free radicals), and, lastly, radical termination [53].

β-Scission and hydrogen abstraction steps often occur together in a chain propagation sequence. That is, a radical abstracts a hydrogen atom from the reactant to form a molecule and a new radical. A bond β is then broken at the radical center (β-scission) to regenerate an abstracting radical and to produce a molecule with a double bond (a molecule with a double bond involving the carbon atom that had been the radical center). Sample size and surface area-to-volume ratio of the melt have a significant influence on the rate and relative importance of the various mechanisms of polymer degradation [53].

Nowadays, interest is growing in developing value-added products such as synthetic lubricants via PE thermal degradation and in development of value-added recycling technologies, as they will increase the economic incentive to recycle polymers. Several methods for chemical recycling are presently in use, such as direct chemical treatment involving gasification, smelting in a blast furnace or coke oven, and degradation by liquefaction. The main advantage of chemical recycling is the possibility of treating heterogeneous and contaminated polymers with only limited pretreatment. As petrochemical plants are much larger in size (6–10 times) than plastic manufacturing plants, it is essential and advisable to utilize petrochemical plants in supplementing their usual feedstock by using PSW-derived feedstock [53].

3.1.6.2 Thermolysis Schemes and Technologies

3.1.6.2.1 Pyrolysis

Thermolysis is the treatment by heat under controlled temperatures without catalysts. Thermolysis processes can be divided into the following:

1. Advanced thermochemical treatment or pyrolysis (thermal cracking in an inert atmosphere)
2. Gasification (in substoichiometric presence of air, usually leading to CO and CO_2 production)
3. Hydrogenation (hydrocracking).

Thermal degradation processes yield a number of constituting molecules, combustible clean gases with high calorific value, and/or energy from a wide variety of solid hydrocarbon-based waste and biomass streams. The reduction of landfilling is an added advantage. The hydrocarbon content of the waste is converted into a gas, which is suitable for utilization in either gas engines, with associated electricity generation, or in boiler applications without the need for flue gas treatment. This gas will typically have a calorific value of 22–30 MJ m^{-3}, depending on the waste material being processed. Solid char is also produced from the process, which contains both carbon and the mineral content of the original feed material. The char can either be further processed onsite to release the energy content of the carbon, or utilized offsite in other thermal processes [53].

Pyrolysis at the microgram scale follows random initiation and intermolecular transfer, while at the milligram scale it follows intermolecular transfer of hydrogen atoms via abstraction by free radicals as the predominant transfer mechanism to produce volatiles.

Pyrolysis provides a number of advantages, such as (i) operational advantages, (ii) environmental advantages, and (iii) financial benefits. Operational advantage is the utilization of residual output of char used as a fuel or as a feedstock for other petrochemical processes. An additional operational benefit of pyrolysis is that it requires no flue gas to clean up the gas produced. Environmentally, pyrolysis provides an alternative to landfilling and reduces GHGs and CO_2 emissions. Financially, pyrolysis produces a high-calorific-value fuel that can be easily marketed and used in gas engines to produce electricity and heat [53]. The main obstacles that exist with pyrolysis are the handling of the char produced and treatment of the final fuel produced if specific products are desired. Besides this, there is no sufficient understanding of the underlying reaction pathways to predict the quantitative distribution of full products [53].

3.1.6.2.2 **Gasification**

In this process, air is used as a gasification agent, which adds a number of advantages, such as simplicity of the process using air instead of O_2 alone and reduction of cost. The only disadvantage associated with this is the presence of (inert) N_2 in air, which causes a reduction in the calorific value of resulting fuels due to the dilution effect on the fuel gases. Hence, steam is introduced in a stoichiometric ratio to reduce the N_2 present. A significant amount of char is always produced in gasification, which needs to be further processed and/or burnt.

An ideal gasification process for PSW is one that can produce a gas with high calorific value, completely burn the char, and easily separate metals from the ash, and does not require any additional installations for air/water pollution abatement [53]. Early gasification attempts were for plastics (PP, PVC, and PET) to get high-calorific-value fuel gas, reported in the 1970s. The need for alternative fuels led for the co-gasification of PSW with other types of waste, mainly biomass [55, 56]. Pinto and his group studied the fluidized bed co-gasification of PE, pine, and coal, and biomass mixed with PE. Xiao *et al.* co-gasified five typical kinds of organic components (wood, paper, kitchen garbage, textile, and plastic (namely PE)) and three representative types of simulated MSW in a fluidized bed (400–800 °C) [57]. It was determined that plastic should be gasified at temperatures greater than 500 °C to reach a lower heating value (LHV) of 10 000 kJ N^{-1} [53].

3.1.6.2.3 **Hydrogenation (Hydrocracking)**

Hydrogenation by definition is the addition of hydrogen by chemical reaction through unit operations. The main technology applied in PSW recycling via hydrogenation technology is the Veba process. Based on the coal liquefaction technology, Veba Oel AG converted coal by this process into naphtha and gas oil [53]. The major technologies are summarized in the review paper by Al-Salem *et al.* [53].

3.1.6.2.4 **Other Chemical Recycling Schemes**

Degradative extrusion provides an optimum engineering solution especially on a small industrial scale (10 kg h^{-1}). The advantages of degradative extrusion are (i) achieving molecular breakdown of thermoplastics and hence low-viscosity polymer melts and (ii) applying a combination of mechanical and chemical recycling schemes, prompting the degradation process by introducing steam, gas, oxygen, or catalysts, if needed. Other advantageous technologies for chemical treatment include catalytic and steam cracking. The concept for both processes is the employment of either steam or a catalyst in a unit operation [53].

3.1.6.3 Reactor Types

3.1.6.3.1 Fluidized-Bed Reactors

Pyrolysis in a fluidized-bed reactor or a similar device is used for large-scale implementation for continuous waste plastic upgradation [58]. *In situ* catalytic pyrolysis is also studied as a relevant research topic [59]. The conical spouted-bed reactor (CSBR) is considered a better reactor for catalytic pyrolysis because of the low bed segregation and lower attrition than the bubbling fluidized bed. The good performance of the CSBR has been proven, as it produces selective waxes [60], fuel-like hydrocarbons, and monomers [61]. This good performance is a consequence of the solid flow pattern, high heat transfer between phases, and the smaller defluidization problems when sticky solids are handled. Defluidization is due to the agglomeration of solid particles (sand) coated with melted plastic, constituting a severe problem in fluidized-bed reactors. In the CSBR, polyolefins melt as they are fed into the reactor, and they uniformly coat the sand and catalyst particles because of their cyclic movement. Vigorous solid flow and the action of the spout avoid the formation of agglomerates. Furthermore, the CSBR has great versatility in terms of gas residence time, which may be reduced to near 20 ms [62], and, consequently, the yield of polyaromatic compounds is minimized. Besides, the smaller attrition of catalyst particles, due to the absence of a distributor plate, is another advantage over the fluidized bed for its use in catalytic processes. This excellent behavior of the CSBR has already been recorded in other processes carried out prior to catalytic pyrolysis, such as catalytic polymerization, where a similar problem of fusion of catalyst particles coated with polymer occurs [63]. The simple design of a CSBR makes its scaling up straightforward. Furthermore, its throughput by reactor volume unit is higher than that of a bubbling fluidized bed due to the lower amount of sand required for fluidization enhancement.

3.1.6.3.2 Batch Reactors

Thermal and catalytic degradation of polyethylene was conducted at atmospheric pressure in a batch type reactor [63]. The reactor was a 1-l circular stainless steel bottle placed in a thermostatic furnace. The reactor system was connected to a nitrogen supply to eliminate air before premixed plastics and catalysts were fed into it. Temperature of the reactor was increased to 450 °C and held for 30 min until the reaction was completely finished.

Van Grieken used another type of laboratory-scale batch reactor with a helicoidal stirrer at 120 rpm for the catalytic degradation of polyethylene [64]. Three temperatures (380, 400, and 420 °C) and different reaction times (0–360 min) under nitrogen flow were studied. The effluent from the reactor was connected to a water-cooled trap in order to condense the liquid products, whereas the effluent gas was finally collected in a Teflon bag.

3.1.6.3.3 Fixed-Bed Reactor

Achilias used a laboratory-scale fixed-bed reactor to study the thermal and catalytic degradation of PE [54]. The reactor was filled with the fluid catalytic cracking (FCC) catalyst, and the piston was filled with the polymer. The experiment lasted for 17 min, and the reaction temperature was 450 °C. Under similar experimental conditions, when an FCC catalyst was used, the liquid fraction increased and the gaseous product yield decreased along with solid residue for all three polymers (LDPE, HDPE, and PP).

3.1.7 Polyolefin Thermal Cracking

Thermal degradation behavior at the laboratory scale enables the assessment of a number of important parameters, such as thermal kinetics, activation energy (energy required to degrade materials treated and product formation), reference temperatures of the half-life of polymers, and the maximum degradation point achievable. Appropriate design and scale (of operation and economy) are of paramount importance in the case of thermal treatment plants. It is also important to perform pilot-scale experiments using a number of rectors and unit operations before upgrading to a full scale. This will also aid in the determination of the mode of the material processing of the thermal plant (i.e., pulsating, continuous, batch, etc.). Pyrolysis (depolymerization in an inert atmosphere) is usually the first process in a thermal plant, which needs appropriate end-product designs [14].

Achilias and his group carried out a number of studies on the thermal cracking of polyolefins in inert (pyrolysis) and/or partially oxidized atmospheres (e.g., step pyrolysis, gasification) [58, 64]. Thermogravimetry is the most commonly used technique for the determination of kinetic parameters, although the experimental conditions used are very different, involving broad ranges of temperature, sample amount, heating rates (in the case of dynamic runs), reaction atmospheres, and pressures [65, 66].

3.1.7.1 Catalytic Degradation

Studies concerning the use of different catalysts in the pyrolysis of polyolefins were conducted by Achilias and his group [58]. Thermogravimetry (TG) and microreactors were used to pyrolyze plastics with zeolite-based acid catalysts [67, 68]. Catalytic pyrolysis of PE samples was carried out in laboratory-scale reactors, such as batch reactors [69, 70], semi-batch reactors (with evacuation of volatile products) [71, 72], and fixed beds [54]. Many catalysts were used for the catalytic degradation of polymeric materials, including amorphous silica–alumina, zeolite Y, mordenite, and ZSM-5, the family of mesoporous MCM-41 materials [73, 74], and a few silicoaluminophosphate (SAPO) molecular sieves [75, 76]. It was found that catalytic activity is closely related to the amount of acid sites, pore size, and also shape of the catalyst [27, 77]. SAPO molecular sieves represent an important class of adsorbents and catalytic materials generated by the introduction of silicon into its aluminophosphate framework.

The medium-pore SAPOs are attractive for catalytic applications due to the presence of specific acid sites in its structure, which can convert the polymer into useful hydrocarbons [71]. The use of $BaCO_3$ as a catalyst for the thermal and catalytic degradation of waste HDPE was also reported [78].

Haung found that the catalysts more frequently employed for the cracking of polyolefins are shape-selective zeolites and mesoporous materials, such as HY, HZSM-5, Hβ, and MCM-41 [79], which undergo inevitable deactivation by coke deposition. Indeed, deactivation is a major hurdle in the implementation and scale-up of the valorization of plastics by cracking [80]. Microporous zeolites have very high thermal stability and customized acid sites. Thus, the selection of the zeolite should be based on target selectivity: HZSM-5 zeolite promotes the production of olefins (original monomers), while Hβ and HY zeolites maximize the production of middle distillates [81].

Marcilla and his group studied the deactivation of zeolites by coke formation and protocols for characterizing the coke deposited on zeolites during the cracking of HDPE, by using mainly a thermobalance as reactor [82]. It should be pointed out that coke formation is strongly affected by the catalyst properties (e.g., shape selectivity, acidity, concentration of acid sites) [79], reactor medium [83], operating conditions, and feedstock properties.

3.1.8 Chemical Recycling of Polypropylene

3.1.8.1 Introduction

Condensation polymers such as PET, Nylon, and so on, can undergo chemolysis with different reagents to produce mainly the monomers from which they have been produced or other oligomers. In contrast, vinyl polymers, such as polyolefins (PP and PE), cannot be degraded with simple chemicals to their monomers due to the random scission of the C–C bonds. Two main chemical recycling routes are the thermal and catalytic degradation of these polymers. In thermal degradation, the process produces a broad range of products and requires high operating temperatures, typically more than 500 °C and even up to 900 °C. Thermal cracking of PE and PP is usually carried out either at high temperatures (>700 °C), to produce an olefin mixture (C1–C4) and aromatic compounds (mainly benzene, toluene, and xylene), or at low temperature (400–500 °C) (thermolysis) where three fractions are obtained:

1. A high-calorific-value gas
2. Condensable hydrocarbon oil
3. Waxes.

In the first case, the objective is to maximize the gas fraction and to recover the olefins, which could be used after separation as monomers for the re-production of the corresponding polyolefins. Cracking in lower temperatures leaves a waxy product in the reactor, which mainly consists of paraffins together with a carbonized char. The gaseous fraction can be used after burning for the supply of the energy required for the pyrolysis. The liquid fraction mainly consists of linear olefins and paraffins with C11–C14 carbon atoms with only traces of aromatic compounds [84]. Thermal cracking of polyolefins proceeds through a random scission mechanism in four steps:

1. Initiation
2. Depropagation
3. Inter or intramolecular hydrogen transfer followed by β-scission
4. Termination.

In general, thermal cracking is more difficult in HDPE followed by LDPE and by PP. Because of the low thermal conductivity of polymers, together with the endotherm of cracking, thermal pyrolysis consumes large amounts of energy. Thus, catalytic technologies have been proposed to promote cracking at lower temperatures, resulting in reduced energy consumption and higher conversion rates. Furthermore, the use of specific catalysts allows the process to be directed toward the formation of a narrower distribution of hydrocarbon products with higher market value. Heterogeneous catalysis has been investigated extensively using solids with acid properties. Zeolites of the kind employed in the catalytic cracking of hydrocarbon feedstock (Y, ZSM-5, Beta), as well as other well-known acid solids like silica–alumina, alumina, and clays, are the most studied. Mixtures of these catalysts, for example, SAHA/ZSM-5 and MCM-41/ZSM-5, have also been used. Cracking with acid catalysts takes place through the formation of carbocations, which requires the presence of strong acidic regions. Acid strength and textural properties are the main parameters dictating the performance of acid solids in the catalytic conversion of polymers. Porosity, surface area characteristics, and particle size determine to a large extent the accessibility of bulky polymeric molecules to the internal catalytic acid sites of the solids. Thus, while the catalyst HZSM-5 presents higher reactivity than HMCM-41 in the cracking of HDPE and LDPE, in the decomposition of the large molecules of PP the transformation is almost the same as with thermal cracking, because the cross section of the polymer is too large to enter the catalysts' micropores [54].

These facts strongly limit their applicability and especially increase the cost of feedstock recycling for waste plastic treatment. Therefore, catalytic degradation provides a means to address these problems. The addition of a catalyst reduces decomposition temperature, promotes decomposition speed, and modifies the products. The catalytic degradation of polymeric materials has been reported for a range of model catalysts centered on the active components, including amorphous silica–aluminas, zeolites Y, mordinite, and ZSM-5 and the family of mesoporous MCM-41 materials. However, even though these catalysts perform well, they are unfeasible from the point of view of practical use due to the cost of manufacturing, which is highly sensitive to the cost of the catalyst. Another option that is attractive for the chemical recycling of polymer wastes is the use of FCC catalysts. Therefore, an alternative way to improve the recycling process via catalytic cracking can be by mixing the polymer waste with commercial FCC catalysts.

Recently, much attention has been paid to the recycling of waste polymers by thermal or catalytic pyrolysis to recover value-added products or energy via the production of high-value petrochemical feedstock or synthetic fuel fractions. The following review is rather selective and not extensive [13, 85].

3.1.8.2
Pyrolysis

Achilias studied the technique of pyrolysis of PP in a laboratory fixed-bed reactor using as raw materials either model PP or waste products based on these polymers [54]. The conclusions are very interesting. The oil and gaseous fractions recovered presented a mainly aliphatic composition consisting of a series of alkanes and alkenes of different carbon numbers with a great potential to be recycled back into the petrochemical industry as a feedstock for the production of new plastics or refined fuels. Details are presented in another section. Hayashi studied the pyrolysis of PP in the presence of oxygen [86]. PP was coated on porous α-alumina particles and then pyrolyzed in a flow of helium or a mixture of helium–oxygen at atmospheric pressure. The mass release from PP was dramatically enhanced in the presence of oxygen at temperatures in the range 200–300 °C. The net mass release rate in the presence of oxygen followed first-order kinetics with respect to the oxygen partial pressure and was controlled by the formation of peroxide on the tertiary carbon of PP. The activation energy was 60–70 kJ mol^{-1}. Oxidative pyrolysis at 2500 °C converted 90% of PP into volatiles, which mainly consisted of CS-soluble oils having a number-average chain length of 10. Dawood studied the influence of γ-irradiation on the thermal degradation of PP by performing thermogravimetric analysis (TGA) at three constant heating rates and at a constant temperature [87]. At all heating rates, it was found that the TG curves of the irradiated samples shifted to lower temperatures in comparison with those of the nonirradiated samples. The shift clearly increased with increasing irradiation dose, which means that the pyrolysis rate was enhanced by irradiation. Since the difference in TG curves between the nonirradiated sample and the samples irradiated at 10 and 30 kGy is quite large, a small dose of irradiation may be enough to cause a significant enhancement of the pyrolysis activity. The samples irradiated with small doses, 10 and 30 kGy, seem to show a pyrolysis behavior different from that of the other irradiated samples. At a small heating rate of 3 K min^{-1}, the TG curves of 10 and 30 kGy samples are close to the TG curve of the 60 kGy sample, whereas the former TG curves are distinctly different from the latter ones at 10 K min^{-1}. These results may suggest that the mechanism of the increase in pyrolysis activity is different between the irradiated samples. A further examination of the influence of irradiation was carried out by pyrolyzing the samples at a constant temperature. Similar to the case of dynamic heating rate, the difference in pyrolysis reactivity between the nonirradiated and the 30 kGy irradiated sample is quite large, while the difference between the irradiated samples is small. This supports the suggestion that a small radiation dose is enough to cause a significant enhancement in the pyrolysis activity of PP.

3.1.8.3 Co-pyrolysis

Assumpcao studied the co-pyrolysis of PP with Brazilian crude oil by varying the temperature (400–500 °C) and the amount of PP fed to the reactor [88]. The co-pyrolysis of plastic waste in an inert atmosphere yielded around 80% oil, of which 50% was diesel oil. This technique is good for recycling of waste PP, as it minimizes the environmental burden caused by inadequate disposal of PP and allows the reuse of a nonrenewable natural resource (petroleum) through the use of diesel oil fractions obtained in this process. It was observed that increase in temperature has a favorable effect on the yield of the pyrolytic liquid and reduces the amount of solid formed [14].

On the other hand, a large increase in the PP amount causes a decrease in the total yield (liquid product). In general, it was observed that with temperature increase, there was a small reduction in yield in the diesel distillation range. Moreover, most part of these liquid distillates was in a range higher than diesel, corresponding to a heavy vacuum gas oil product (GOP). This product (GOP) can still be cracked in an FCC, generating more profitable products (naphtha and LPG), or can be used as fuel oil. The increase of PP in the reaction favors a yield increase in the diesel distillation range compared to pyrolysis of pure heavy oil, also forming a significant amount of compounds with a distillation range lower than that of diesel [14, 88].

Ballice investigated the temperature-programmed co-pyrolysis of Soma lignites from Turkey with PP [89]. A series co-pyrolysis operations were performed with lignites and PP using lignite/plastic total carbon ratios of 1 : 3, 1 : 1, and 3 : 1. A fixed-bed reactor was used to pyrolyze small samples of lignites and PP mixture under an inert gas flow (argon). In addition, the performance of the experimental apparatus was investigated by establishing a carbon balance, and the degree of recovery of total organic carbon of the samples as aliphatic hydrocarbons and in solid residue was determined. Conversion into volatile hydrocarbons was found to be higher with increasing PP ratio in the lignite–PP system, while C16+ hydrocarbons and the amount of coke deposit were lower in the presence of PP. The maximum product release temperature was found to be approximately 440 °C for the co-pyrolysis of lignite and PP.

Straight and branched-chain paraffins and olefins from methane to C26, diene, and simple aromatic hydrocarbons were determined in co-pyrolysis products. The fraction of *n*-paraffins was higher than that of 1-olefins at a high proportion of lignite in the mixture. Co-pyrolysis of lignite with PP has been found to give less C16+ *n*-paraffins and 1-olefins than pyrolysis of lignite by increasing PP ratios.

Coke deposits in co-processing decreased by increasing the ratio of PP. The *n*-paraffins were found to consist of mainly C1–C9 and relatively small amount of C10–C15 and C16 fractions. The evolution of 1-olefins decreased in co-pyrolysis operation because of the higher hydrogen content in feed by increasing ratios of PP. A slightly synergistic effect was seen in the co-pyrolysis operation, and the experimental results indicated that the pyrolysis products of PP were highly aliphatic in character and, during the initial stages of pyrolysis, these pyrolysis products of PP were relatively poor solvents for the structures of lignite. In addition, relative to liquefaction source materials such as coals, the dominant components of MSWs (mainly PE, PS, PET, and PP) are hydrogen rich so that co-processing of coal with waste plastics could be a good way to recycle waste plastics into useful products.

Hajekova and Bajus investigated the thermal decomposition of polyalkenes as a recycling route for the production of petrochemical feedstock [90]. PP was thermally decomposed individually in a batch reactor at 450 °C, thus forming oil/wax products. Then the product was dissolved in primary heavy naphtha to obtain steam-cracking feedstock. The selectivity and kinetics of co-pyrolysis for 10 mass% solutions of oil/waxes from PP with naphtha in the temperature range from 740 to 820 °C at residence times from 0.09 to 0.54 s were studied.

The decomposition of polyalkene oil/waxes during copyrolysis was confirmed. It was shown that the yields of the desired alkene (propene) increased or slightly decreased compared to the yield from naphtha. In addition to the primary reactions, the secondary reactions leading to coke formation have also been studied. Slightly higher formation of coke was obtained from PP wax solution at the beginning of the measurements, on the clean surface of the reactor. After a thin layer of coke covered the walls, the production was the same as that from naphtha. The results confirmed the possibility of polyalkene recycling via the co-pyrolysis of polyalkene oils and waxes with conventional liquid steam cracking feedstocks on already existing industrial ethylene units.

3.1.8.4 Catalytic Cracking

A large number of laboratory studies have been conducted on the direct catalytic cracking of different type of plastics. A large variety of catalysts have been used; though performing well, they can be unrealistic from the point of view of practical use due to the cost of manufacturing and the high sensitivity of the process to the cost of the catalyst. Some studies reported in the literature are reviewed below.

Zhao studied the effects of different zeolites such as H-Y, Na-Y, L, H-mordenite, and Na-mordenite on the catalytic degradation of PP by thermogravimetry under nitrogen flow [91]. It was found that the degradation temperature of PP strongly depended on the type of zeolite used and the amount added. One type of H-Y zeolite (320HOA) was found be a very effective catalyst. Pyrolysis products, which were identified by using coupled gas chromatography-mass spectrometry (GC/MS), were also affected by the type of zeolites. Some zeolites did not change the structure of the products but narrowed the product distribution to a smaller molecule region, while the H-Y zeolite led to hydrocarbons concentrated at those containing four to nine carbons. Furthermore, some new compounds with cyclic structures were found in the presence of the H-Y zeolite [91].

Zhao studied the effect of irradiation on the pyrolysis of PP in the presence of zeolites. The results revealed that the thermal degradation temperature of PP was significantly reduced when PP was irradiated in the presence of a zeolite. The irradiation-induced temperature reduction depended on the zeolite structure and composition, as well as on the morphology of the mixture. Identification of the pyrolysis products indicated that, in the absence of zeolite, irradiation resulted only in a change of the product distribution but no formation of new compounds. In the presence of the zeolite, however, a series of oxidized products were formed. In addition, the pyrolysis could be performed at a much lower temperature. Irradiation was able to render PP much more susceptible to thermal degradation when carried out in the presence of zeolite. However, this effect was closely related to the type of zeolite, mixing methods, and irradiation conditions. Furthermore, in the pyrolysis of properly irradiated PP–zeolite mixtures, new chemicals such as acetone, acetic acid, and so on, could possibly be obtained in addition to traditional hydrocarbons.

Ishihara investigated the catalytic degradation of PP by silica–alumina at temperatures between 180 and 300 °C in a semi-batch reactor under nitrogen flow [92]. The production of gas precursors was found essential to decomposition. The most important elementary reaction was the intramolecular rearrangement of chain-end secondary carbonium ions in the liquid fraction to inner tertiary carbon atoms. The catalytic decomposition of PP proceeds as follows: polymer to degraded polymer + oligomer + liquid + gas. Gas is produced from the chain-ends of the liquid fraction, and its components are primarily isobutene and isopentane. The most important elementary reaction in the decomposition is intramolecular rearrangement taking place via a six-membered transition state to inner tertiary carbon atoms (back-biting reactions). The main gas components are produced by the decomposition of the C fraction formed by the back-biting reaction.

Durmus studied thermal–catalytic degradation kinetics of PP over BEA, ZSM-5, and MOR zeolites [93]. The degradation rate of the PP over zeolites was studied by TGA employing four different heating rates, and the apparent activation energies of the processes were determined by the Kissinger equation. The catalytic activity of zeolites decreases as BEA > ZSM-5a (Si/Al = 12.5) > ZSM-5b (Si/Al = 25) > MOR, depending on pore size and acidity of the catalysts. On the other hand, initial degradation is faster over MOR and BEA than over both ZSM-5 catalysts, depending on the apparent activation energy. It can be concluded that acidity of the catalyst is the most important parameter determining the activity for the polymer degradation process, apart from other structural parameters such as the pore structure and size.

Lin investigated the catalytic cracking of PP in a fluidized-bed reactor using H-ZSM-5, H-USY, H-mordenite, silica–alumina, and MCM-41, with nitrogen as fluidizing gas [24] (Figure 3.1.2). PP was pyrolyzed over various catalysts using

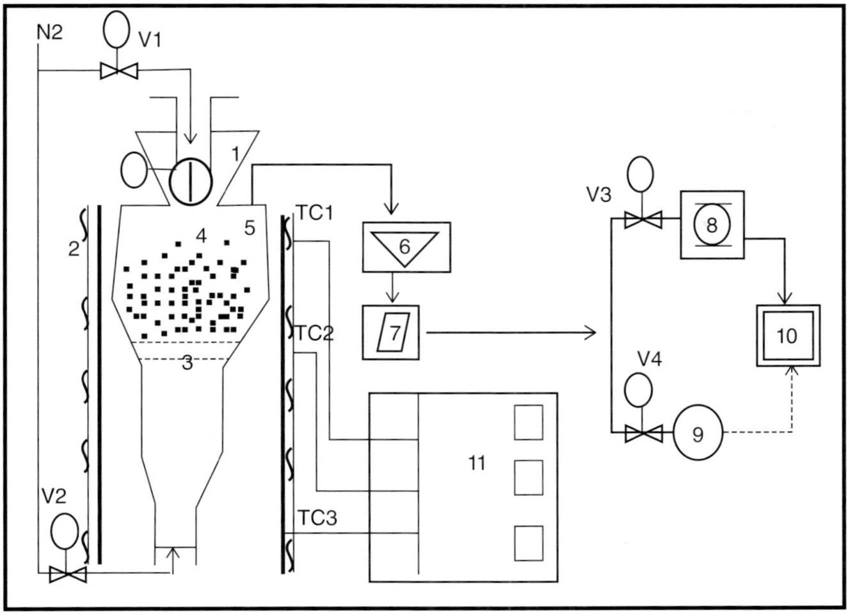

Figure 3.1.2 Schematic diagram of catalytic fluidized-bed reactor system [94].

a laboratory fluidized-bed reactor operating isothermally at ambient pressure. The yield of volatile hydrocarbons for zeolite catalysts was higher than that for non-zeolite catalysts. Product distributions with HZSM-5 contained more olefinic materials with about 60 wt% in the range C3–C5. However, both HMOR and HUSY produced more paraffin streams with large amounts of isobutene (i-C4), and both catalysts were deactivated during the course of the degradation. SAHA and MCM-41 showed the lowest conversion and generated an olefin-rich product with the broadest carbon range C3–C7.

Greater product selectivity was observed with HZSM-5 and HMOR as catalysts, with about 60% of the product in the C3–C5 range and HMOR generating the highest yield of i-C4 among all catalysts studied. The larger pore zeolites (HUSY and HMOR) showed deactivation in contrast to the more restrictive HZSM-5. The observed differences in product yields and product distributions under identical reaction conditions can be attributed to the microstructure of catalysts. Valuable hydrocarbons of olefins and iso-olefins were produced by low temperatures and short contact times. Lin pyrolyzed PP over a spent FCC commercial catalyst (FCC-s1) using a laboratory fluidized-bed reactor operating isothermally at ambient pressure [94]. The yield of gaseous and liquid hydrocarbon products at 390 °C for the spent FCC commercial catalyst (87.8 wt%) was much higher than with silicate (only 17.1 wt%). Greater product selectivity was observed with FCC-s1 as a post-use catalyst with about 61 wt% olefins products in the C3–C7 range [94]. The use of a fluidized-bed reaction system coupled with a spent FCC equilibrium catalyst can be a better option from the cost point of view since it can gives a good conversion with a relatively short reaction time. Though its activity is lower than that of the zeolites (ZSM-5 and HUSY) and silica–aluminas (SAHA), this can be compensated by increasing the catalyst to PP ratio. Product distributions with the FCC-s1 catalyst contained more olefinic materials in the range C3–C7 (about 56 wt% at 390 °C). It was concluded that the use of a spent FCC commercial catalyst and under appropriate reaction conditions has the ability to control both the product yield and product distribution from polymer degradation, potentially leading to a cheaper process with more valuable products. Dawood studied the effect of exposing PP to γ-irradiation prior to catalytic pyrolysis over an H-Y zeolite using a thermobalance and a semi-batch reactor [87]. A significant increase in the rate of the catalytic pyrolysis was realized when PP was exposed to a small irradiation dose of 10 kGy. The high reactivity of the irradiated PP was combined with low yields of residue and coke, in addition to enhanced selectivity for light distillate (C7–C10). Examination of the effect of pyrolysis temperature revealed that catalytic pyrolysis preferred high temperature in the investigated temperature range of 325–375 °C. The results presented above clarify that a significant increase in the rate of the catalytic pyrolysis with enhanced selectivity of C7–C10 compounds can be obtained by exposing PP to the ionizing irradiation prior to catalytic pyrolysis. The results suggest the applicability of the proposed pyrolysis method for enhancing the catalytic conversion of plastic waste into useful hydrocarbons. Uemichi investigated the degradation of PP to aromatic hydrocarbons over activated carbon catalysts containing Pt and Fe. The results obtained were compared with those of the degradation of polyethylene [87]. The addition of Pt or Fe to activated carbon resulted in an increase in the yield of aromatics from PP. However, the increase was less than that from PE. Pt metal was more effective than Fe only when the reaction conditions involved a longer contact time. The formation of aromatics was explained by essentially the same mechanism as the case of PE, in which an influence of methyl branching of PP on the aromatization yield and a difference in catalytic activity of the catalysts containing Pt and Fe for a ring-expansion reaction were considered.

Park reported the pyrolysis of PP over mesoporous MCM-48 material,which were employed as catalysts for the degradation of PP [71]. The catalytic activity of Al–MCM-48 was much higher than that of Si–MCM-48. Al–MCM-48 mainly generated C7–C10 hydrocarbons, while Si–MCM-48 gave a broader distribution of oil products (C7–C14). Al–MCM-48 showed high catalytic stability for the degradation of PP. In view of these facts, Al–MCM-48 can be considered a promising catalyst for the degradation of other waste plastics. Cardona and Corma studied the tertiary recycling of PP by catalytic cracking in a semi-batch stirred reactor. A semi-continuous reactor has been presented that allows carrying out efficiently the catalytic cracking of PP [95, 96]. By working with USY zeolites with different unit cell sizes, it has been proven that neither the total amount nor the strength of the acid sites is the most determininf factor for cracking PP. The first cracking event of PP occurs at or close to the external surface. Then the formation of mesopores in the zeolite strongly improves the cracking activity. This has been supported by the results obtained with a Y zeolite synthesized with smaller crystallite sizes. Finally, it has been shown that amorphous or ordered silica–aluminas are very active catalysts. However, an FCC equilibrium catalyst can be a better option from the cost point of view since it gives very good selectivity, and even though its activity is lower than that of the silica–aluminas, this can be compensated by increasing the catalyst to PP ratio. Xie reported the catalytic cracking of PP over MCM-41modified by Zr and Mo [97]. The relationship between the structure, acidity, and catalytic activity of Zr–Mo–MCM-41 was studied. The results showed that Zr–Mo–MCM-41 exhibited high activity for the cracking of PP and good selectivity for producing liquid hydrocarbons of higher carbon numbers. The results were compared with those obtained over HZSM-5, $SiO_2-Al_2O_3$, and other MCM-41 mesoporous molecular sieves. For the catalytic cracking of PP, Mo enhances the selectivity to hydrocarbons of high carbon number and Zr enhances the acidity of catalyst and results in the increasing the cracking conversion of PP. Zr–Mo–MCM-41 using $Zr(SO_4)_2$ as Zr source has the best catalytic activity and selectivity to high carbon number hydrocarbon, which means that Zr–Mo–MCM-41 has good potential as catalyst for the cracking of PP.

Panda investigated the catalytic performances of kaolin and silica–alumina in the thermal degradation of PP [98]. PP was cracked thermally and catalytically in the presence of kaolin and silica–alumina in a semi-batch reactor in the temperature range 400–550 °C in order to obtain suitable liquid fuels. It was observed that up to 450 °C thermal cracking temperature, the major product of pyrolysis was liquid oil, whereas the major product at higher temperatures (475–550 °C) were viscous liquid or wax and the highest yield of pyrolysis product was 82.85% by weight at 500 °C. Use of kaolin and silica–alumina decreased the reaction time and increased the yield of the liquid fraction. Again, the major pyrolysis product in catalytic pyrolysis at all temperatures was low-viscosity liquid oil. Silica–alumina was found to be better than kaolin in liquid yield and in reducing the reaction temperature. The maximum oil yield using silica–alumina and kaolin catalysts was 91% and 89.5%, respectively. On the basis of the obtained results, hypothetical continuous process of waste PP plastics processing for engine fuel production can be envisaged.

In conclusion, catalytic pyrolysis of PP reduces environmental impacts and also the time of recycling, and results in very useful products with potential use as fuel replacements.

3.1.9
Chemical Recycling of Polystyrene

3.1.9.1
Introduction

PS is widely used for packaging various electronics and other fragile items and in the manufacturing of many products because of its favorable properties such as good strength, light weight, and durability. In the past several years, PS has received much public and media attention, as it accounts 9–10% of the plastic waste in MSW. It is boycotted by several localities because of various misconceptions, such as nondegradable, non-recyclable, toxic when burned, choking landfills, depleting ozone, killing wildlife, and even carcinogenic. Actually, PS comprises less than 0.5% of the solid waste going to landfills [14].

PS is used in the solid form (coffee cups, trays) and expanded forms, both of which can be recycled. Solid PS can be recycled back into alternative applications such as videocassette cases, office equipment, and so on, whereas expanded polystyrene (EPS) foam waste loses its foam characteristics during the recovery process. The recovered material can be re-gassed, but the product becomes more expensive than virgin material. Both expanded and solid PS wastes have been successfully recycled in extruded plastic timber–lumber. Recycled PS is used to produce plant pots and desk items such as pens, pencils, and so on. As with other types of plastic materials, PS recycling takes place after consideration by the industry of a number of issues including eco-efficiency, availability, corporate social responsibility, product quality, hygiene aspects, and traceability [14]. The boom in electronic products has sharply increased the quantities of waste from electrical and electronic equipment (WEEE), amplifying the problem of their disposal. More than 1000 t of PS foam worldwide is disposed off into the environment as MSW, and this amount is increasing every year. Modern design for environment (DFE) with special attention to recycling and disassembly can be the only solution of this problem.

The following types of PS materials are accepted for recycling:

1. EPS foam used for packaging, which is the familiar white material, custom-molded to cushion, insulate, and protect all types of products during transportation.
2. PS resin foamed used for EPS insulation boards in houses and commercial construction, foodservice products like cups, plates, trays, and so on, which provide unique insulating quality and loose-filled packaging.
3. Non-foam PS products, that is, high-impact polystyrene (HIPS)
4. Oriented polystyrene (OPS), post-consumer products, post-industrial products, and Styrofoam (a Dow Chemical Company brand trademark for a PS foam thermal insulation product) [99].

3.1.9.2
Recycling Methods for Polystyrene Products

A large amount of expanded PS is collected from wholesale markets, supermarkets, department stores, restaurants, and shops (such as electrical appliances stores), and from factories of machinery manufacturers. It is collected in-house by the companies or by resource recycling agents, and becomes a recycled resource. Then it is rinsed off for the removal of any food or dirt particles. Oversized materials like cartons, milk jugs, and so on, are crushed so that they can fit into the bin and into the truck more easily. The caps of the plastic bottles and glass jars are thrown away. The volume of EPS is reduced by methods such as solvent volume reduction (dissolved using solvent), heating volume reduction, and pulverizing volume reduction (powdering). The processed EPS is used in its reduced state as an ingredient for recycled products, or it is burnt to generate heat energy. There are four methods commonly used for recycling of PS.

3.1.9.2.1 The Dissolution Technique

In this method, an appropriate solvent is used to selectively dissolve the polymer, and another solvent is used to recover the polymers from the mixture of different plastics. In this sense, a naturally occurring compound, that is, limonene (occurring in citrus fruits), has been successfully used to dissolve EPS [100]. This solvent can dissolve EPS in large amounts safely and with negligible degradation of the performance of the polymers. The dissolved PS can be precipitated through the addition of a non-solvent in the mixture. The solvent is vacuum-evaporated and reused. It is to be noted that conventional melt separation methods cause a large drop in the polymer's molecular weight due to thermal degradation.

3.1.9.2.2 Chemical Recycling of PS: Catalytic Degradation

Modified Fe-based catalysts are employed for the catalytic degradation of EPS waste. This process gives styrene monomer (S) at a relatively low temperature with a high selectivity [101]. The yield of oil (YOil) and yield of S (YS) are increased in the presence of Fe-based catalysts and with increasing reaction temperature. YOil and YS, obtained over $Fe-K/Al_2O_3$ at a relatively low reaction temperature (400 °C), are 92.2 and 65.8 wt%, respectively. The value of E_a (activation energy) is obtained as 194 and 138 kJ mol^{-1} for the thermal degradation of EPS degradation in the presence of the catalysts ($Fe-K/Al_2O_3$). Bajdur synthesized sulfonated derivatives of expanded PS wastes by means of known methods, which may be used as polyelectrolytes. The products have different contents of sulfogroups in the polymer chain [102]. It was observed that polyelectrolytes have good flocculation properties similar to those of anionic commercial polyelectrolytes. The effect of a base catalyst, MgO, on the decomposition of PS was studied through degradation of both a mono-disperse polymer and a PS mimic 1,3,5-triphenylhexane (TPH) [103]. It was observed that the presence of the catalyst increased the decomposition rate of the model compound but decreased the degradation rate of PS, as measured by the evolution of low molecular weight products. Results of the model compound suggested that the rate of initiation was enhanced in both cases by the addition of catalyst but decreased the "zip length" during depropagation due to termination reactions facilitated by the catalyst.. A decrease in the selectivity to styrene monomer in the presence of MgO was observed for both PS and TPH. Degradation of PS into styrene, including monomer and dimer, was studied by Ukei *et al.* [104] using solid acids and bases, such as MgO, CaO, BaO, K_2O, SiO_2/Al_2O_3, and HZSM5, and active carbon as catalyst. It was found that solid bases were more effective catalysts than solid acids for the degradation of PS into styrene. This was attributed to the differences in the degradation mechanisms of PS over solid acids and bases. Among the solid bases employed, BaO was found to be the most effective catalyst, and about 90 wt% of PS was converted into styrene when thermally degraded PS was subjected to BaO powder at 350 °C. Koji tried to obtain PS foam by mixing PS with a basic metal oxide as catalyst and passing an inert blowing agent [105]. When this foam is wasted, it could be recycled into styrene by decomposing it by heating to 300–450 °C in a non-oxidizing atmosphere. The basic oxides Na_2O, MgO, CaO can be used, and among them CaO is more desirable. The blowing agent used may be nitrogen gas, chlorofluorocarbon, propane, or any similar gas. Several solid acids, such as silica–alumina, HZSM-5, HY, mordenite, and clinoptilolite, have been studied as catalysts and screened for their performances in the catalytic degradation of PS [106]. More than 99% styrene and ethylbenzene were the major products in the liquid fraction when clinoptilolite catalysts (HNZ, HSCLZ) was used. The increase of acidity favored the production of ethylbenzene by promoting the hydrogenation reaction of styrene. Higher selectivity to styrene is observed at higher temperatures, whereas an increase of contact time by reducing the nitrogen gas flow rate enhanced the selectivity to ethylbenzene. Thus product distribution between styrene monomer and ethylbenzene is adjusted by design parameters such as acidity of catalyst, reaction temperature, and contact time.

The degradation of PS in various supercritical solvents like benzene, toluene, xylene, and so on, at 310–370 °C and 6.0 MPa pressure was studied by Ke *et al.* [107]. It was found that PS can be successfully depolymerized into monomer, dimmer, and other products in a very short reaction time with high conversion. Toluene as supercritical solvent showed more effectiveness than other solvents such as benzene, ethylbenzene, and *p*-xylene for the recovery of styrene from PS. The highest yield of styrene (77 wt%) was obtained from PS in supercritical toluene at 360 °C for 20 min. Subcritical water treatment in the presence of an amino alcohol showed de-crosslinking of unsaturated polyesters crosslinked with styrene [108]. A linear PS derivative bearing hydroxy-terminated side chains was recovered. After modification of these hydroxy groups with maleic anhydride, the PS derivative was re-crosslinked with styrene to form a networked structure again. The resulting solid was degraded by subcritical water treatment in the presence of the amino alcohol and another PS derivative bearing hydroxy groups. This is a novel recycling method of thermosetting resins and can be successfully repeated. The PS derivative was re-crosslinked again on heating with an alternative copolymer of styrene and maleic anhydride as a result of the formation of a linkage between the hydroxy groups and carboxylic anhydride moieties.

If oligo-styrene (of several thousands of Da molecular weight) is produced (instead of monomer), which can be used as a kind of fuel oil, it will save energy. For this, methylstyrene as a chain-transfer agent was used for the thermal decomposition of EPS at 200 °C [109]. Three kinds of organic peroxides were tested as radical accelerators. It was found that the addition of dicumyl peroxide (DCP) enhanced the thermal decomposition of EPS even at a lower temperature, about 140 °C, but the addition of *tert*-butylcumyl peroxide was less effective than

DCP. On the other hand, di-*tert*-butyl peroxide showed almost no effect on the thermal decomposition of EPS.

3.1.9.2.3 **Mechanism of Polystyrene Cracking**

Radical depolymerization of neat PS samples produces large quantities of monomer (styrene), and chain-end backbiting yields substantial amounts of dimer and trimer. Polymer decomposition proceeds by entirely different processes when a catalyst is present. The formation of the primary PS catalytic cracking volatile products can be explained by the initial electrophilic attack on the polymer's aromatic rings by protons. Protons preferentially attack the ortho and para ring positions because the aliphatic polymer backbone is an electron-releasing group for the aromatic rings. Most volatile products can be derived from mechanisms beginning with ring protonation. Thermal decomposition of ortho-protonated aromatic rings in the polymer chain (**1**) can lead directly to the liberation of benzene, the primary catalytic cracking product, or may result in chain shortening. Benzene cannot be obtained directly from para-protonated aromatic rings in the polymer. However, para-protonated rings can react with neighboring polymer chains to yield the same chain scission products that are formed by ortho-protonation. The macro cation remaining after benzene evolution (**2**) may undergo chain-shortening β-scission to produce (**3**) and an unsaturated chain end, rearrange to form an internal double bond and protonate a neighboring aromatic ring (either by intra or intermolecular proton transfer), and cyclize to form an indane structure or abstract a hydride to produce a saturated chain segment. The substantial quantities of indane obtained by PS catalytic cracking suggest that cyclization of (**2**) to form indane structures is a favored process. A consequence of chain unsaturation resulting from (**2**) might be the formation of conjugated polyene segments, which may subsequently cyclize to form naphthalenes. Decomposition of (**3**), which might be formed from (**1**) or (**2**), can result in the formation of styrene or may lead to chain-end unsaturation and neighboring ring protonation (Figure 3.1.3).

Hydride abstraction by (**3**) would result in a saturated chain end. The lack of significant styrene production from any of the PS catalyst samples suggests that β-scission of (**3**) to form styrene is not a dominant decomposition pathway at low temperatures. Chain-end unsaturation derived from (**3**) may result in the formation of indenes, which were detected in substantial amounts only when HZSM-5 catalyst was present. The restricted volume of the HZSM-5 channels apparently inhibits hydride abstraction pathways for (**3**), which results in increased production of indenes and styrene for PS-HZSM-5. Protonation of the aromatic rings adjacent to methyl-terminated chain ends (**4**) can result in the formation of alkyl benzenes, propene, and benzene, depending on how the macro cation decomposes.

3.1.9.2.4 **Thermochemical Recycling of PS**

Thermochemical recycling techniques such as pyrolysis are usually applied. In pyrolysis, PS is thermally depolymerized at relatively low temperatures in order

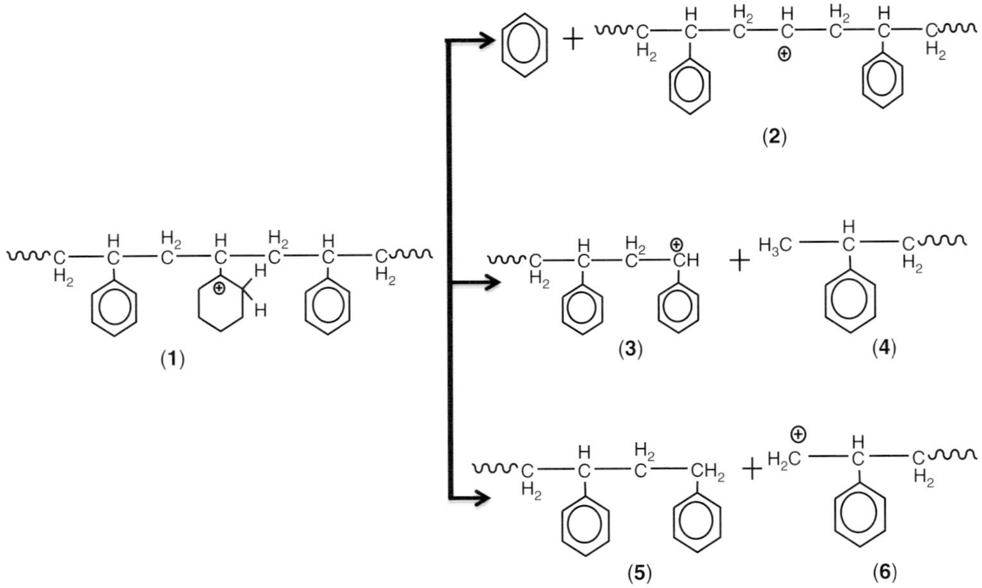

Figure 3.1.3 Mechanism of PS cracking.

to obtain the monomer styrene with a high selectivity. The thermal cracking of PS and poly(styrene–butadiene) (PSBD) on mesoporous silica (not measurably acidic) was studied by Arandes *et al.* [110]. As the content of PS in domestic plastic wastes is approximately 10 wt%, less attention has been paid to the cracking of dissolved PS than to the cracking of dissolved polyolefins. The design of the reactor also plays an important role in the cracking of PS. The plastics are fed in the solid state, and sand is used for helping fluidization in the fluidized bed for thermal cracking [85, 111, 112]. The design of fluidized beds used at the laboratory or pilot-plant scale is based on the kinetics of pyrolysis of plastics. As it involves a great level of uncertainty caused by factors such as heterogeneity of the material, synergy in the cracking of different constituents, and limitations to heat and mass transfer, these factors prevent obtaining kinetics that is reliable for the design of the reactor at temperatures of industrial interest (above 450 °C) [113]. The main disadvantage in the cracking of PS is the rapid deactivation of the catalyst caused by the coke formed on the acid sites, which is favored by the aromatic nature of styrene and its high C/H ratio. If a CSBR is used for the kinetic study of PS pyrolysis in the 450–550 °C range, the high heat transfer rate between solid and gas avoids agglomeration of particles [114]. A swirling fluidized-bed reactor (0.0508 m ID and 1.5 m high) was developed to recover the styrene monomer and valuable chemicals effectively from the PS waste, since it can control the residence time of the feed materials and enhance the uniformity of the temperature distribution [115]. The swirling fluidization mode makes the temperature fluctuations more periodic and persistent, which increases the uniformity of temperature distribution by reducing the temperature gradient in the reactor. The yields of styrene monomer as well as oil products can be increased by increasing the ratio of the swirling gas, but they exhibit their maximum values with increasing total volume flow rate of the gas.

Fe_2O_3, BaO, or HZSM-5 is used as catalyst to increase the selectivity and yield of styrene monomer in the product. It was observed that the reaction time and temperature could be reduced considerably by adding the solid catalyst.

Achilias conducted catalytic and noncatalytic pyrolysis experiments in a fixed-bed reactor using both a model polymer and commercial waste products as the feedstock [54]. The liquid fraction produced from all the pyrolysis experiments consisted mainly of the styrene monomer, and this could be repolymerized without any further purification in differential scanning calorimetry (DSC) with azobisisobutyronitrile (AIBN) initiator. A basic (BaO) and an acidic commercial FCC catalyst were examined in relation to the yield and composition of gaseous and liquid products. Aromatic compounds were identified in the liquid fraction. Products of the thermal and catalytic pyrolysis of model PS and commercial waste were aromatic compounds, which could be polymerized to produce a polymer similar to the original PS. Other aromatic compounds of this fraction act as chain-transfer agents, lowering the average molecular weight of the polymer produced and contributing to apolymer with a lower T_g. Therefore, the polymer can be reproduced but with inferior properties compared to a polymer prepared from neat styrene.

Sterling developed a general model for polymer degradation by concurrent random and chain-end processes using continuous distribution kinetics. Population balance equations based on fundamental, mechanistic free-radical reactions were solved analytically by the moment method [116]. The model was applicable for all molecular weight distributions (MWDs), and reduces to the cases of independent random or chain-end scission.

PS degradation experiments in a mineral oil solution at 275–350 °C supported the model and determined the reaction rate parameters. The degradation proceeded to moderate extents requiring an MW-dependent random scission rate coefficient. PS random scission activation energy was 7.0 kcal mol^{-1} similar to thermolysis, but was lower than that found by pyrolysis due to fundamental differences between the processes.

Thermal and thermo-oxidative degradation of PS in the presence of ammonium sulfate was studied with thermogravimetry and Fourier transform infrared (FTIR). TGA results indicated that ammonium sulfate accelerated thermal degradation in nitrogen but delayed thermo-oxidative degradation of PS in air [117]. Infrared (IR) analysis of tetrahydrofuran extracts, from the samples degraded at 340 °C and of residues after thermal treatment at 340 °C in a furnace, showed that the acceleration of thermal degradation and the suppression of thermo-oxidative degradation were due to sulfonation and oxidation of ammonium sulfate and its decomposition products and formation of unsaturated structures in the PS chain.

Thermal degradation of PS was investigated in the presence of water under subcritical conditions (hydrous pyrolysis) in closed systems under an inert atmosphere, in temperature range 300–350 °C and pressures up to 18 MPa, for 1–120 h. The products obtained were separated as gases, volatiles, and heavy compounds [118]. The results showed that the presence of water increased the yields of the volatile products, mainly in the first steps of the pyrolytic process and led to higher yields of the monomer. This latter observation suggests a lowering of the secondary reaction extent. The catalytic degradation of waste plastics such as HDPE, LDPE, PP, and PS over a spent FCC catalyst was carried out at atmospheric pressure with a stirred semi-batch operation at 400 °C [106, 119]. The objective of the investigaton was to test the influence of plastic types on the yield, liquid product rate, and liquid product distribution for catalytic degradation. The catalytic degradation of waste PE and PP with polyolefinic structure resulted in the liquid yield of 80–85% and the solid yield below 1%, whereas that of waste PS with polycyclic structure produced much more liquid and solid products and much less gas products. Accumulative liquid product weight by catalytic degradation strongly depended on the degradation temperature of the plastics.

In accordance with the option of recycling plastics into fuels by dissolving them in standard feedstock for the process of catalytic cracking of hydrocarbons, FCC and various acidic catalysts (zeolites ZSM-5, mordenite, Y, and a sulfur-promoted zirconia) were tested in the conversion of PS dissolved in inert benzene at 550 °C in a fluidized-bed batch reactor [120]. Experiments were performed with very short contact times of up to 12 s. The main products were in the gasoline range, including benzene, toluene, ethylbenzene, styrene, and minor amounts of C9–12 aromatics and light C5 compounds. Coke was always produced in very significant amounts. Even though sulfur-promoted zirconia is highly acidic, the low proportion of Brönsted-type acid sites does not allow the occurrence of secondary styrene reactions. It was shown that the most favorable product distributions (higher yields of desirable products) are obtained on equilibrium commercial FCC catalysts.

PS can be recycled into styrene monomer in association with some other aromatics, from which styrene can be converted to biodegradable plastic such as poly(hydroxy alkanoate)s (PHAs). Ten percen yield of PHA from PS has been achieved [121]. The chain length of PHA produced was 10. The yield and composition of oils and gases derived from the pyrolysis and catalytic pyrolysis of PS have been investigated. The pyrolysis and catalytic pyrolysis was carried out in a fixed-bed reactor. Two catalysts were used, zeolite ZSM-5 and Y-zeolite, and the influence of the temperature of the catalyst, the amount of catalyst loading, and the use of a mixture of the two catalysts were investigated. The main product from the uncatalyzed pyrolysis of PS was oil consisting mostly of styrene and other aromatic hydrocarbons such as toluene and ethylbenzene. In the presence of either catalyst, an increase in the yield of gas and decrease in the amount of oil production were found, but there was significant formation of carbonaceous coke on the catalyst. Increasing the temperature of the Y-zeolite catalyst and also the amount of catalyst in the catalyst bed resulted in a decrease in YOil and an increase in the yield of gas. Oil derived from the catalytic pyrolysis of PS contained aromatic compounds such as single-ring compounds like benzene, toluene, styrene, *m*-xylene, *o*-xylene, *p*-xylene, ethylmethylbenzene, propenylbenzene, and methylstyrene; two-ring compounds like indene, methylindene, naphthalene, 2-methylnaphthalene, 1-methylnaphthalene, biphenyl, methylbiphenyl, dimethylnaphthalene, trimethylnaphthalene, tetramethylnaphthalene, and ethylbiphenyl; three-ring compounds like phenanthrenes; and four-ring compounds like pyrenes and chrysenes.

3.1.9.3 Future Prospects

Some future prospects of PS recycling includethe following:

1. Thermal recycling of PS yields higher percentage of styrene monomer, which can be fermented by bacteria to produce PHAs – the starting material for the synthesis of biodegradable polymers.
2. The waste PS can be blended with biodegradable polymers to produce biodegradable polymers.
3. Styrene monomer produced by recycling can be grafted onto biodegradable polymers to give biodegradable polymers [122].

3.1.10 Chemical Recycling of Poly(vinyl chloride)

3.1.10.1 Introduction

The main applications of PVC include food packaging, shoes, flooring, pipes, clothing (leather-like material), ceiling tiles, and multilayered flooring and window frames. Approximate life of a PVCitem is 5 years, and it has the same density as PET. This makes the separation of PVC difficult prior to recycling of plastic wastes containing both polymers. The technique used for the efficient separation is X-ray fluorescence and IR sorting. The chlorine atoms of PVC are detected, and the wastes are sent for separation [123].

In general, PVC waste occurs in two ways:

1. MPW fraction (with a rather low PVC content)
2. PVC-rich plastics fraction.

PVC incineration is associated with formation of the high chlorine content of this polymer in the form of hydrochloric acid (HCl) and the formation of persistent and toxic compounds such as dioxines and furans [124–126]. In addition, when PVC wastes are fired in an incinerator, HCl corrodes the boiler tubes of the incinerator and other equipment. Therefore, the steam pressure must be kept relatively low to prevent corrosion of the heat recovery boiler [127].

Neutralization of HCl with calcium carbonate (lime) and/or sodium hydroxide (caustic soda) to convert the released HCl to the respective salts is the best solution of this problem, and special filters can also be used to prevent problems related to atmospheric emissions during the combustion process [128].

Complete combustion of PVC-rich waste occurs at high temperatures (>1700 K) because PVC is inherently noncombustible. The net energy recovered by incineration of PVC-rich waste is also not high enough to make it highly economical. The calorific value of the incineration of PVC is about 64 MJ kg^{-1} (under ideal conditions), while for paper it is 17 and for wood 16 MJ kg^{-1} [129]. Therefore, mechanical and/or chemical recycling of PVC plastic wastes does not seem a logical solution. The usual approach for chemical recycling of PVC wastes is currently "thermal cracking" via hydrogenation, pyrolysis, or gasification [130–134].

The main intermediate product of the thermal cracking is a polyene material, which continuously degrades by the evolution of aromatics and converts to products, the composition of which is strongly determined by the processing variables such as the type of atmosphere, temperature, and residence time. In an inert atmosphere, the degradation products are HCl, gaseous and liquid hydrocarbons, and char. HCl is the main product and can be reused either in vinyl chloride production or in other chemical processes [135].

In the recycling process of vinyl chloride, a gas purification unit is also incorporated to get high-purity hydrogen chloride gas. In a steam atmosphere and at high temperatures, the hydrocarbon fraction is converted into other products such as carbon monoxide, carbon dioxide, and hydrogen.

In a reported process, a bench-scale bubbling fluidized bed is used to investigate some processing parameters on the product outcome. The choice of the type of bed material influences the product outcome; for example, the use of catalytic inactive solid quartz as bed material yields large amounts of char and tar, whereas the application of a catalytically active material such as porous alumina results in a high conversion of PVC into syngas. In addition, temperature has a large impact on the composition of the products, so that the carbon to gas conversion improves from about 70% at 1150 K to approximately 100% by increasing the reactor temperature to 1250 K [14].

For chemical recycling of PVC, an increase in efficiency of dehydrochlorination process usually leads to successful recycling [136]. Besides, the emission of hydrogen chloride changes significantly with the oxides used, indicating the chlorine-fixing ability of oxides. Also, the use of poly(ethylene glycol) (PEG) can accelerate dehydrochlorination of PVC, so that at 210 °C for 1 h, the dehydrochlorination degree was as high as 74% for poly(vinyl chloride)/poly(ethylene glycol) (PVC/PEG) while for PVC it was only 50%. Moreover, it was demonstrated that for PVC/PEG the decomposition of PVC shifted to lower temperatures compared with that of pure PVC, suggesting some interactions exist between PEG and PVC that caused the increased dehydrochlorination rate. According to the results, during this process no waste byproducts such as KCl were produced, and satisfactory recyclability of PEG (10 cycles) could be obtained [137].

An alternative method to the thermal process of dehydrochlorination is the rather easy process of dehydrochlorination under the influence of alkaline media to recover HCl with a possibility that the degradation of PVC by oxygen oxidation in an aqueous alkaline solution produces various carboxylic acids.

Some researchers have demonstrated that dehydrochlorination of flexible PVC occurred first, followed by oxidation. The major products reported were oxalic acid, a mixture of benzenecarboxylic acids, and CO_2. However, the chlorine content could also be recovered in the form of HCl by adjusting the reaction conditions such as the alkali concentration [138].

Among the various methods of thermal cracking, pyrolysis is a better known procedure in the chemical recycling of PVC. The process of pyrolysis, which takes places at 500–900 °C without any oxygen, is a very suitable recycling method especially in the case of mixed plastic wastes (PVC recycling, 2005). In a typical process, a PVC-rich waste can be pyrolyzed to hydrocarbons (oil), soot, HCl, chlorinated hydrocarbons, and so on. HCl needs to be removed from the pyrolysis gas, although this removal process can result in the formation of toxic dioxins in some stages. The main end product of pyrolysis is, however, oil [123].

Corrosion of the process equipment (e.g., pyrolysis reactor and piping) is the main problem associated with the pyrolysis of PVC and mixed plastics containing PVC materials through the formation of the acid gas (HCl). Moreover, many petrochemical specifications limit the amounts of halogens (appeared in the form of hydrogen chloride and chloro-organic compounds) to a very low range in the gas and oil derived from plastic waste. Therefore, in the case of mixed plastic wastes (uneconomical to separate into a single polymer) with a low PVC content, the conventional chemical recycling is frequently used only for a waste stream in which the PVC content is less than 30% (e.g., the multiple material products) [139].

Milling of PVC with CaO is an effective way to extract Cl from the waste, among the several solutions proposed to this problem [140]. An attempt has also been made to develop a process for recovering metals from alloy wastes by using a mechanochemical reaction consisting of a co-grinding alloy and PVC waste, followed by washing with water and filtration [141]. Currently, the NKT-Watech pyrolysis process in Europe uses another two-step pyrolysis of PVC wastes in a stirred vessel. Calcium carbonate and a filler are used to react with the liberated HCl and produce calcium chloride. Then at the increased temperature, the polymer chains break down, producing a solid coke residue. Finally, the residual calcium chloride can be treated to make it suitable for the market [142].

In an alternative approach, PVC and other halogenated plastics are removed from the feed during pretreatment of mixed plastic waste. Such pretreatment methods include dilution of the wastes with excessive chlorine content with less chlorine-containing or chlorine-free polymer mixture. It is also common to dilute the chlorine-containing hydrocarbon feed with chlorine-free petroleum fractions coming from refineries. Another approach that is less expensive and more acceptable is thermal dehalogenation, which takes place either in a liquid or in a fluidized bed.

A new method involving co-pyrolysis of PVC with nitrogen compounds (biowaste) was carried out to reduce the corrosive effects of the generated HCl. The researchers used cattle manure with PVC during pyrolysis using a statistical method, and optimized the conditions to provide the highest HCl reduction during PVC pyrolysis. They compared results under same optimized conditions with a plastic mixture without cattle manure, and then determined the quality of the obtained products. It was concluded that the lowest heating rate, the highest reaction temperature (450 °C), and the PVC/cattle manure ratio of 1 : 5 were the most suitable conditions providing the highest HCl reduction. However, according to their results the presence of manure decreases the oil yield of pyrolysis by about 17% [139].

Nucleophilic substitution of chlorine atoms of rigid PVC can chemically modify its structure. Reactions of rigid PVC with various nucleophiles (Nu's), such as iodide, hydroxide, azide, and thiocyanate, in ethylene glycol as solvent were used to modify its structure. Such reactions lead to the substitution of Cl by Nu and finally elimination of HCl, resulting in the dehydrochlorination of the rigid PVC. According to the results, the dehydrochlorination yield increased with increasing Nu concentration, resulting in maximum substitution at high Nu amounts. Moreover, when ethylene glycol was replaced by diethylene glycol, the dehydrochlorination was found to accelerate, which may be due to the higher compatibility of diethylene glycol with PVC, making it easier to penetrate the rigid PVC particles [143].

Other technologies based on depolymerization and repolymerization have been developed for chemical recycling of PVC, most of which unfortunately are more expensive than mechanical recycling [144].

PVC waste was used in a study that was carried out for recycling electric arc furnace dust by heat treatment with PVC. The entire process was aimed at recovering zinc, lead, and cadmium from the dust, and was adjusted so that the residual dust can be injected into the electric arc furnace [145].

3.1.10.2 Mixed Plastic Recycling Processes

The company Recovinyl in UK deals with post-consumer PVC to re-produce two grades via mechanical recycling. Because of its structure and composition, PVC can easily be mechanically recycled in order to obtain good-quality recycled material. Careful and proper sorting is of crucial importance for the optimal recycling of PVC [146]. A pyrolytic process, which has proven to be successful for PSW rich in PVC, is the Akzo process (Netherlands). With a capacity of 30 kg h^{-1}, this fast pyrolysis process is based on a circulating fluidized-bed system (two reactors) with subsequent combustion. Input to the process is shredded mixed waste including a high percentage of PVC waste. The main output consists of HCl, CO, H_2, CH_4, and, depending on the feedstock composition, other hydrocarbons and fly ash [147].

The NRC process is another successful pyrolysis scheme. This process is based on the pyrolysis with subsequent metal extraction technology. The aim is to produce purified calcium chloride instead of HCl. The input to the process is PVC waste (cables, flooring, profiles, etc.). No other PSW type is fed to the processing system, which results in calcium chloride, coke, organic condensate (for use as fuels), and heavy metals for metal recycling as products [53].

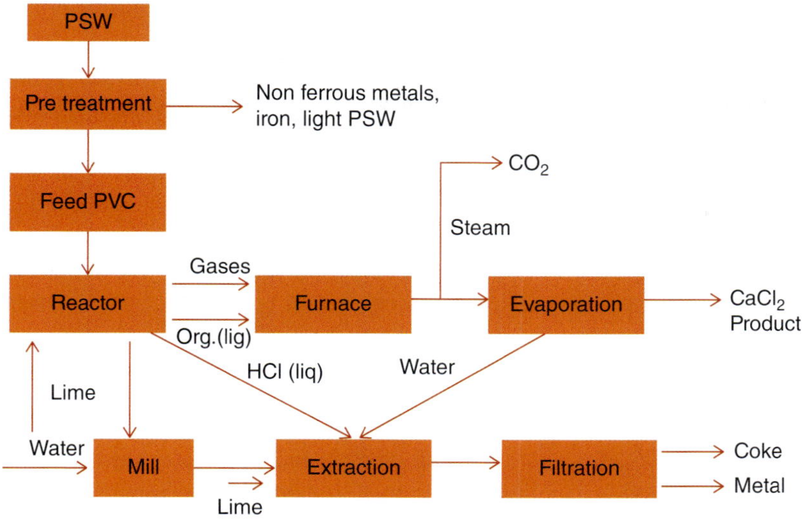

Figure 3.1.4 Schematic diagrams showing the NTK process diagram.

The NTK process, depicted in (Figure 3.1.4), is a very successful recycling process. The process is based on an initial pretreatment step that involves separating light plastics (PP, PE, etc.) and other materials, for example, wood, sand, iron, steel, brass, copper, and other metallic pollutants.

PSW waste is then fed to a reactor at a low pressure (2–3 bar) and a moderate temperature (375 °C). The process emits no dioxins, chlorine, metals, or plasticizers. Also, there are no liquid waste streams in the process since all streams are recycled within the system. There is a small volume of carbon dioxide gas formed by the reaction between lime/limestone and hydrogen chloride. Mixed PVC building waste containing metals, sand, soil, PE, PP, wood, and rubber waste has been successfully treated [53]. The gasification into high calorific value fuel gas obtained from PVC was also reported by Borgianni *et al.* [134].

Chemical recycling of PVC has also been attempted. Most of the proposed processes use the rather easy dehydrochlorination of PVC either under the influence of heat or in alkaline media. The oxidative degradation of PVC by molecular oxygen in an aqueous alkaline solution at temperatures between 150 and 260 °C with oxygen pressures of 1–10 MPa has been observed. The main products are oxalic acid and carbon dioxide, their yield depending on the reaction conditions and the alkali concentration. The maximum yield of oxalic acid was 45%, and 42% of the chlorine content could be recovered in the form of HCl [148].

Oxidative degradation of rigid PVC (R-PVC) pellets with oxygen was carried out in 1–25 mol kg^{-1}-H_2O (m) NaOH solutions, at 150–260 °C and PO_2 of 1–10 MPa, in order to investigate the chemical recycling of PVC materials. The apparent rate of oxidative degradation of R-PVC progressed as a zero-order reaction, and the apparent activation energy was 38.5 kJ mol^{-1}. The major products were oxalic acid, a mixture of benzene and carboxylic acids, and CO_2. The tin in R-PVC was extracted completely. The possibility of converting PVC materials into raw materials such as carboxylic acids by chemical recycling has been reported [127].

3.1.10.3
Mixed PVC Wastes World Initiatives

Regarding the chemical recycling of mixed plastic wastes with PVC content of up to several percent, some elementary preparation of the waste plastics feed is required, including size reduction and removal of most non-plastics. This prepared feed is fed directly into the heated fluidized-bed reactor, which forms the heart of the polymer cracking process. The reactor operates at approximately 500 °C and in the absence of air. The plastics crack thermally under these conditions to hydrocarbons, which vaporize and leave the bed with the fluidizing gas. Solid impurities, including metals, PVC stabilizers, and some coke, are either accumulated in the bed or carried away in the hot gas as fine particles for capture by the cyclone. The decomposition of PVC leads to the formation of HCl, which is neutralized by bringing the hot gas into contact with a solid lime absorbent. This results in a $CaCl_2$ fraction that has to be sent to a landfill. The purified gas is cooled to condense most of the hydrocarbons as valuable distillate feedstock. This is then stored and tested against agreed specifications before transfer to the downstream user plant. The remaining light hydrocarbon gas is compressed, reheated, and returned to the reactor as fluidizing gas. Part of the stream could be used as fuel gas for heating the cracking reactor, but as it is olefin-rich, recovery options are being considered. The process shows very good results with regard to the removal of elements such as chlorine. With an input of 10 000 ppm (or 1%) Cl, the products will contain around 10 ppm Cl. This is somewhat higher than the specification of 5 ppm typical for refinery use. However, in view of the high dilution likely in any refinery or petrochemical application, BP thinks that this is acceptable. Also, metals such as Pb, Cd, and Sb can be removed to very low levels in the products. Tests have shown that all the hydrocarbon products can be used for further treatment in refineries [149].

3.1.10.4
The BASF Feedstock Recycling Process

It was designed to handle the recycling of mixed plastic waste supplied by the direct store delivery (DSD) collection system. The process is as follows: Before the waste plastics can be fed to the process, a pretreatment is necessary (Figure 3.1.5). In this step, the plastics are ground, separated from other materials such as metals, and agglomerated. The conversion of the pretreated, mixed plastic into petrochemical raw materials takes place in a multistage melting and process. In the first stage, the plastic is melted and dehalogenized to protect the subsequent plant

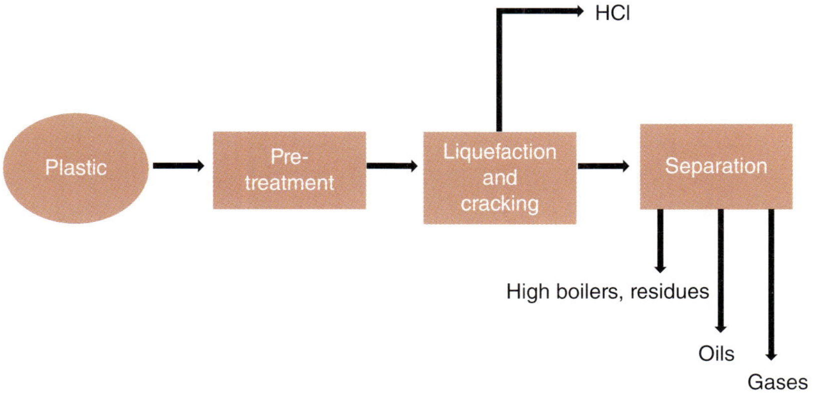

Figure 3.1.5 BASF processes [150].

segments from corrosion. The hydrogen chloride separated out in this process is absorbed and processed in the HCl production plant. Hence, the major part of the chlorine present in the input (e.g., from PVC) is converted into saleable HCl. Minor amounts come as NaCl or $CaCl_2$ effluent [150]. Gaseous organic products are compressed and can be used as feedstock in a cracker. In the subsequent stages, the liquefied plastic waste is heated to over 400 °C and cracked into components of different chain lengths. About 20–30% of gases and 60–70% of oils are produced, which are subsequently separated in a distillation column. Naphtha produced by the feedstock process is treated in a steam cracker, and the monomers (e.g., ethylene, propylene) are recovered. These raw materials are used for the production of virgin plastic materials. High-boiling oils can be processed into syngas or conversion coke and then be transferred for further use. The residues consist of 5% minerals at most, for example, pigments or aluminum lids. It seems likely that metals present in PVC formulations mainly end up in this outlet. The process is carried out under atmospheric pressure in a closed system and, therefore, no other residues or emissions occur.

Mixed PVC waste can be used as reduction agent in blast furnaces. For the production of pig iron for steel production, iron ore (Fe_2O_3) has to be reduced to Fe. This process takes place in a blast furnace. Coke, coal, and heavy oil are normally used as reducing agents in this process. Iron and steel companies try to lower the consumption of coke by partly replacing it with coal, gas, or fuel oil (30% in weight seems to be the maximum) via coal injection technology. Recently, new developments have started to replace the conventional reducing agents by plastics waste. The pioneer in this field is Stahlwerke Bremen, Germany, though others like British Steel (UK) have done trials as well. Stahlwerke Bremen is a large German steel manufacturer that operates two blast furnaces to produce over 7000 t day^{-1} or some 3 million tpa pig iron. Currently, Germany is the only country where blast furnace use waste in this way [151].

3.1.10.5
Veba Combi Cracking Process

The plant configuration includes a depolymerization section and the Veba Combi cracking (VCC) section. Depolymerization is required to allow further processing in the VCC section. In the depolymerization section, the agglomerated plastic waste is kept between 350 and 400 °C to effect depolymerization and dechlorination. The overhead product of the depolymerization is partially condensed. The main part (80%) of the chlorine introduced with PVC is present as HCl in the light gases. It is washed out in the following gas purification process, yielding technical HCl. The condensate, containing 18% of the chlorine input, is fed into a hydro treater. The HCl is eliminated with the formed water. The resulting Cl-free condensate and gas are mixed with the depolymerized material for treatment in the VCC section. The depolymerized material is hydrogenated in the VCC section at 400–450 °C under high pressure (about 100 bar) in a liquid-phase reactor with no internals. Separation yields a product, which, after treatment in a fixed-bed hydrotreater, is a synthetic crude oil, a valuable product that may be processed in any refinery. From the separation, a hydrogenated residue stream also results, which comprises heavy hydrocarbons contaminated with ashes, metals, and inert salts. This hydrogenation bitumen is a byproduct, which is blended with the coal for coke production (2 wt%). It is likely that the major part of any metals present in a PVC formulation ends up in this residue flow. Light cracking products end up in off-gas (E-gas), which is sent to a treatment section for H_2S and ammonia removal. As indicated previously, the main part of the chlorine present in the input (i.e., from PVC) is converted into usable HCl. Some 2% of the chlorine input is bound to $CaCl_2$ in the process, which is 4 times leaner than the stochiometric amount of CaO.

3.1.11
Chemical Recycling of Poly(methyl methacrylate)

3.1.11.1
Introduction

PMMA is a thermoplastic used throughout the world in such applications as transparent all-weather sheets, electrical insulation, bathroom units, automotive parts, surface coating, ion exchange resins, and so on. The plastics made from PMMA are widely used under the commercial trade names PLEXIGLAS or PERSPEX.

In contrast to condensation polymers (e.g., PET), addition polymers such as PMMA cannot be easily recycled to the monomer by simple chemical methods. Pyrolysis is applied instead thermochemical recycling techniques. Among the various methods of depolymerization, the most prominent ones are the molten metal bath process and fluidized-bed pyrolysis [23, 150–153].

The effect of temperature, addition of the filler, and amount of feed on the amount and distribution of pyrolytic products was investigated by Kaminsky and coworkers in a fluidized-bed reactor [23, 154–156]. Furthermore, PMMA's thermal pyrolysis results in nearly 97% recovery of the monomer methyl methacrylate (MMA) at relatively low temperatures (400–500 °C) [152]. It has been reported that the liquid pyrolysis product was so pure that it could be polymerized again without any further treatment [154]. Achilias investigated the chemical recycling of PMMA using pyrolysis, aiming at the recovery of the pure monomer to be repolymerized back to the polymer [54].

The chemical recycling of PMMA can be done by two different methods.

3.1.11.2 Dissolution/Reprecipitation

The traditional method of dissolution/reprecipitation belongs to mechanical recycling [157, 158]. In this technique, the polymer is separated and recycled using a solvent/non-solvent system. For this purpose, different systems are examined at different weight percents and temperatures. Solvent-based processes include stages of treating plastic waste with solvents so that the polymeric materials are dissolved and then recovered by reprecipitation. In this approach, a model PMMA is used together with the commercial product containing PMMA. Xylene and toluene are used as solvents, and *n*-hexane as the non-solvent. Some other parameters are adjusted according to requirement of the process, such as the concentration of the polymer, solvent/non-solvent volume ratio, and the dissolution temperature below the boiling point for each solvent (140 °C for xylene and 110 °C for toluene).

To start with, the polymer and the solvent are added into a flask equipped with a vertical condenser and a magnetic stirrer. The system is heated for 30 min to the desired temperature. Then, the flask is cooled and the solution of the polymer is properly poured into the non-solvent. The polymer is reprecipitated, washed, filtrated, and dried in an oven at 80 °C for 10 h. The recycled polymer is obtained in the form of a powder or grains.

These processes have the advantage that they are able to deal with mixtures of polymers, based on the principle of selective dissolution. Moreover, the dissolution/reprecipitation technique seems to have several advantages:

1. The plastic waste is eventually converted into a form acceptable to the fabrication equipment (powder or small grains).
2. Additives and insoluble contaminants can be removed by filtration, leaving the pure material.
3. Except heating for dissolving, no further degradation, due to the recycling process itself, is anticipated.
4. The value added during the polymerization stage is maintained intact, and the recycled polymers, free of any contaminants, can be used for any type of applications, since the final product is of a quality comparable to the virgin

material. This method has already been studied in the recovery of a variety of (mainly model) polymers including PVC, PS, LDPE, and HDPE [155, 156].

This method is widely used in several countries, but it has several serious disadvantages: for example, the raw condensate MMA may be contaminated by the metal used (usually lead) or other byproducts [153].

3.1.11.3
Chemical/Feedstock Recycling

Thermal cracking, or pyrolysis, involves the degradation of polymeric materials by heating in the absence of oxygen (usually in a nitrogen atmosphere). During pyrolysis at high temperatures, depending on polymer type, either end-chain or random scission of the macromolecules occurs. In the first case (occurring in PMMA), the monomer is produced in large amounts. Conventional (thermal) pyrolysis can be carried out using either a model using the raw polymer or a commercial product as feedstock. The experiments are carried out in a laboratory fixed-bed reactor at 450 °C, which is the optimum temperature for the maximization of the monomer MMA [155].

The liquid product obtained from both the model and the commercial samples is very high, 99% and 98%, respectively. Monomer recovery is higher when feeding pure PMMA (98.3 wt%) compared to the commercial sample (94.9 wt%). In both cases, the gas fraction is very small. Only 0.6 or 1.5 wt% of gases is produced. Also, the residue obtained is very low, 0.1 and 0.4 wt%. Furthermore, the gas composition of both samples is approximately the same, with large amounts of CO_2 followed by CO and methane. The first two are degradation products of PMMA and MMA due to the existence of oxygen in the macromolecular chain. The liquid fraction mostly consists of the monomer MMA in a large amount (99 and 97 wt% for the model polymer and the commercial product, respectively) with a small percentage of some other organic compounds, mainly esters. The potential use of the liquid pyrolysis fraction as a raw material for the reproduction of PMMA by polymerization was investigated using DSC, and the results were compared with those of virgin monomer. The results for the direct polymer reproduction were not very encouraging [54].

Furthermore, thermal pyrolysis was carried out using either the model polymer as raw material or a commercial product (PERSPEX). The experiments were carried out in a laboratory fixed-bed reactor. The potential use of the oil produced as a raw material for the reproduction of PMMA by polymerization was investigated using DSC, and the results were compared with those from the virgin monomer. The results for the direct polymer reproduction were not very encouraging [13].

All experiments were carried out in the laboratory Lf Environmental Fuels and Hydrocarbons, situated at CPERI, Thessaloniki, Greece. The reactor was filled with 0.7 g glass beads, and the piston was filled with the polymer (1.5 g). The system was always heated in the presence of N_2 (30 ml min^{-1}). As soon as the reaction temperatures were achieved, the polymer entered the reactor and the experiment started. The time of the experiment was 17 min, and the reaction temperature was 450 °C. At the end of the experiment, purging (30 min) was performed. Both the experiment (100 ml min^{-1}) and purging (30 ml min^{-1}) were performed in the presence of N_2. Details on the procedure can be found in Refs. [159, 160].

The liquid products were collected in a liquid bath (−17 °C) and quantitatively measured in a preweighed glass receiver. The gaseous products were collected and measured by water displacement. The liquid samples were analyzed by GC/MS in an HP 5989 MS engine. Gaseous products were analyzed in an HP 6890 GC, equipped with four columns and two detectors (TCD and FID). The chromatograph was standardized with gases at known concentrations as standard mixtures [13].

References

1. Williams, P.T. and Bagri, R. (2004) Hydrocarbon gases and oils from the recycling of polystyrene waste by catalytic pyrolysis. *Int. J. Energy Res.*, **28** (1), 31–44.
2. Yamada-Onodera, K., Mukumoto, H., Katsuyaya, Y., Saiganji, A., and Tani, Y. (2001) Degradation of polyethylene by a fungus, *Penicillium simplicissimum* YK. *Polym. Degrad. Stab.*, **72**, 323–327.
3. Zheng, Y., Yanful, E.K., and Bassi, A.S. (2005) A review of plastic waste biodegradation. *Crit. Rev. Biotechnol.*, **25**, 243–250.
4. Marqués-Calvo, M.S., Cerdà-Cuéllar, M., Kint, D.P.R., Bou, J.J., and Muñoz-Guerra, S. (2006) Enzymatic and microbial biodegradability of poly(ethylene terephthalate) copolymers containing nitrated units. *Polym. Degrad. Stab.*, **91**, 663–671.
5. Bonhomme, S., Cuer, A., Delort, A.-M., Lemaire, J., Sancelme, M., and Scott, G. (2003) Environmental degradation of polyethylene. *Polym. Degrad. Stab.*, **81**, 441–452.
6. Albertsson, A.C., Andersson, S.O., and Karlsson, S. (1987) The mechanism of biodegradation of polyethylene. *Polym. Degrad. Stab.*, **18**, 73–87.
7. Jayasekara, R., Harding, I., Bowater, I., and Lornergan, G. (2005) Biodegradability of selected range of polymers and polymer blends and standard methods for assessment of biodegradation. *J. Polym. Environ.*, **13**, 231–251.
8. Andrady, A.L. (2011) Microplastics in the marine environment. *Mar. Pollut. Bull.*, **62**, 1596–1605.
9. Raquez, J.-M., Bourgeois, A., Jacobs, H., Degée, P., Alexandre, M., and Dubois, P. (2011) Oxidative degradations of oxodegradable LDPE enhanced with thermoplastic pea starch: thermo-mechanical properties, morphology, and UV-ageing studies. *J. Appl. Polym. Sci.*, **122**, 489–496.
10. Müller, R.-J., Kleeberg, I., and Deckwer, W.-D. (2001) Biodegradation of polyesters containing aromatic constituents. *J. Biotechnol.*, **86**, 87–95.
11. Seyhan, N. Ege, (1999) *Organic Chemistry–Structure and Reactivity*, Houghton Mifflin Company, pp. 1024–1064.
12. Francis, R. and Sethi, B. (2012) Catalytic feedstock recycling of polymers: a green approach towards sustainable environment. *Sci. Res. Adv. Mater. Phys. Chem.*, **2**, 263–266.
13. Achilias, D.S. (2006) Chemical recycling of polymers: the case of poly(methyl methacrylate). Proceedings of the 2006 IASME/WSEAS International Conference on Energy & Environmental Systems, pp. 271–276.
14. Achilias, D.S. *et al.* (2012) Recent advances in the: chemical recycling of polymers (PP, PS, LDPE, HDPE, PVC, PC, Nylon, PMMA), in *Material Recycling – Trends and Perspectives*, InTech: Croatia.

15. Karayannidis, G.P. and Achilias, D.S. (2007) Chemical recycling of poly(ethylene terephthalate). *Macromol. Mater. Eng.*, **292** (2), 128–146.
16. Scheirs, J. (1998) *Polymer Recycling, Science, Technology and Applications*, John Wiley & Sons, Inc., New York.
17. Butkutė, R. and Miknius, L. (2015) Properties of residual marine fuel produced by thermolysis from polypropylene waste. *Mater. Sci.-Medziagotyra*, **21** (2), 249–254.
18. Jagtap, M.D., Khatavkar, S.S., and Quazi, T.Z. (2015) Methods for waste plastic recycling. *Int. J. Recent Technol. Mech. Electr. Eng.*, **2** (6), 120–122.
19. Stelmachowski, M. (2010) Thermal conversion of waste polyolefins to the mixture of hydrocarbons in the reactor with molten metal bed. *Energy Convers. Manage.*, **51**, 2016–2024.
20. Arabiourrutia, M., Elordi, G., Lopez, G., Borsella, E., Bilbao, J., and Olazar, M. (2012) Characterization of the waxes obtained by the pyrolysis of polyolefin plastics in a conical spouted bed reactor. *J. Anal. Appl. Pyrolysis*, **94**, 230–237.
21. Stelmachowski, M. and Slowinski, K. (2012) Thermal and thermo-catalytic conversion of waste polyolefins to fuel-like mixture of hydrocarbons. *Chem. Process Eng.*, **33** (1), 185–198.
22. Jung, S.H., Cho, M.H., Kang, B.S., and Kim, J.S. (2010) Pyrolysis of a fraction of waste polypropylene and polyethylene for the recovery of BTX aromatics using a fluidized bed reactor. *Fuel Process. Technol.*, **91**, 277–284.
23. Kaminsky, W., Predel, M., and Sadiki, A. (2004) Feedstock recycling of polymers by pyrolysis in a fluidised bed. *Polym. Degrad. Stab.*, **85**, 1045–1050.
24. Lin, Y.H. and Yen, H.Y. (2005) Fluidised bed pyrolysis of polypropylene over cracking catalysts for producing hydrocarbons. *Polym. Degrad. Stab.*, **89**, 101–108.
25. Shah, S.H., Khan, Z.M., Raja, I.A., Mahmood, Q., Bhatti, Z.A., Khan, J., Farooq, A., Rashid, N., and Wu, D. (2010) Low temperature conversion of plastic waste into light hydrocarbons. *J. Hazard. Mater.*, **179**, 15–20.
26. Na, J.G., Jeong, B.H., Kim, S.S., and Chung, S.H. (2006) Pyrolysis of low-density polyethylene using synthetic catalysts produced from fly ash. *J. Mater. Cycles Waste Manage.*, **8**, 126–132.
27. Park, H.J., Yim, J.H., Jeon, J.K., Kim, J.M., Yoo, K.S., and Park, Y.K. (2008) Pyrolysis of polypropylene over mesoporous MCM-48 material. *J. Phys. Chem. Solids*, **69**, 1125–1128.
28. Tekin, K., Akalin, M.K., Kadi, C., and Karagoz, S. (2012) Catalytic degradation of waste polypropylene by pyrolysis. *J. Energy Inst.*, **85** (3), 150–155.
29. Mehrdad, S.A.A., Mehdi, N.H., and Yeganeh, H. (2013) Evaluation of pyrolysis product of virgin high density polyethylene degradation using different process parameters in a stirred reactor. *Fuel Process. Technol.*, **109**, 90–95.
30. Demirbas, A. (2004) Pyrolysis of municipal plastic wastes for recovery of gasoline-range hydrocarbons. *J. Anal. Appl. Pyrolysis*, **72**, 97–102.
31. Aboulkas, A., Makayssi, T., Bilali, L., Elharfi, K., Nadifiyine, M., and Benchanaa, M. (2012) Co-pyrolysis of oil shale and plastics: influence of pyrolysis parameters on the product yields. *Fuel Process. Technol.*, **96**, 209–213.
32. Adrados, A., Marco, I., Caballero, B.M., López, A., Laresgoiti, M.F., and Torres, A. (2012) Pyrolysis of plastic packaging waste: a comparison of plastic residuals from material recovery facilities with simulated plastic waste. *Waste Manage.*, **32**, 826–832.
33. Williams, E.A. and Williams, P.T. (1997) Analysis of products derived from the fast pyrolysis of plastic waste. *J. Anal. Appl. Pyrolysis*, 347–363.
34. Domınguez, A., Blanco, C.G., Barriocanal, C., Alvarez, R., and Dıez, M.A. (2001) Gas chromatographic study of the volatile products from co-pyrolysis of coal and polyethylene wastes. *J. Chromatogr. A*, **918**, 135–144.
35. Singh, S., Wu, C., and Williams, P.T. (2012) Pyrolysis of waste materials using TGA-MS and TGA-FTIR as complementary characterisation techniques. *J. Anal. Appl. Pyrolysis*, **94**, 99–107.

36. Kiran, N., Ekinci, E., and Snape, C.E. (2000) Recycling of plastic wastes via pyrolysis. *Resour. Conserv. Recycl.*, **29**, 273–283.
37. Nikles, D.E. and Farahat, M.S. (2005) New motivation for the depolymerization products derived from PET waste: a review. *Macromol. Mater. Eng.*, **290**, 13–30.
38. Achilias, D.S. and Karayannidis, G.P. (2004) The chemical recycling of PET in the framework of sustainable development. *Water Air Soil Pollut. Focus*, **4**, 385–396.
39. Zhang, J., Wang, X., Gong, J., and Gu, Z. (2004) A study on the biodegradability of polyethylene terephthalate fiber and diethylene glycol terephthalate. *J. Appl. Polym. Sci.*, **93**, 1089–1096.
40. Tansel, B. and Yildiz, B.S. (2011) Goal-based waste management strategy to reduce persistence of contaminants in leachate at municipal solid waste landfills. *Environ. Dev. Sustainability*, **13**, 821–831.
41. Massardier-Nageotte, V., Pestre, C., Cruard-Pardet, T., and Bayard, R. (2006) Aerobic and anerobic biodegradability of polymer films and phsyico-chemical characterization. *Polym. Degrad. Stab.*, **91**, 620–627.
42. Tollner, E.W., Annis, P.A., and Das, K.C. (2011) Evaluation of strength properties of polypropylene-based polymers in simulated landfill and oven conditions. *J. Environ. Eng.*, **137**, 291–296.
43. Urase, T., Okumura, H., Panyosaranya, S., and Inamura, A. (2008) Emission of volatile organic compounds from solid waste disposal sites and importance of heat management. *Waste Manage. Res.*, **26**, 534–538.
44. Xu, S.-Y., Zhang, H., He, P.-J., and Shao, L.-M. (2011) Leaching behaviour of bisphenol A from municipal solid waste under landfill environment. *Environ. Technol.*, **32**, 1269–1277.
45. Svenson, A., Sjöholm, S., Allard, A.-S., and Kaj, L. (2009) Antiestrogenicity and estrogenicity in leachates from solid waste deposits. *Environ. Toxicol.*, **26**, 233–239.
46. Tsuchida, D., Kajihara, Y., Shimidzu, N., Hamamura, K., and Nagase, M. (2011) Hydrogen sulphide production by sulfate-reducing bacteria utilizing additives eluted from plastic resins. *Waste Manage. Res.*, **29**, 594–601.
47. Sinha, V., Patel, M.R., and Patel, J.V. (2010) PET waste management by chemical recycling: a review. *J. Polym. Environ.*, **18**, 8–25.
48. Astrup, T., Møller, J., and Fruergaard, T. (2009) Incineration and co-combustion of waste: accounting of greenhouse gases and global warming contributions. *Waste Manage. Res.*, **27**, 789–799.
49. Khoo, H.H. and Tan, R.B.H. (2010) Environmental impacts of conventional plastic and bio-based carrier bags: part 2: end-of-life options. *Int. J. Life Cycle Assess.*, **15**, 338–345.
50. Shen, C., Tang, X., Yao, J., Shi, D., Fang, J., Khan, M.I., Cheema, S.A., and Chen, Y. (2010) Levels and patterns of polycyclic aromatic hydrocarbons and polychlorinated biphenyls in municipal waste incinerator bottom ash in Zhejiang province, China. *J. Hazard. Mater.*, **179**, 197–202.
51. Simoneit, B.R.T., Medeiros, P.M., and Didyk, B.M. (2005) Combustion products of plastics as indicators for refuse burning in the atmosphere. *Environ. Sci. Technol.*, **39**, 6961–6970.
52. Valavanidis, A., Iliopoulos, N., Gotsis, G., and Fiotakis, K. (2008) Persistent free-radicals, heavy metals and PAHs generated in particulate soot emissions and residue ash from controlled combustion of common types of plastics. *J. Hazard. Mater.*, **156**, 277–284.
53. Al-Salem, S.M., Lettieri, P., and Baeyens, J. (2009) Recycling and recovery routes of plastic solid waste (PSW): a review. *Waste Manage.*, **29**, 2625–2643.
54. Achilias, D.S. (2007) Chemical recycling of poly(methyl methacrylate) by pyrolysis. Potential use of the liquid fraction as a raw material for the reproduction of the polymer. *Eur. Polym. J.*, **43** (6), 2564–2575.
55. Pinto, F., Franco, C., Andre, R.N., Miranda, M., Gulyurtlu, I., and Cabrita,

I. (2002) Co-gasification study of biomass mixed with plastic wastes. *Fuel*, **81** (3), 291–297.

56. Pinto, F., Franco, C., Andre, R.N., Tavares, C., Dias, M., Gulyurtlu, I., and Cabrita, I. (2003) Effect of experimental conditions on co-gasification of coal, biomass and plastics wastes with air/steam mixtures in a fluidized bed system. *Fuel*, **82** (15–17), 1967–1976.
57. Xiao, G., Ni, M.-J., Chi, Y., Jin, B.-S., Xiao, R., Zhong, Z.-P., and Huang, Y.-J. (2009) Gasification characteristics of MSW and an ANN prediction model. *Waste Manage.*, **29** (1), 240–244.
58. Predel, M. and Kaminsky, W. (2000) Thermal & thermo- cataytic conversion of waste polyolefins to fuel. *Polym. Degrad. Stab.*, **70**, 373.
59. Mastral, J.F., Berrueco, C., Gea, M., and Ceamanos, J. (2006) Catalytic degradation of high density polyethylene over nanocrystalline HZSM-5 zeolite. *Polym. Degrad. Stab.*, **91**, 3330.
60. Aguado, R., Olazar, M., San José, M.J., Gaisán, B., and Bilbao, J. (2002) Wax formation in the pyrolysis of polyolefins in a conical spouted bed reactor. *Energy Fuels*, **16**, 1429.
61. Elordi, G., López, G., Aguado, R., Olazar, M., and Bilbao, J. (2007) Catalytic pyrolysis of high density polyethylene on a HZSM-5 zeolite catalyst in a conical spouted bed reactor. *Int. J. Chem. Reactor Eng.*, **5**, A72.
62. Olazar, M., San José, M.J., Aguado, R., Gaisán, B., and Bilbao, J. (1999) Bed voidage in conical sawdust beds in the transition regime between spouting and jet spouting. *Ind. Eng. Chem. Res.*, **38**, 4120.
63. Olazar, M., San José, M.J., Zabala, G., and Bilbao, J. (1994) New reactor in jet spouted bed regime for catalytic polymerizations. *Chem. Eng. Sci.*, **49**, 4579.
64. Achilias, D.S. (2006) Chemical recycling of poly(methyl methacrylate). *WSEAS Trans. Environ. Dev.*, **2** (2), 85–91.
65. Al-Salem, S.M., Lettieri, P., and Baeyens, J. (2010) The valorization of plastic solid waste (PSW) by primary to quaternary routes: from re-use to energy and chemicals. *Prog. Energy Combust. Sci.*, **36**, 103–129.
66. Al-Salem, S.M. and Lettieri, P. (2010) Kinetic study of high density polyethylene (HDPE) pyrolysis. *Chem. Eng. Res. Des.*, **88**, 1599–1606.
67. Marcilla, A., Beltran, M., and Conesa, J.A. (2001) Catalyst addition in polyethylene pyrolysis thermogravimetric study. *J. Anal. Appl. Pyrolysis*, **58**, 117.
68. Marcilla, A., Beltran, M.I., Hernandez, F., and Navarro, R. (2004) HZSM5 and HUSY deactivation during the catalytic pyrolysis of polyethylene. *Appl. Catal., A*, **278**, 37–43.
69. Seo, Y.-H., Lee, K.-H., and Shin, D.-H. (2003) Investigation of catalytic degradation of high density polyethylene by hydrocarbon group type analysis. *J. Anal. Appl. Pyrolysis*, **70** (2), 383–398.
70. Van Grieken, R., Serrano, D.P., Aguado, J., García, R., and Rojo, C. (2001) Thermal and catalytic cracking of polyethylene under mild conditions. *J. Anal. Appl. Pyrolysis*, **58**, 127.
71. Akpanudoh, N.S., Gobin, K., and Manos, G. (2005) Catalytic degradation of plastic waste to liquid fuel over commercial cracking catalysts. *J. Mol. Catal. A: Chem.*, **235**, 67.
72. Cardona, S.C. and Corma, A. (2002) Kinetic study of the catalytic cracking of polypropylene in a semibatch stirred reactor. *Catal. Today*, **75**, 239.
73. Marcilla, A., Gómez, A., García, A.N., and Olaya, M.M. (2002) Kinetic study of the catalytic decomposition of different commercial polyethylenes over an MCM-41 catalyst. *J. Anal. Appl. Pyrolysis*, **64**, 85–101.
74. Marcilla, A., Gomez, A., Reyes-Labarta, J.A., Giner, A., and Hernandez, F. (2003) Kinetic study of polypropylene pyrolysis using ZSM-5 and an equilibrium fluid catalytic cracking catalyst. *J. Anal. Appl. Pyrolysis*, **68–69**, 46–480.
75. Araujo, A.S., Fernandes, V.J. Jr., and Fernandes, G.J.T. (2002) Thermogravimetric kinetics of polyethelyne degredation over silicoaluminophosphate. *Thermochim. Acta*, **392**, 55–61.
76. Fernandes, G.J.T., Fernandes, V.J. Jr., and Araujo, A.S. (2002) Catalytic

degradation of polyethylene over SAPO-37 molecular sieve. *Catal. Today*, **75**, 233–238.

77. Serrano, D.P., Aguado, J., Escola, J.M., Rodriguez, J.M., Morselli, L., and Orsi, R. (2003) Thermal and catalytic cracking of a LDPE–EVA copolymer mixture. *J. Anal. Appl. Pyrolysis*, **68–69**, 481–494.
78. Jan, M.R., Shah, J., and Gulab, H. (2010) Catalytic degradation of waste high-density polyethylene into fuel products using BaCO3 as a catalyst. *Fuel Process. Technol.*, **91**, 1428–1437.
79. Huang, J., Jiang, Y., Marthala, V.R.R., Bressel, A., Frey, J., and Hunger, M. (2009) Effect of pore size and acidity on the coke formation during ethylbenzene conversion on zeolite catalysts. *J. Catal.*, **263**, 277–283.
80. Marcilla, A., Beltran, M.I., and Navarro, R. (2007) Study of the deactivation process of HZSM5 zeolite during polyethylene pyrolysis. *Appl. Catal., A*, **333**, 57–66.
81. Elordi, G., Olazar, M., Lopez, G., Amutio, M., Artetxe, M., Aguado, R., and Bilbao, J. (2009) Catalytic pyrolysis of HDPE in continuous mode over zeolite catalysts in a conical spouted bed reactor. *J. Anal. Appl. Pyrolysis*, **85**, 345–351.
82. Cerqueira, H.S., Caeiro, G., Costa, L., and Ramoa Ribeiro, F. (2008) Deactivation of FCC catalysts. *J. Mol. Catal. A: Chem.*, **292**, 1–13.
83. Aguayo, A.T., Gayubo, A.G., Ortega, J., Olazar, M., and Bilbao, J. (1997) Catalyst deactivation by coking in the MTG process in fixed and fluidized bed reactors. *Catal. Today*, **37**, 239–248.
84. Aguado, J. and Serrano, D.P. (eds) (1999) *Feedstock Recycling of Plastic Wastes*, The Royal Society of Chemistry, Cambridge.
85. Scheirs, J. and Kaminsky, W. (eds) (2006) *Feedstock Recycling and Pyrolysis of Waste Plastics. Converting Waste Plastics into Diesel and Other Fuels*, Wiley-VCH Verlag GmbH.
86. Hayashi, J.-I. (1998) Pyrolysis of polypropylene in the presence of oxygen. *Fuel Process. Technol.*, **55**, 265–275.
87. Dawood, A. and Miura, K. (2001) Pyrolysis kinetics of g-irradiated polypropylene. *Polym. Degrad. Stab.*, **73**, 347–354.
88. Assumpção, L.C.F.N., Carbonell, M.M., and Marques, M.R.C. (2011) Co-pyrolysis of polypropylene waste with Brazilian heavy oil. *J. Environ. Sci. Health Part A*, **46**, 461–464.
89. Ballice, L. and Reimert, R. (2002) Temperature-programmed co-pyrolysis of Turkish lignite with polypropylene. *J. Anal. Appl. Pyrolysis*, **65**, 207–219.
90. Hájeková, E. and Bajus, M. (2005) Recycling of low-density polyethylene and polypropylene via copyrolysis of polyalkene oil/waxes with naphtha: product distribution and coke formation. *J. Anal. Appl. Pyrolysis*, **74** (1-2), 270–281.
91. Zhao, W. *et al.* (1996) Effects of zeolites on the pyrolysis polypropylene. *Polym. Degrad. Stab.*, **53**, 129–135.
92. Ishihara, Y., Nanbu, H., Saido, K., Ikemura, T., Takesue, T., and Kuroki, T. (1993) Mechanism of gas formation in catalytic decomposition of polypropylene. *Fuel*, **72** (8), 1115–1119.
93. Durmus, A., Koc, S.N., Pozan, G.S., and Kasgoz, A. (2005) Thermal-catalytic degradation kinetics of polypropylene over BEA, ZSM-5 and MOR zeolites. *Appl. Catal., B*, **61**, 316–322.
94. Lin, Y.-H. and Yang, M.-H. (2007) Chemical catalysed recycling of waste polymers: catalytic conversion of polypropylene into fuels and chemicals over spent FCC catalyst in a fluidised-bed reactor. *Polym. Degrad. Stab.*, **92**, 813–821.
95. Uemichi, Y., Makino, Y., and Kanazuka, T. (1989) Degradation of polypropylene to aromatic hydrocarbons over Pt- and Fe- containing activated carbon. *J. Anal. Appl. Pyrolysis*, **16**, 229–238.
96. Cardona, S.C. and Corma, A. (2000) Tertiary recycling of polypropylene by catalytic cracking in a semibatch stirred reactor use of spent equilibrium FCC commercial catalyst. *Appl. Catal., B*, **25**, 151–162.
97. Xie, C. *et al.* (2008) Catalytic cracking of polypropylene into liquid hydrocarbons over Zr and Mo modified

MCM-41 mesoporous molecular sieve. *Catal. Commun.*, **10**, 79–82.
98. Panda, A.K. and Singh, R.K. (2011) Catalytic performances of kaoline and silica alumina in the thermal degradation of polypropylene. *J. Fuel Chem. Technol.*, **39** (3), 198–202.
99. Vilaplana, F., Ribes-Greus, A., and Karlsson, S. (2006) Degradation of recycled high-impact polystyrene: simulation by reprocessing and thermooxidation. *Polym. Degrad. Stab.*, **91**, 2163–2170.
100. Achilias, D.S., Giannoulis, A., and Papageorgiou, G.Z. (2009) Recycling of polymers from plastic packaging materials using the dissolution/reprecipitation technique. *Polym. Bull.*, **63** (3), 449–465.
101. Kim, J.S., Lee, W.Y., Lee, S.B., Kim, S.B., and Choi, M.J. (2003) Degradation of polystyrene waste over base promoted Fe catalysts. *Catal. Today*, **87**, 59–68.
102. Bajdur, W., Pajaczkowska, J., Makarucha, B., Sulkowska, A., and Sulkowski, W.W. (2002) Effective polyelectrolytes synthesized from expanded polystyrene wastes. *Eur. Polym. J.*, **38**, 299–304.
103. Woo, O.S., Ayala, N., and Broadbelt, L.J. (2000) Mechanistic interpretation of base-catalyzed depolymerization of polystyrene. *Catal. Today*, **55**, 161–171.
104. Ukei, H., Hirose, T., Horikawa, S., Takai, Y., Taka, M., Azuma, N., and Ueno, A. (2000) Catalytic degradation of polystyrene into styrene and a design of recyclable polystyrene with dispersed catalysts. *Catal. Today*, **62**, 67–75.
105. Koji, U., Yoshio, M., Naoto, A., Tamaki, H., Sanae, H., and Mitsunori, O. (1998) Recycling of Polystyrene. Patent JP10130418.
106. Lee, K.H., Noh, N.S., Shin, D.H., and Seo, Y. (2002) Comparison of plastic types for catalytic degradation of waste plastics into liquid product with spent FCC catalyst. *Polym. Degrad. Stab.*, **78**, 539–544.
107. Ke, H., Li-hua, T., Zi-Bin, Z., and Cheng-Fang, Z. (2005) Reaction mechanism of styrene monomer recovery from waste polystyrene by supercritical solvents. *Polym. Degrad. Stab.*, **89**, 312–316.
108. Suyama, K., Masafumi, K., Masamitsu, S., and Hiroyuki, Y. (2010) Chemical recycling of networked polystyrene derivatives using subcritical water in the presence of an aminoalchocol. *Polym. Degrad. Stab.*, **95**, 1588–1592.
109. Xue, F., Takeda, D., Kimura, T., and Minabe, M. (2004) Effect of organic peroxides on the thermal decomposition of expanded polystyrene with the addition of a -methylstyrene. *Polym. Degrad. Stab.*, **83**, 461–466.
110. Arandes, J.M., Abajo, I., Lopez-Valerio, D., Fernández, I., Azkoiti, M.J., Olazar, M., and Bilbao, J. (1997) Transformation of several plastic wastes into fuels by catalytic cracking. *Ind. Eng. Chem. Res.*, **36**, 4523–4529.
111. Milne, B.J., Behie, L.A., and Berruti, F. (1999) Recycling of waste plastics by ultrapyrolysis using an internally circulating fluidized bed reactor. *J. Anal. Appl. Pyrolysis*, **51**, 157–166.
112. Westerhout, R.W.J., Waanders, J., Kuipers, J.A.M., and van Swaaij, W.P.M. (1997) Kinetics of the low temperature pyrolysis of polyethene, polypropene, and polystyrene modeling, experimental determination, and comparison with literature models and data. *Ind. Eng. Chem. Res.*, **36**, 1955–1964.
113. Mehta, S., Biederman, S., and Shivkumar, S. (1995) Thermal degradation of foamed polystyrene. *J. Mater. Sci.*, **30**, 2944–2949.
114. Aguado, R., Olazar, M., Gaisan, B., Prieto, R., and Bilbao, J. (2003) Kinetics of polystyrene pyrolysis in a conical spouted bed reactor. *Chem. Eng. J.*, **92**, 91–99.
115. Lee, C.G., Cho, Y.J., Song, P.S., Kang, Y., Kim, J.S., and Choi, M.J. (2003) Effects of temperature distribution on the catalytic pyrolysis of polystyrene waste in a swirling fluidizedbed reactor. *Catal. Today*, **79–80**, 453–464.
116. Sterling, W.J., Walline, K.S., and McCoy, B.J. (2001) Experimental study of polystyrene thermolysis to moderate conversion. *Polym. Degrad. Stab.*, **73**, 75–82.

117. Zhu, X., Elomaa, M., Sundholma, F., and Lochmuller, C.H. (1998) Infrared and thermogravimetric studies of thermal degradation of polystyrene in the presence of ammonium sulfate. *Polym. Degrad. Stab.*, **62**, 487–494.
118. Beltrame, P.L., Bergamasco, L., Carniti, P., Castelli, A., Bertini, F., and Audisio, G. (1997) Hydrous pyrolysis of polystyrene. *J. Anal. Appl. Pyrolysis*, **40–41**, 451–461.
119. Lee, S.Y., Yoon, J.H., Kim, J.R., and Park, D.W. (2002) Degradation of polystyrene using clinoptilolite catalysts. *J. Anal. Appl. Pyrolysis*, **64**, 71–83.
120. Puente, D.L.G. and Sedran, U. (1998) Recycling polystyrene into fuels by means of FCC: performance of various acidic catalysts. *Appl. Catal., B*, **19**, 305–311.
121. Ward, P.G., Goff, M., Donner, M., Kaminsky, W., and O'Connor, K.E. (2006) A two step chemobiotechnological conversion of polystyrene to a biodegradable thermoplastic. *Environ. Sci. Technol.*, **40**, 2433–2437.
122. Maharana, T., Negi, Y.S., and Mohanty, B. (2007) Review article: recycling of polystyrene. *Plast. Technol. Eng.*, **46** (7), 729–736.
123. Sadat-Shojai, M. and Bakhshandeh, G.-R. (2011) Recycling of PVC wastes. *Polym. Degrad. Stab.*, **96**, 404–415.
124. Garcia, D., Balart, R., Sanchez, L., and Lopez, J. (2007) Compatibility of recycled PVC/ABS blends effect of previous degradation. *Polym. Eng. Sci.*, **47**, 789–796.
125. Ulutan, S. (1998) A recycling assessment of PVC bottles by means of heat impact evaluation on its reprocessing. *J. Appl. Polym. Sci.*, **69**, 865–869.
126. Siddiqui, M.N., Achilias, D.S., Redhwi, H.H., Bikiaris, D.N., Katsogiannis, K.-A.G., and Karayannidis, G.P. (2010) Hydrolytic depolymerization of PET in a microwave reactor. *Macromol. Mater. Eng.*, **295**, 575–584.
127. Yoshioka, T., Furukawa, K., and Okuwaki, A. (2000) Chemical recycling of rigid-PVC by oxygen oxidation in NaOH solutions at elevated temperatures. *Polym. Degrad. Stab.*, **67**, 285–290.
128. Machado, H.M.A.M.M.S., Rodrigues Filho, G., De Assunc, R.M.N., Soares, H.M., Cangani, A.P., Cerqueira, D.A. *et al.* (2010) Chemical recycling of poly(vinyl chloride): application of partially dehydrochlorinated poly(vinyl chloride) for producing a chemically modified polymer. *J. Appl. Polym. Sci.*, **115**, 1474–1479.
129. Xiong L. (2010) PVC: an Enemy of Public Health and Environment, http://cdn.intechopen.com/pdfs/32560.pdf.
130. Ryu, C., Sharifi, V.N., and Swithenbank, J. (2007) Waste pyrolysis and generation of storable char. *Int. J. Energy Res.*, **31** (2), 177–191.
131. Williams, P.T. and Williams, E.A. (1998) Recycling plastic waste by pyrolysis. *J. Inst. Energy*, **71**, 81–89.
132. De Marco, I., Caballero, B., Torres, A., Laresgoiti, M.F., Chomón, M.J., and Cabrero, M.A. (2002) Recycling polymeric wastes by means of pyrolysis. *J. Chem. Technol. Biotechnol.*, **77** (7), 817–824.
133. Kaminsky, W. and Kim, J.-S. (1999) Pyrolysis of mixed plastics into aromatics. *J. Anal. Appl. Pyrolysis*, **51** (1), 127–134.
134. Borgianni, C., Filippis, P.D., Pochetti, F., and Paolucci, M. (2002) Gasification process of wastes containing PVC. *Fuel*, **14** (81), 1827–1833.
135. Slapak, M.J.P., van Kasteren, J.M.N., and Drinkenburg, B.A.A.H. (1999) Hydrothermal recycling of PVC in a bubbling fluidized bed reactor: the influence of bed material and temperature. *Polym. Adv. Technol.*, **10**, 596–602.
136. Wu, Y.-H., Zhou, Q., Zhao, T., Deng, M.-L., Zhang, J., and Wang, Y.-Z. (2009) Poly (ethylene glycol) enhanced dehydrochlorination of poly(vinyl chloride). *J. Hazard. Mater.*, **163**, 1408–1411.
137. Brown, T. (2001) Infinity nylon – A never-ending cycle of renewal. Presentation at 6th Annual Conference on Recycling of Fibrous Textile and Carpet Waste, Dalton, Georgia, 2001.

138. Yoshioka, T., Furukawa, K., Sato, T., and Okuwaki, A. (1998) Chemical recycling of flexible PVC by oxygen oxidation in NaOH solutions at elevated temperatures. *J. Appl. Polym. Sci.*, **70**, 129–135.
139. Duangchan, A. and Samart, C. (2008) Tertiary recycling of PVC-containing plastic waste by copyrolysis with cattle manure. *Waste Manage.*, **28**, 2415–2421.
140. Tongamp, W., Kano, J., Zhang, Q., and Saito, F. (2008) Simultaneous treatment of PVC and oyster-shell wastes by mechanochemical means. *Waste Manage.*, **28** (3), 484–488.
141. Zhang, Q., Saeki, S., Tanaka, Y., Kano, J., and Saito, F. (2007) A soft-solution process for recovering rare metals from metal/alloy-wastes by grinding and washing with water. *J. Hazard. Mater.*, **139**, 438–442.
142. Scheirs, J. (2010) End-of-Life Environmental Issues with PVC in Australia.
143. Kameda, T., Fukuda, Y., Grause, G., and Yoshioka, T. (2010) Chemical modification of rigid poly(vinylchloride) by the substitution with nucleophiles. *J. Appl. Polym. Sci.*, **116**, 36–44.
144. La Mantia, F.P. (1996) *Recycling of PVC and Mixed Plastic Waste*, ChemTec Publishing, Toronto.
145. Lee, G.-S. and Song, Y.J. (2007) Recycling EAF dust by heat treatment with PVC. *Miner. Eng.*, **20**, 739–746.
146. Recovinyl (2008) Application Covered by Recovinyl, http://www.recovinyl.com/certification/recyclers/applications-covered.
147. Tukker, A., de Groot, H., Simons, L., and Wiegersma, S. (1999) Chemical Recycling of Plastic Waste: PVC and other Resins. European Commission, DG III, Final Report, STB-99-55 Final. Delft, the Netherlands.
148. Shin, S.-M., Yoshioka, T., and Okuwaki, A. (1998) Dehydrochlorination behavior of rigid PVC pellet in NaOH solutions at elevated temperature. *Polym. Degrad. Stab.*, **61** (2), 349–353.
149. Brophy, J.H., Hardmann, S., and Wilson, D.C. (1997) Polymer cracking for feedstock recycling of mixed plastic wastes, in *Chemical Aspects of Plastics Recycling* (eds W. Hoyle and D.R. Karsa), The Royal Society of Chemistry Information Services, Cambridge.
150. Heyde, M. and Kremer, S. (1999) LCA Packaging Plastics Waste. LCA Documents, Vol. 2, No. 5, EcoMed, Landsberg, Germany.
151. DKR/DSD (1999) Mit Altkunststoffen Ressourcen schonen. Leaflet with information on chemical recycling, DKR, Germany.
152. Smolders, K. and Baeyens, J. (2004) Thermal degradation of PMMA in fluidized beds. *Waste Manage.*, **24**, 849–857.
153. Sasse, F. and Emig, G. (1998) Chemical recycling of polymer materials. *Chem. Eng. Technol.*, **21**, 777–789.
154. Kaminsky, W. and Frank, J. (1991) Monomer recovery by pyrolysis of poly(methyl methacrylate). *J. Anal. Appl. Pyrolysis*, **19**, 311–318.
155. Kaminsky, W. and Eger, C. (2001) Pyrolysis of filled PMMA for monomer recovery. *J. Anal. Appl. Pyrolysis*, **58–59**, 781–787.
156. Grause, G., Predel, M., and Kaminsky, W. (2006) Monomer recovery from aluminum hydroxide high filled poly(methyl methacrylate) in a fluidized bed reactor. *J. Anal. Appl. Pyrolysis*, **75**, 236–239.
157. Papaspyrides, C.D., Gouli, S., and Poulakis, J.G. (1994) Recovery of poly(methyl ethacrylate) by the dissolution/reprecipitation technique. *Adv. Polym. Technol.*, **13** (3), 213–218.
158. Gouli, S., Poulakis, J.G., and Papaspyrides, C.D. (1994) Solvent recycling of poly (methyl methacrylate) decorative sheets. *Adv. Polym. Technol.*, **13** (3), 207–211.
159. Billmeyer, F.W. Jr., (1971) *Textbook of Polymer Science*, 2nd edn, John Wiley & Sons, Inc.
160. Achilias, D.S., Kanellopoulou, I., Megalokonomos, P., Lappas, A., and Antonakou, E. (2006) Chemical Recycling of Polystyrene, Protection and Restoration of the Environment VII, Chania, Greece, July 3–7, 2006.

3.2 Methods of Recycling of Polymers: Condensation Polymers

Beena Sethi

3.2.1 Introduction

Condensation polymerization is the combination of simple dissimilar molecules, with the formation of small molecules as byproducts such as water or ammonia. Condensation polymers are usually formed by the stepwise intermolecular condensation of reactive groups. By understanding the structures of the different types of polymers found in municipal solid waste MSW, their mechanisms of degradation can be studied (and perhaps altered by the presence of catalysts) in order to achieve the most useful decomposition products [1]. A notable exception occurs with the synthesis of polyurethanes, which are formed by reaction of isocyanates with hydroxyl compounds and follow step kinetics, without elimination of small molecules from the respective units [2]. This chapter reviews the methods of recycling mainly Nylon and polycarbonate (PC) polymers.

3.2.2 Chemical Recycling of Nylon

3.2.2.1 Introduction

Nylon is one of the most commonly used polymers, known as *polyamides*, and can be produced by any one of the following reactions: reaction of a diamine with a dicarboxylic acid, condensation of the appropriate amino acid, ring opening of a lactam, reaction of a diamine with a diacid chloride, or the reaction of a diisocyanate with a dicarboxylic acid [3]. Nylon is a crystalline polymer with high modulus, strength, impact properties, low coefficient of friction, and resistance to abrasion. A variety of commercial nylons are available including nylon 6, nylon 11, nylon 12, nylon 6,6, nylon 6,10, and nylon 6,12. The most widely used nylons are nylon 6,6 and nylon 6. Polyamides are used most often in the form of fibers, primarily nylon 6,6 and nylon 6, although engineering applications are also of importance [3].

3.2.2.2 Recycling Methods

According to the U.S. Department of Energy, about 3.5 billion pounds of waste carpets are discarded each year in the United States, with about 30% of them made from nylon 6. Nylon recycling has increased substantially in the last several years. Therefore, most recycling efforts have focused on the recovery of carpets. Processing of recyclables is necessary to transform the collected materials into raw materials for the manufacture of new products. In general, there are two categories for nylon recycling:

1. *Chemical recycling.* Chemicals are used to break down the molecular structure of the polymer. The products of the reaction are purified and used again to produce either the same or a related polymer [3].
2. *Thermal recycling (pyrolysis).* Heat is used to break down the chemical structure of the polymer. In pyrolysis, the polymer (or mixture of polymers) is subjected to high heat in the absence of sufficient oxygen for combustion. At elevated temperatures, the polymeric structure breaks down [3].

3.2.3 Chemical Recycling Involving Depolymerization of Nylons Which Can Be Carried Out by Hydrolysis or Ammonolysis of Nylon 6,6 and Nylon 6

The waste carpets are collected, sorted, and then subjected to a mechanical shredding process before depolymerization. Because of the higher value of nylons in comparison with other polymers used in carpets, nylon carpet has been looked at as a resource for making virgin nylon via depolymerization. Most of polyamides used commercially are nylon 6,6 or nylon 6, and the largest supply of waste for recycling of nylons comes from used carpets [4].

3.2.3.1 Hydrolysis of Nylon 6

In this process, ground scrap is dissolved in high-pressure steam at 125–130 psig (963–997 kPa) and 175–180 °C for 0.5 h in a batch reactor and then continuously hydrolyzed with superheated steam at 350 °C and 100 psig (790 kPa) to form ε-caprolactam. Almost 98% of caprolactam can be recovered. The recovered monomer is repolymerized without additional purification [5, 6]. Braun reported the depolymerization of nylon 6 carpets in a small laboratory apparatus with steam at 340 °C and 1500 kPa (200 psig) for 3 h. Ninety-five percent yield of crude ε-caprolactam of purity 94.4% was obtained [7, 8]. Acid hydrolysis of nylon 6 wastes [9] in the presence of superheated steam has been used to produce aminocaproic acid, which under acid conditions is converted to ε-caprolactam [10]. For this purpose, inorganic or organic acids such as nitric, formic, benzoic, and hydrochloric acids are used [11]. Orthophosphoric acid and boric acid are typically used as catalysts at temperatures 250–350 °C. In a typical process, superheated steam is passed through the molten nylon 6 waste at 250–300 °C in the presence of phosphoric acid. The resulting solution is purified before concentrating to 70% liquor, which is fractionally distilled in the presence of a base to recover pure ε-caprolactam. Boric acid (1%) may be used to depolymerize nylon 6 at 400 °C under ambient pressure. A recovery of 93–95% ε-caprolactam was obtained by passing superheated steam through molten nylon 6 at 250–350 °C [11].

3.2.3.2
Hydrolysis of Nylon 6,6 and Nylon 4,6

The depolymerization of nylon 6,6 and nylon 4,6 involves the hydrolysis of the amide linkages in both acid- and base-catalyzed hydrolysis. An effective phase-transfer catalyst, benzyltrimethylammonium bromide in aqueous 50% sodium hydroxide solution, is used for the conversion of nylon 6,6 and nylon 4,6 to oligomers [12]. The depolymerization efficiency (% weight loss) and the molecular weight of the reclaimed oligomers are dependent on the amount and concentration of the aqueous sodium hydroxide and the reaction time. Nylon 4,6 fibers ($M_v = 41\,400\ \text{g mol}^{-1}$) do not undergo depolymerization on exposure to 100 ml of 25 wt% sodium hydroxide solution at 165 °C. If 6.0 g of nylon fibers is fed for depolymerization, 5.95 g remains unaffected. When the concentration of sodium hydroxide is increased to 50 wt%, the depolymerization process results in the formation of low molecular weight oligomers [12]. The feasibility of alkaline hydrolysis in respect to recycling of nylon 4,6, is established only after the determination whether the recovered oligomers can be repolymerized to form nylon 4,6. For this purpose, solid-state polymerization was performed on nylon 4,6 oligomers formed via alkaline hydrolysis with 50 wt% NaOH at 165 °C for 24 h. In order to isolate adipic acid, nylon 6,6 fibers were depolymerized under reflux with a 50% NaOH solution in the presence of catalytic amounts of benzyltrimethylammonium bromide. The oligomers formed in successive steps were depolymerized under similar conditions. However, hexamethylene diamine was not isolated. The overall yield of adipic acid was 59.6% [12].

3.2.3.3
Ammonolysis of Nylon 6,6

In this reaction, ammonia is added to nylon 6,6 and nylon 6,6/nylon 6 mixtures at temperatures between 300 and 350 °C and a pressure of about 68 atmospheres in the presence of an ammonium phosphate catalyst to yield a mixture of the following monomeric products: HMDA (hexamethylene diamine), adiponitrile (ADN), and 5-cyanovaleramide from nylon 6,6 and ε-caprolactam, 6-aminocapronitrile, and 6-aminocaproamide from nylon 6 [13, 14]. The equilibrium is shifted toward products by continuous removal of the water formed. Most of HMDA was formed hydrogenation of monomers. The mechanism of ammonolysis includes the amide bond breakage and amide end dehydration (nitrilation) reactions, plus ring addition and ring opening for cyclic lactams present in nylon 6 [15]. Bordrero utilized a two-step ami/ammonolysis process to depolymerize nylon 6,6. The first step was based on an aminolysis treatment of nylon-6,6 by *n*-butylamine at a temperature of 300 °C and a pressure of 45 atm. Free HMDA and *N*,*N*′-dibutyl adipamide were

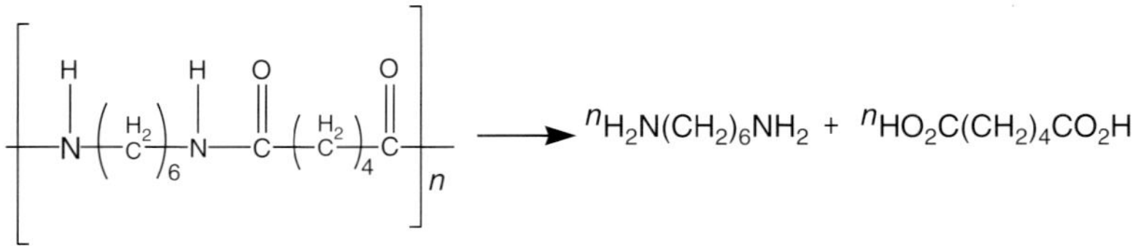

Figure 3.2.1 Depolymerization of nylon 6,6 by hydrolysis.

generated [16]. The second step was ammonolysis of *N*,*N*′-dibutyl adipamide at a temperature of 285 °C and a pressure of 50 atm. The end product was ADN. It was estimated that the yield was about 48% for ADN and about 100% for HMDA at optimized reaction conditions [16].

3.2.3.4 Recovery of Nylon 6,6 Monomers

Adipic acid and HMDA can be obtained from nylon 6,6 by the hydrolysis of the polymer in concentrated sulfuric acid. Adipic acid is obtained by recrystallization, and the HMDA is recovered by distillation after neutralizing the acid. This process is slow and inefficient for treating large amounts of waste because recrystallization of adipic acid requires a long time [16]. In a continuous process, 30–70% concentration of mineral acid is used to hydrolyze nylon 6,6 wastes, and the resulting hydrolysate is fed to a crystallization zone. The adipic acid crystallizes and the crystals are continuously removed from the hydrolysate. Addition of calcium hydroxide to neutralize the mother liquor liberates HMDA after subsequent distillation (Figure 3.2.1).

In this process, hot hydrolysate containing 10–20% adipic acid is continuously introduced into an agitated crystallization vessel maintained at an average temperature of 20–30 °C. The slurry obtained from the crystallization vessel is filtered to collect the adipic acid crystals. The filtrate, which contains the HMDA acid salt, is continuously neutralized with calcium hydroxide. The calcium salt formed is removed by filtration and the HMDA in the filtrate is isolated by distillation.

In the case of nylon 6,6 waste recycled by ammonolysis, nylon is treated with ammonia in the presence of a phosphate catalyst at 330 °C and 7 MPa. Distillation of the reaction mixture produces ammonia, which is recycled, and three fractions containing (i) caprolactam, (ii) HMDA and aminocapronitrile, and (iii) ADN. Aminocapronitrile and ADN are hydrogenated to yield pure HMDA, and the caprolactam is either converted to aminocapronitrile by further ammonolysis or distilled to produce pure caprolactam. The HMDA produced by this process is extremely pure (>99.8%). The main impurities are aminomethylcyclopentylamine and tetrahydroazepine, which are expected to be removed more effectively in the larger distillation columns employed in larger plants [16].

3.2.3.5 Catalytic Pyrolysis

Catalytic pyrolysis a hybrid process involving chemical recycling and high-temperature pyrolysis. Chemical recycling converts nylon 6 into caprolactam, and thermal pyrolysis of caprolactam yields a synthetic natural gas. Czernik investigated the catalysis of the thermal degradation of nylon 6 with an α-alumina-supported KOH catalyst in a fluidized-bed reactor [17]. At 330–360 °C, the yield of caprolactam exceeded 85%. Bockhorn used a liquid catalyst composed of a eutectic mixture of 60 mol% NaOH and 40 mol% KOH, which melted at 185 °C. At 290 °C, the caprolactam yield exceeded 95% [18].

3.2.3.6 Applications of Depolymerized Nylon 6

Fibers are developed from chemical recycling of nylon 6 carpets into a closed-loop recycling process [11, 19]. The recovered nylon 6 face fibers are depolymerized in a depolymerization reactor and treated with superheated steam in the presence of a catalyst to produce a distillate containing caprolactam. The crude caprolactam is distilled and repolymerized to form nylon 6. The caprolactam obtained is comparable to virgin caprolactam in purity. The repolymerized nylon 6 is converted into yarn and tufted into carpet. The carpets obtained from this process are very similar in physical properties to those obtained from virgin caprolactam. Its process involves collection of used nylon 6 carpet, shredding and separation of face fibers, pelletizing the face fiber for depolymerization and chemical distillation to obtain a purified caprolactam monomer, and repolymerization of caprolactam into nylon polymer [20].

In the two-stage selective pyrolysis process, the ground nylon scrap is dissolved with high-pressure steam and then continuously hydrolyzed with superheated steam to form caprolactam. Over 100 000 t of post-consumer carpets are recycled every year to produced virgin-quality caprolactam, thus saving landfill sites [19].

3.2.4 Chemical Recycling of Polycarbonate

3.2.4.1 Introduction

PC plastics, $C_{16}H_{14}O_3$, are polyesters have excellent mechanical properties, high impact resistance, UV resistance, and flame retardancy, as well as excellent electrical resistance. PCs do not have their own recycling identification code and therefore fall under the seven “other” classifications [3]. Because of its excellent properties, PC is used to make compact disks (CDs), bullet-proof windows, food packaging, and soft-drink bottles. With the rapid increase in the production and consumption of PC, the chemical recycling of waste PC to obtain valuable products has received greater attention in recent years.

3.2.4.2 Recycling Techniques

Various methods for the chemical recycling of waste PC to recover raw materials have been reported. These methods include thermal pyrolysis [21–24], alcoholysis [25], and hydrolysis [23, 26]. It is difficult to recover pure bisphenol A (BPA) using thermal pyrolysis, and it can be only obtained using hydrolysis [27]. For the purpose of recycling of PC, a number of depolymerization methods have been reported, for example, poly[2,2-bis(4-hydroxyphenyl) propane carbonate], in the form of an essential monomer BPA [28].

3.2.4.2.1 Methanolysis

It is one of the most important methods to recover pure monomer BPA and dimethyl carbonate (DMC). Methanolysis requires high temperature, pressure, and presence of a large amount of concentrated bases or acids, as PC is insoluble in methanol. The traditional acid or base catalysts are not used because these cannot be reused and also have other disadvantages such as equipment corrosion, tedious work-up procedures, and environmental problems. Even though the supercritical method can overcome some of above-mentioned shortcomings, it has its own disadvantages such as severe conditions, so its application is limited. At high pressure (not atmospheric pressure) and high-temperature steam (300 °C), PC can be completely decomposed into its monomer (BPA) in 5 min of reaction time. It is known that PC can be decomposed into the monomer in alkaline alcohol or aqueous solutions. However, the monomer BPA yield is relatively low because of BPA's instability under these conditions. To develop a highly effective process of PC recycling, a reactive atmosphere is required that can preserve the stability of BPA as well as high reactivity for PC. To determine the optimum conditions for recycling PC, it is important to know the stability or reactivity of BPA, as well as the decomposition rate of PC. Alkali-catalyzed depolymerization of PC wastes by alcoholysis in supercritical or near-critical conditions is carried out in order to recover the essential monomer BPA and DMC as a valuable byproducts [29]. In a continuous process, PC plastic waste is decomposed using methanol as solvent/reagent and NaOH as alkali catalyst. Total depolymerization of PC can be achieved working in the temperature range 75–180 °C and pressures 2–25 MPa.

Methanolysis in the Presence of Ionic Liquids Methanolysis of PC in the presence of the ionic liquid [Bmim][Ac] can be carried out under moderate conditions without an acid or base catalyst, and the starting monomers BPA and DMC are obtained. Almost 100% conversion of PC and over 95% yield of BPA are observed under the following conditions: m([Bmim][Ac]):m(PC) = 0.75 : 1, m(methanol) : m(PC) = 0.75 : 1, reaction temperature = 90 °C, and total reaction time = 2.5 h. The ionic liquid [Bmim][Ac] can be reused up to six times without any apparent decrease in the conversion of PC and yield of BPA.

The methanolysis of PC using the ionic liquid [Bmim][Ac] as a catalyst was studied recently by Liu *et al.* [27]. The effects of temperature, time, methanol dosage, and [Bmim][Ac] dosage on the methanolysis reaction were examined. According to the authors, this strategy can overcome the shortcomings associated with traditional methods such as the infeasibility of reusing the catalyst, equipment corrosion, tedious work-up procedures, and environmental problems. Moreover, investigation on the kinetics indicated that the methanolysis of PC in [Bmim][Ac] was a first-order reaction and the activation energy was 167 kJ mol^{-1}. In this process, PC is dissolved in the ionic liquid [Bmim][Ac]; after PC is dissolved or swelled in the ionic liquid, it reacts with methanol to form oligomers under ionic liquid catalysis. Then, the resulting oligomers react with methanol further to produce the final products BPA and DMC.

Alkali-Catalyzed Methanolysis Oku studied the alkali-catalyzed methanolysis of poly[2,2-bis(4-hydroxyphenyl)propane carbonate] in a mixed solvent of methanol and toluene or dioxane [30]. In this process, PC pellets were treated in MeOH with a catalytic amount of NaOH at 60 °C for 330 min, and about 7% yield of BPA was obtained. For a mixed solvent of MeOH and toluene and 70 min reaction time, the complete depolymerization of PC was observed (96%), and free BPA (96%) in the solid form and DMC (100%) in solution were obtained. The characteristic feature of this methanolysis is that PC can be depolymerized to its starting monomer components BPA and DMC by the use of a catalytic amount of alkali metal hydroxide under mild reaction conditions. The monomers can be obtained almost quantitatively in very pure states, and they can be recycled as the monomers of PC and epoxy resins. Liu studied the alkali-catalyzed methanolysis but in a reactor with a stirrer and a refluxing condenser [29]. The results were almost same because of the use of a refluxing condenser. The temperature, on the other hand, affected the efficiency of methanolysis in both studies, with a temperature of 60 °C providing a higher rate of BPA formation [27].

3.2.4.2.2 Hydrolysis with High-Temperature Steam

Watanabe observed that PC gets rapidly hydrolyzed in high-pressure high-temperature steam around the saturated pressure of water at 573 K. In 300 s (5 min) of reaction time, PC completely gets decomposed into BPA with a maximum yield of around 80% [31]. In liquid water phase at 573 K, PC remains even for 3000 s (50 min).

The high yield of BPA in high-pressure steam is due to its high stability. The amount of water required for degradation is drastically reduced, and therefore the high-pressure high-temperature steam process is energetically and economically preferable [31]. Pyrolytic hydrolysis in the presence of alkali-earth oxides and hydroxides (MgO, CaO, $Mg(OH)_2$, or $Ca(OH)_2$ as catalysts is carried out in a steam atmosphere. These catalysts accelerated the hydrolysis of PC drastically, with MgO and $Mg(OH)_2$ being more effective than their Ca counterparts. There is little difference in the reaction rates when using oxides and hydroxides, which shows that the same mechanism operates irrespective of nature of catalyst used. Temperature of the reaction has an effect on the yield and nature of the product; BPA is the main product (yield 78%) at 300 °C in the presence of MgO, but at 500 °C BPA gets degraded to phenol and isopropenyl phenol (IPP). It was shown that a combined process involving PC hydrolysis at 300 °C and BPA fission at 500 °C leads to high yields of phenol and IPP and a drastic decrease of residue [26]. The reaction mechanism of PC hydrolysis in high-pressure high-temperature steam seems to be both surface and bulk erosion.

3.2.4.2.3 Hydrolysis in Other Solvents

Solvents such as *N*-methyl-2-pyrrolidone, 1,4-dioxane, tetrahydrofuran, and dimethylformamide (DMF) are used for alkali-catalyzed hydrolysis of PC in which it can substantially dissolve [29]. In these solvents, hydrolysis of PC is carried out under moderate conditions. If the reaction is carried out in water without co-solvents, no BPA is detected. However, when the hydrolysis is carried out under the same conditions in presence of a co-solvent, accelerated rate of hydrolysis is observed. Because of the insolubility of PC in water, the aqueous depolymerizations require severe conditions such as long reaction times as well as high temperatures and pressures. Therefore, instead of aqueous systems, organic solvent systems such as methylene chloride in combination with ammonia are used [32]. A mixed solvent of phenol and methylene chloride in combination with an alkali catalyst has been reported [33]. With organic solvents, however, a tedious product separation process is generally required. The yield of BPA increases with the amount of H_2O, and the maximum yield is obtained when a ratio of PC/H_2O close to 1.5 : 1 is used. Above this concentration of water, the yield of BPA decreased. Almost 100% conversion of PC is observed. Moreover, under the conditions of reaction temperature = 100 °C, m(PC):m(H_2O) = 1 : 0.7, m(PC):m(NaOH) = 10 : 1, and reaction time = 8 h, and using 1,4-dioxane as solvent, the hydrolysis conversion of PC was almost 100% and the yield of BPA was over 94%.

3.2.4.2.4 NO-Catalyzed Glycolysis of PC in Ethylene Glycol

This is a green process because it involves use of neither toxic solvents nor an alkali catalyst. Using this method, the depolymerization of PC waste by glycolysis using ethylene glycol (EG) was explored without any catalyst in order to get the monomer BPA [31–33]. The maximum yield of BPA of (95.6%) achieved at a reaction temperature 220 °C for 85 min with an EG/PC weight. The reaction mechanism is illustrated in Figure 3.2.2.

PC polymer particles get swollen as EG penetrates into the particles. Then PC gets depolymerized in the solid state by the diffusion of EG. Random scissions of PC take place to lower the average molecular weight until the resulting oligomer can be dissolved in the bulk EG solution but retaining the solid state. The size of the PC particle shrinks as the dissolution proceeds, and finally the oligomers get

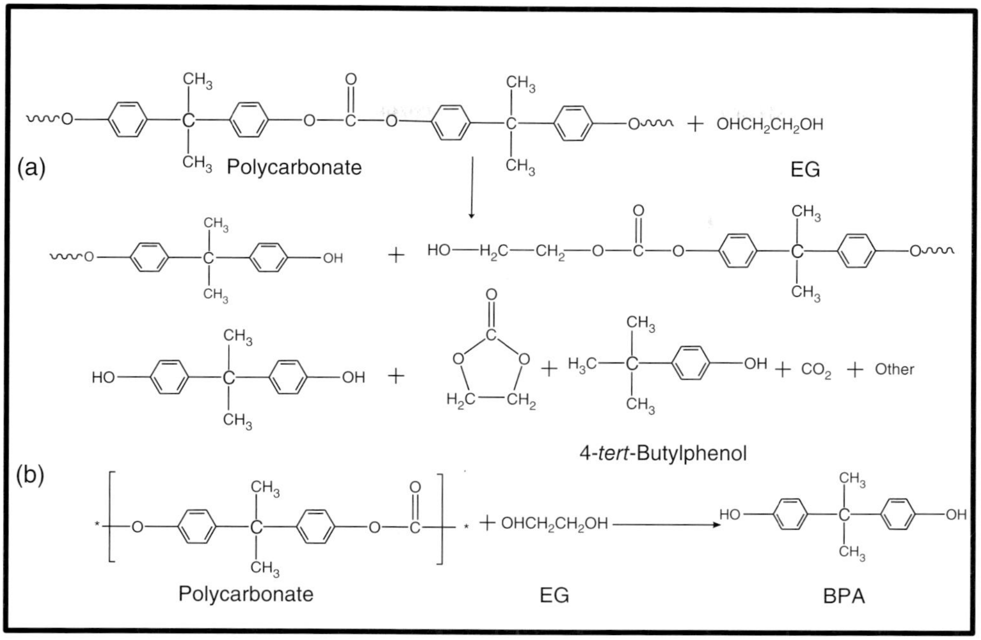

Figure 3.2.2 PC glycolysis reaction pathway [34].

dissolved in EG, which is a heterogeneous reaction. The dissolved oligomer continues to be depolymerized with EG in the bulk solution to produce its monomer BPA, which is a homogeneous reaction.

3.2.4.2.5 **Pyrolysis of PC-Based Polymers**

Pyrolysis of PC and PC-based waste electric and electronic equipment as a means of chemical recycling of this polymer was investigated by Achilias *et al.* [21]. For this, a laboratory-scale fixed-bed reactor was used, and the appropriate pyrolysis temperature was selected after measuring the thermal degradation of model PC by thermogravimetric analysis. After pyrolysis, a large amount of oil, together with a smaller amounts of gaseous products, was obtained, leaving little solid residue. For both samples (model PC and a CD based on PC), the gaseous fraction consisted mainly of CO_2 and CO, whereas in the liquid fraction a large amount of different phenolic compounds, including the monomer BPA, was obtained. It suggests that recycling of used CDs by pyrolysis is a very promising technique with the potential of producing useful high-value chemicals, which may find applications in the petrochemical industry.

3.2.4.2.6 **Problems of PC Recycling**

The addition of fillers and pigments decrease the plastic's resilience. Therefore, recycled PC is usually less resilient and has decreased impact resistance compared to newly manufactured PC. This problem can be overcome by the use of chemicals to modify the impact resistance in recycled PC. It has been observed that up to 15% recycled material can safely be added to the virgin resin without significantly altering properties of the latter. The nature of CDs does not allow them to be easily recycled. The disk is a multilayer product consisting of a PC substrate and three coatings. These coatings, namely aluminum, lacquer, and printing, respectively, make up only a small portion of the entire disk. These materials should be separated or recovered in order to recycle the PC. There are a variety of methods for the removal of paint or plating from engineering plastics, ranging from chemical to physicomechanical procedures. Such techniques include chemical stripping or chemical recovery (high-temperature alkaline treatment), melt filtration, mechanical abrasion, hydrolysis, liquid cyclone, compressed vibration, cryogenic grinding, dry crushing, and roller pressing. The disadvantages of PC include high melt viscosity and notch sensitivity. Used PC usually would have from crazing due to light, radiation, and chemicals present in the service environment, which makes the problem of notch sensitivity even worse. As with other thermoplastics, the level of mechanical and physical properties of PC depends on the molecular weight. However, production waste, recyclates, and so on, frequently do not, or no longer, possess the required molecular weights. Direct material recycling of production waste or recyclates is possible only to a very limited extent. When recycling PC residues, production wastes, and remainders, the recyclates have similar PC compositions; it is therefore desirable and essential to increase the molecular weight to a sufficiently high level for the projected new use. So, for example, low molecular production scrap from PC production for CDs could be increased to the molecular weight range required for injection molding; or the average molecular weight of the PC recyclate from the de-lamination of CDs should be increased sufficiently to allow the material to be used as a component in the production of polycarbonate/acrylonitrile/butadiene/styrene (PC/ABS) blends. It was found that, surprisingly, it is possible to condense PCs from waste by simply melting in a vacuum, optionally with bisphenols or suitable oligocarbonates with OH terminal groups, to produce directly PCs of higher molecular weights.

3.2.5
Advantages of Recycling and Reuse of Polymers

In can be concluded, therefore, that recycling and reuse are approaches for end-of-life waste management of plastic products. These make increasing economic as well as environmental sense. Recent trends demonstrate a substantial increase in the rate of recovery and recycling of plastic wastes. These processes have advantages and disadvantages. The main advantages are the following:

1. Recycling helps in protecting the environment in the most balanced manner.
2. It helps in conserving important raw materials.

3. It protects natural habitats for future generations. Protecting natural resources such as wood, water, and minerals ensures their optimal use.
4. The government and various environmental organizations regularly emphasize the many benefits of recycling.
5. Recycling reduces the amount of wastes that must be sent to landfills or incinerated.
6. The recycling of metals, glass, and other materials reduces pollution that would be caused by the manufacturing of products from virgin materials.
7. Using recycled materials also saves energy because it takes less energy to use recycled materials than to make a product from virgin materials.
8. Reusing plastic is preferable to recycling as it uses less energy and fewer resources. Some plastic items end up in the waste stream after a single use only (single-life or cycle) or a short time after purchase (e.g., food packaging).
9. Reusing plastics conserves fossil fuels since plastic production uses 4–8% of global oil production (4% as feedstock and 4% during conversion).
10. Recycling reduces energy usage and the production of MSW.
11. Recycling leads to the reduction of carbon dioxide (CO_2), nitrogen oxides (NO_2), and sulfur dioxide (SO_2) emission [35].

The disposal of waste plastic is a well-known problem worldwide. Many researchers have proposed many solutions that are energetically viable and environmentally sound. Still there is need of more research to develop more accurate methods by understanding the structures of the many different types of polymers used today and the mechanisms of their degradation. It will make it possible to optimize the decomposition products of this waste stream to give useful products. In the presence of catalysts, the degradation pathways of a polymer may change from a high-energy free-radical mechanism to a more stable carbocation mechanism that results in higher proportions of high RON (Research Octane Number) fuel, so energy saving may also be achieved. In the recycling industry, sorting and identification must be attempted within a short time to positively affect a recycler's finances. Both fast and accurate identification of the primary plastic contained in a particular item and some type of manual or automated sorting are essential. In the case of plastic bottles, the following sorting methods are used:

1. Automated techniques (not always applicable due difference in shape and size, or the existence of paint and coating that delays the analysis technique, and so on.
2. Density sorting methods, which are not particularly helpful because most plastics have nearly the same density (e.g.,$q_{HDPE} = 0.941$, $q_{MDPE} = 0.926–0.940$, $q_{LDPE} = 0.915–0.925$, $q_{LLDPE} = 0.91–0.94$, $q_{PP} = 0.96\ g\ cc^{-1}$).
3. Heavy medium separation in the case of rigid plastic solid waste (PSW) coming from electronic parts, which can be done by adding a modifier to water or by using tetrabromoethane (TBE). However, this is a costly process and can lead to contamination of the recovered plastic.

4. Hydrocyclones to enhance the effectiveness of density separation. These use centrifugal force to enhance material wettability. Some of the factors affecting liquid separation of a given material are its wettability, its variation in density (from porosity, fillers, pigments, etc.), shape factors of size-reduced particles, and its level of liberation from other materials. Even surface air bubbles, which can attach to plastics as the result of poor wetting or surface contamination, can cause an individual flake of material to float in a solution less dense than that of bulk material [38].

Along with sorting, the most important step in every recycling loop is the removal of the paint on plastics, which remains the main issue facing recyclers. Properties of recycled plastics can be compromised because of the stress concentration created by these coating materials.

1. Grinding is used to remove coatings: for example, chrome from plated plastics can be removed by simple grinding, sometimes assisted with cryogenic methods to enhance the liberation process and to prevent the plating materials from being embedded in the plastic granules. These cryogenic methods provide good liberation, but the actual separation of plastic particles from the paint is still problematic.
2. Abrasion is another method to remove coatings, best applied to whole parts of significant size. Solvent stripping is also used by recyclers, which involves the dipping of the coated plastic into a solvent, liberating coatings from the plastic. This method is applicable for the removal of coating from CDs [38].

In conclusion, we suggest that research should continue to develop proper methods of recycling of polymers along with appropriate sorting and identification methods.

References

1. Billmeyer, F.W. Jr., (1971) *Textbook of Polymer Science*, 2nd edn, John Wiley & Sons, Inc.
2. Nicholson, J.W. (1991) *The Chemistry of Polymers*, Royal Society of Chemistry.
3. Achilias, D.S. *et al* (2012) Recent advances in the: chemical recycling of polymers (PP, PS, LDPE, HDPE, PVC, PC, Nylon, PMMA), in *Material Recycling – Trends and Perspectives*, InTech, www.intechopen.com (accessed 11 May 2016).
4. Sifniades, S., Levy, A., and Hendrix, J. (1999) Process for depolymerizing nylon-containing waste to form caprolactam. US Patent 5,932,724.
5. Braun, D. (2002) Recycling of PVC. *Prog. Polym. Sci.*, **27**, 2171–2195.
6. Braun, M., Levy, A.B., and Sifniades, S. (1999) Recycling nylon 6 carpet to caprolactam. *Polym. Plast. Technol. Eng.*, **38** (3), 471–484.
7. Chaupart, N., Serpe, G., and Verdu, J. (1998) Molecular weight distribution and mass changes during polyamide hydrolysis. *Polymer*, **39**, 1375.
8. Corbin, T., Handermann, A., Kotek, R., Porter, W., Dellinger, J., and Davis, E. (1999) Reclaiming epsiloncaprolactam from nylon 6 carpet. US Patent 5,977,193.
9. Bajaj, P. and Sharma, N.D. (1997) Reuse of polymer and fibre waste, in *Manufactured Fibre Technology* (eds V.B. Gupta and V.K. Kothari), Chapman & Hall, New York, p. 615.

10. Polk, M.B., LeBoeuf, L.L., Shah, M., Won, C.-Y., Hu, X., and Ding, Y. (1999) Nylon 66, nylon 46, and pet phase-transfer-catalyzed alkaline depolymerization at atmospheric pressure. *Polym. Plast. Technol. Eng.*, **38** (3), 459.
11. Kasserra, H.P. (1998) in *Science and Technology of Polymers and Advanced Materials* (eds P.N. Prasad *et al.*), Plenum Press, New York, pp. 629–635.
12. McKinney, R.J. (1994) Recycling of carpet and textile fibres. US Patent 5,302,756.
13. Kalfas, G.A. (1998) Mathematical modeling of the depolymerization of polyamide mixtures part 1 : Kinetic Mechanism and parameteric studies in batch reactor. *Polym. React. Eng.*, **6** (1), 41.
14. Bodrero, S., Canivenc, E., and Cansell, F. (1999) Chemical recycling of polyamide 6.6 and polyamide 6 through a two step ami-/ammonolysis process. Presentation at 4th Annual Conference on Recycling of Fibrous Textile and Carpet Waste, Dalton.
15. Czernik, S., Elam, C., Evans, R., Meglen, R., Moens, L., and Tatsumoto, K. (1998) Catalytic pyrolysis of Nylon 6 to recover caprolactam. *J. Anal. Appl. Pyrolysis*, **56**, 51–64.
16. Bockhorn, H., Donner, S., Gernsbeck, M., Hornung, A., and Hornung, U. (2001) Pyrolysis of polyamide 6 under catalytic conditions and its application to reutilization of carpets. *J. Anal. Appl. Pyrolysis*, **58-59**, 79–84.
17. BASF Corp. (2001) BASF 6ix Again Program, www.nylon6ix.com.
18. Yoshioka, T., Sugawara, K., Mizoguchi, T., and Okuwaki, A. (2005) Chemical recycling of polycarbonate to raw materials by thermal decomposition with calcium hydroxide/steam. *Chem. Lett.*, **34**, 282–283.
19. Duangchan, A. and Samart, C. (2008) Tertiary recycling of PVC-containing plastic waste by copyrolysis with cattle manure. *Waste Manage.*, **28**, 2415–2421.
20. Yoshioka, T., Furukawa, K., and Okuwaki, A. (2000) Chemical recycling of rigid-PVC by oxygen oxidation in NaOH solutions at elevated temperatures. *Polym. Degrad. Stab.*, **67**, 285–290.
21. Achilias, D.S., Giannoulis, A., and Papageorgiou, G.Z. (2009) Recycling of polymers from plastic packaging materials using the dissolution/reprecipitation technique. *Polym. Bull.*, **63** (3), 449–465.
22. Chiu, S.J., Chen, S.H., and Tsai, C.T. (2006) Effect of metal chlorides on thermal degradation of (waste) polycarbonate. *Waste Manage.*, **26**, 252–259.
23. Hu, L.C., Oku, A., and Yamada, E. (1998) Alkali-catalyzed methanolysis of polycarbonate. A study on recycling of bisphenol A and dimethyl carbonate. *Polymer*, **39**, 3841–3845.
24. Grause, G., Sugawara, K., Mizoguchi, T., and Yoshioka, T. (2009) Pyrolytic hydrolysis of polycarbonate in the presence of earth-alkali oxides and hydroxides. *Polym. Degrad. Stab.*, **94**, 1119–1124.
25. Ikeda, A., Katoh, K., and Tagaya, H. (2008) Monomer recovery of waste plastics by liquid phase decomposition and polymer synthesis. *J. Mater. Sci.*, **43**, 2437–2441.
26. Liu, F.-S., Li, L., Yu, S., Lv, Z., and Ge, X. (2011) Methanolysis of polycarbonate catalysed by ionic liquid [Bmim][Ac]. *J. Hazard. Mater.*, **189**, 249–254.
27. Liu, F.-S., Li, Z., Yu, S.T., Cui, X., Xie, C.X., and Ge, X.P. (2009) Methanolysis and hydrolysis of polycarbonate under moderate conditions. *J. Polym. Environ.*, **17**, 208–211.
28. Oku, A., Tanaka, S., and Hata, S. (2000) Chemical conversion of poly(carbonate) to bis(hydroxyethyl) ether of bisphenol A. An approach to the chemical recycling of plastic wastes as monomers. *Polymer*, **41**, 6749–6753.
29. Watanabe, M., Matsuo, Y., Matsushita, T., Inomata, H., Miyake, T., and Hironaka, K. (2009) Chemical recycling of polycarbonate in high pressure high temperature steam at 573 K. *Polym. Degrad. Stab.*, **94**, 2157–2162.
30. Fox, D.W. and Peters, E.N. (1989) Methods of recycling a dihydric phenol from a scrap polyester. US Patent 4,885,407.

31. Buysch, H.J., Schoen, N., and Kuehling, S. (1994) Polymerization of polycarbonate waste by ethylene glycol. Ger. Offen., DE 4,220,412.
32. Shafer, S.J. (1994) Introduction to green chemistry : second edition. US Patent 5,336,814.
33. Kim, D., Kim, B., Cho, Y., Han, M., and Kim, B.S. (2009) Kinetics of polycarbonate glycolysis in ethylene glycol. *Ind. Eng. Chem. Res.*, **48**, 685–691.
34. Kim, D., Kim, B.-K., Cho, Y., Han, M., and Kim, B.S. (2009) Kinetics of polycarbonate methanolysis by a consecutive reaction model. *Ind. Eng. Chem. Res.*, **48** (14), 6591–6599.
35. Al-Salem, S.M., Lettieri, P., and Baeyens, J. (2009) Recycling and recovery routes of plastic solid waste (PSW): a review. *Waste Manage. (Oxford)*, **29**, 2625–2643.

4 Recycling of Plastics

Preetha Balakrishnan and Meyyappallil Sadasivan Sreekala

4.1 Introduction

There has been concern about the environmental and health aspects of plastics for many years. Starting in about 2000, there have been developments that signaled the possibility of a major impact of plastics' acceptability to the public. These include emphasis on disposal and recycling and health concerns. Common plastics such as PE (polyethylene), PP (polypropylene), PS (polystyrene), PVC (poly(vinyl chloride)), and PC (polycarbonate) do not degrade by biological means when discarded. Incineration has its opponents, partly because of the production of air pollutants such as mercury, sulfur oxides, and dioxins. Landfills are objectionable as a bad way to utilize space and due to possible health effects of leachates. Plastics are used widely in many types of products. The success of plastics and their growth in diverse applications have resulted in environmental and health concerns, which have to be dealt with and their growth reversed through banning of some plastics. Packaging accounts for 35% of plastics use and is a major source of disposal and recycling problems. About 7% of household waste in the UK is plastics, of which 3% constitutes plastic films. Much of that is not recycled and ends up in landfills or incinerators. Materials have become an inevitable part of our consuming society, and it would be hard to imagine a modern society today without plastics, which have found many uses in fields as diverse as household appliances, packaging, construction, medicine, electronics, and automotive and aerospace components. Plastics can be classified as thermoplastics and thermosets. Examples of thermoplastics are PE, poly(ethylene terephthalate) (PET), poly(vinyl chloride) (PVC), PP, PS, and expanded polystyrene (EPS). Examples of thermosets are epoxide (EP), phenol-formaldehyde (PF), and polyurethane (PUR). Thermoplastics can be shaped by heating and will maintain their shape after cooling, whereas the melting point of thermosets is high enough that they will burn before they melt, which makes them hard to recycle [1]. The aforementioned plastics are derived from crude oil through the distillation of naphtha. Some plastics may also be produced from

Recycling of Polymers: Methods, Characterization and Applications, First Edition. Edited by Raju Francis.

natural gas by cracking it into ethylene. Additional monomeric chemicals are added depending on the type of plastic produced [2]. Plastic recycling can be done either by feedstock recycling or mechanical recycling. Feedstock recycling breaks down the plastic monomers, which can then be used in refineries and chemical production. Mechanical recycling includes shredding, cleaning, melting, or granulation [1], and it produces quality products only when the materials are devoid of contaminants, sorted into a single type of polymer, and sufficiently segregated according to the color required for the end use. For example, when containers are recycled into black plastics, sorting by color is not important. Australia produces approximately 1 million tons of plastic resins annually. Of this, 450 000 tons are discarded in landfill waste sites throughout the country, and represents between 5% and 15% of the total weight of the landfill waste stream [1]. Fifty million tons of plastic resins is produced per annum in the world, but the total amount recycled is difficult to quantify. The high volume-to-weight ratio of plastics means that on a volume basis, plastics represent between 10% and 30% of waste going to landfill. The volume of plastics in landfill is difficult to quantify since the degree of compaction has a definite impact on the figures quoted in the literature. Targets have been set by various organizations to reduce waste problems. At the end of this decade, the Commonwealth Environmental Protection Agency (CEPA) wishes to reduce the total municipal solid waste (MSW) by 50%; the Australian and New Zealand Environment and Conservation Council (ANZECC) aims to reduce packaging waste by 50% in a similar time; and KEMCOR, the supplier of 90% of the high-density polyethylene (HDPE) used for milk packaging in Australia, has set a target of reducing the amount of HDPE sent to landfill by 50% [3]. This follows the worldwide trend of various organizations that have set similar goals. Plastics recycling, as an industry, face a number of hurdles before establishing itself as an effective means of reducing the amount of plastic waste contributing to landfill and litter problems. Currently, the level of recycling occurring in the household waste sector is not high enough to provide an appreciable effect on the problem of plastic waste. Solid markets for the recycled and reprocessed products are yet to be secured because of economic, technological, and institutional barriers. How effective is recycling plastic waste, and how effective can it be? This chapter attempts to address the feasibility of plastics recycling, ways of recycling, and mechanism of recycling of various plastic products.

Natural biofiber composites are emerging as a viable alternative to glass-fiber-reinforced plastics, especially in automotive applications. Thermoplastic biocomposites, composed of PP, PE, PS, or biopolymers such as poly(lactic acid) (PLA) and reinforced with natural fibers have especially attracted great attention during the past few years. Such thermoplastic biocomposites, in contrast to those made using a thermoset matrix, can be processed into different shapes and have the potential to be mechanically recycled. Eureka, a European research and development organization, announced the completion of a research project in 2007 to produce new recyclable compounds for consumer products and audio components that involves the combination of natural fibers such as flax, wood, jute, and hemp with thermoplastics. The key measurement tool to assess a product's environmental impact is life-cycle assessment (LCA). Through LCA it is possible to account for all of the environmental impacts associated with a product, covering all stages in a product's life, from the extraction of resources to waste disposal. However, using renewable feedstock does not guarantee that a plastic will remain environmentally friendly over its entire life cycle. The sustainability benefits of using renewable feedstock may not be sufficient if the material cannot be recycled. To study the recyclability of the polymeric materials including bioplastics and their derivatives, it is a well-tried practice to simulate the mechanical recycling by doing multiple extrusions and to find the durability or service life by accelerated thermal and hydrothermal aging. These methods allow the assessment of the effects of thermal, hydrothermal, and thermomechanical degradation [4, 5].

Recycling plastics encompasses four phases of activity, namely collection, separation, processing, and manufacturing and marketing. Because using clean, homogeneous resins alone can produce the highest quality recycled plastic products in the existing secondary process (material recycling) and high-value chemical products in the existing tertiary process (feedstock recycling), an effective separation of mixed plastics waste is necessary. Although several separation technologies, including automatic sorting, gravity separation, and electrostatic separation, can be applied to separate mixed plastics waste, they have their limitations. Froth flotation is a major and efficient separation method in mineral processing engineering, which could prove useful for mixed plastics separation as well.

4.2 Plastic Waste Management Scenario

Plastics have become an integral part of our lives. The amount of plastics consumed annually has been growing steadily. Its low density, strength, user-friendly design and fabrication capabilities, and low cost are the drivers of such growth. Besides their wide use in packaging as well as automotive and industrial applications, they are extensively used in medical delivery systems, artificial implants and other health care applications, water desalination, removal of bacteria, and many others. Use of plastics in the preservation and distribution of food, housing, and appliances are too many to mention here. Specially designed plastics have been an integral part of the communication and electronics industry – be it in the manufacturing of chips or printed circuit boards, or housings for computers. Durable plastics, as opposed to most packaging and convenience goods which are discarded after a single use, tend to have a life of 3 or more years. Automobiles, computers, household appliances, carpets, and fabrics fall into this category. The use of plastics in durable applications continues to grow as design engineers, manufacturers, and consumers continue to rely on its performance, low cost, and design benefits. The recovery of plastics from such durable goods is complex. Often, they are integrated with several other plastic and non-plastic components.

Their separation, recovery, and purification require several steps and, generally, the volumes of such materials available for recovery are limited. Nevertheless, several efforts are under way exploring the recycling of such products after their lifetime. Manufacturers of such products have committed to using recycled materials, wherever possible, as a part of their total material needs. Rigamonti *et al.* [6] suggested a general approach in which plastic waste management scenarios were defined by taking into account the peculiarity of these materials and some of the most common practices in Europe. When compared to other packaging materials (iron, aluminum, paper, etc.), what is commonly referred to as "plastic" is in fact still a very heterogeneous fraction. As a consequence, an important sorting step is required prior to recycling, which is aimed at the following:

Removing non-plastics fractions
Sorting by different polymers (PET, PE)
Sorting by different colors (PET only).

After sorting, the different plastic flows are sent to the recycling process, which yields some further residues.

The short one-trip lifetime of plastic packaging makes it the major contributor to the problem of plastic waste and litter, and consequently plastic packaging is the focus of most household recycling efforts. However, when plastic packaging is recycled, it is reprocessed into items other than for direct food contact applications due to stringent regulations. In effect, this means that an entirely new market is created for this type of recycled and reprocessed plastic. Products currently processed from recycled plastics include garbage bins, crates, piping, piping joints, and structural items for timber and concrete alternatives (these products are not necessarily made from recycled packaging).

Laws and regulations for recycling plastic waste have been enacted, and an efficient collecting system has formed with the economic development of China. There are many parks and residential areas in China acting as collection points for waste plastics. The recycling of plastic wastes is becoming a new attractive field in that it can save resources and protect the environment. The National Committee of Recycling and Utilization of Plastic Wastes was formed as early as 1992 in China. Its duty is to guide the recycling and utilization of waste plastics. The technology for converting plastic wastes to oil can adapt itself to the model of China. In China, further tasks are on to formulate the standards for process and products of post-consumer recycled plastics and to adopt the more advanced pyrolysis technologies for waste plastics, such as fixed-bed catalytic reactors or fluidized-bed reactors, following to the experiences of developed countries such as Germany and Japan [7–10]. In addition, the type of pyrolysis reactor must be designed to suit mixed waste plastics and small-scale and middle-scale production. Scientists should pay more attention to explore the recycling and pyrolysis of waste PVC and the optimization of operational conditions of pyrolysis for waste plastics.

4.3 Ways of Recycling

There is growing concern about the recycling of plastics, as it generates severe environmental and health hazards. As most of these plastics are good to use and cheap, they face a major problem of disposal. There are mainly three types of recycling: mechanical and chemical recycling and reuse. Suitable waste handling and recycling methods are in the focus of our modern society. The main reason for this is that the consumption and production of, for example, metals, plastics, and so on, are increasing monotonously and wastes from them pose difficult energetic and environmental problems. On the other hand, the growth of the gross domestic product (GDP) and standard of living is unimaginable without the application of these materials. Chemical and mechanical recycling of plastic wastes is of growing importance. During chemical recycling, waste plastics are converted into fuels or feedstock (gases, oils, and residue) for the petrochemical industry. Information is abundant on the degradation of polymers that are widely used for packaging and other domestic purposes (e.g., polyolefins, PS, PET, etc.). Yields and characteristics of the fractions obtained by cracking can be considerably affected by the structure of the raw materials and by the cracking parameters (e.g., temperature, catalysts, etc.). Catalysts have some positive effect on degradation. The yields of valuable lighter hydrocarbons are increased in the presence of catalysts and, furthermore, products of better quality are obtained [11–23]. The commonly used catalysts are acidic zeolites (silica–alumina, β-zeolite, Y-zeolite, mordenite, clinoptilolite, etc.). The yields and structure of the products are mainly affected by the characteristics of catalysts (e.g., Si/Al ratio, micro or macroporous surface area, etc.) [24, 25]. The cracking behavior of C–C bonds and the probability of β-scission are mainly determined by the acidity of the catalyst. Also, the greater the pore size, the longer the main chain of the formed hydrocarbons. Landfill dominates solid waste management in the United States, presently absorbing 62% (by weight) of the MSW. However, it is estimated that many existing landfills will close by 2005 either because of capacity limits or because of the failure to meet the stringent environmental regulations. Public opposition has prevented large-scale use of incineration to dispose of MSW largely because of potential health hazards. Over the next 10 years, the amount of MSW will grow mainly because of increases in population and material standards of living. With severe restrictions on landfill and incineration, reduction of the volume of MSW has become an increasingly important objective of solid waste management. This can be achieved through source reduction and recycling. Source reduction activities include measures that reduce material uses for the production of goods or packaging such as the following:

1. Designing products or packages to reduce the quantity of materials or the toxicity of the materials used, for example, concentrated laundry detergent;
2. Lengthening the life of products to postpone disposal and reducing the availability of single-use products like disposable diapers;
3. Reusing products or packages, for example, refillable containers.

Recycling is a necessary component of solid-waste management for much of the country, but it may not be viable for remote, sparsely populated areas where landfill is cheap and abundant and wastes need not be transported over long distances. To facilitate recycling, manufacturers can take a number of actions to make it technically easier and cheaper to recycle plastic, such as the following:

1. Redesign products and packaging to reduce the number of resins.
2. Identify parts made of different resins for easy disassembly and sorting.
3. Reduce the use of additives, adhesives, coatings, and labels which contaminate or weaken recycled resins, or replace them with standardized ones that can easily be separated from plastics.

Government action is needed to achieve a broader range of objectives affecting consumers as well as the industry. These include policies to promote household cooperation, which include the following:

1. Levy a fee based on the amount and type of solid waste collected.
2. Collect a deposit for items (as is being done for soda bottles, for example) of packaging and some durable items like automobiles.

Other policies that can help provide greater certainty regarding the long-term demand for recycled plastic include the following:

1. Taxes on packaging that cannot be recycled or on virgin materials to cover the costs of disposal;
2. A procurement policy that requires, or is prepared to pay, a higher price for recycled content;
3. Recycled-content requirements in the private sector, analogous to local government requirements that newspapers should contain a certain percentage of recycled material.

4.3.1 Reuse

Plastics can be classified as thermoplastics or thermosets. The mainly used thermoplastics are PE, PET, PVC, PP, PS, and EPS. Examples of thermosets are EP, PF, and PUR. Thermoplastics can be shaped by heating and will maintain their shape after cooling, whereas the melting point of thermosets is high enough that they will burn before they melt, which makes them hard to recycle [26]. The aforementioned plastics are derived from crude oil through the distillation of naphtha. Some plastics may also be produced from natural gas by cracking it into ethylene. Additional monomeric chemicals are added, depending on the type of plastic produced [2, 27]. Plastic recycling can be done either by feedstock recycling or by mechanical recycling. Feedstock recycling breaks down the plastic monomers, which can then be used in refineries and chemical production. Mechanical recycling includes shredding, cleaning, melting, or granulation [26], and it produces quality products only when the materials are free of contaminants, sorted into a single type of polymer, and sufficiently segregated according to the color required for end use. For example, when containers are recycled into black plastic items, sorting by color is not important. The most common examples of reuse are with glass containers, where milk and drinks bottles are returned to be cleaned and used again. Reuse is not widely practiced in relation to plastic packaging. Plastic products in general tend to be discarded after their first use. However, there are examples of reuse in the marketplace. For example, a number of detergent manufacturers market refill sachets for bottled washing liquids and fabric softeners. Consumers can refill and hence reuse their plastic bottles at home, but in all of these cases the reusing of the plastic bottles and containers do not continue for long time epically in the food applications.

4.3.2 Mechanical Recycling

This is also known as *physical recycling*. The plastic is ground down and then reprocessed and compounded to produce a new component that may or may not be meet its original use [28]. Vallim *et al.* [29] recycled HDPE waste by blending with virgin polyamide (PA6) using a twin-screw extruder. The characterization of the prepared blend revealed that the mechanical properties and thermal stability of the blend were improved by using PA6, which was attributed to the decrease of the size of domains of the recycled HDPE. The recyclability of other PP composites was also investigated by Bahlouli *et al.* [30], where the effect of recycling on the properties of the PP-based composite (ethylene propylene diene monomer (EPDM)/PP and talc/PP) using extrusion was examined. The rheological, mechanical, and structural properties of the composites were determined and compared after each extrusion cycle. The results showed that the melt viscosity of the composites decreases with processing number in the same way as with pure PP [31]; also, the mechanical properties of the composites decreased with the number of processing cycles. All noted changes in the composites' properties were attributed to the changes in their structural properties during the processing cycle. The obtained results from this work were useful for optimizing the recycling process and for a better use of the recycled materials in components design.

4.3.3 Chemical Recycling

The polymer waste is turned back into its oil/hydrocarbon component in the cases of polyolefins, and monomers in the case of polyesters and PAs, which can be used either as raw materials for new polymer production and petrochemical industry or into pure polymers using suitable chemical solvents [32]. Thermal cracking in the presence of phenol as a solvent was also used for the chemical recycling process of HDPE in a work reported by Vicente *et al.* [33], who investigated the effect of phenol on the thermal cracking process. Their results showed that the main products in the cracking reaction were olefins, which are very important for the petrochemical industry. Also, the presence of phenol as a solvent can promote the cracking reaction due to its role in favoring random scissions and chain reactions, which can explain the plastic conversions obtained as well as the yields and selectivity of each hydrocarbon product. Achilias *et al.* [34] reported catalytic and noncatalytic pyrolysis of PS waste in a fixed-bed reactor using either a model polymer or commercial waste products as the feedstock. It was found that the pyrolysis oil fraction could be repolymerized again to produce virgin PS. However, aromatic compounds included in this fraction may act as chain-transfer agents, resulting in alterations in the shape of the reaction rate curve and lowering significantly the average molecular weight and the glass transition temperature of the PS prepared form the pyrolysis process.

4.4 Poly(Lactic Acid)

PLA is one of the most important biodegradable polyesters derived from renewable sources (mainly starch and sugar). Until the last few decades, the main uses of PLA had been limited to biomedical and pharmaceutical applications because of its high cost and low molecular weight. Because of the presence of both a hydroxyl and carboxyl group in lactic acid, it can be converted directly into polyesters via a poly condensation reaction. Considerable interest has been generated in academic research associated with PLA and its copolymers [35–38]. Duigou *et al.* [39] studied the effect of recycling on the mechanical behavior of biocompostable flax/poly(L-lactide) composites, which were used as an alternative to glass-fiber-reinforced petrochemical polymers. The composites were fabricated using a single-screw extruder and then molded using an injection molding machine. In order to investigate the recyclability of the material, the fabricated composites were subjected to six injection cycles, and the effect of the number of the injection cycle on the rheological, mechanical, and morphological properties was determined. The results showed that the stiffness of PLA improved after the addition of the fiber, but the stress at break and strain at break decreased dramatically. Also, it was found that the stiffness of the composites was not affected with the number of injections in spite of the molecular weight reduction associated with the processing cycles, which was attributed to the increase in crystallinity in the PLA phase after the processing cycles [40], whereas the stress at break and strain at break decreased after the injection cycles.

Mechanical recycling as a favorable recycling method has been under consideration for bioplastics such as PLA. For cradle-to-gate and cradle-to-grave LCAs, Piemonte [41] showed, for both PLA and starch-based bioplastics and using four different recycling methods, that compared with the mechanical recycling process, the other methods, namely incineration, composting, and anaerobic digestion, were clearly inferior from the environmental point of view. Although PLA is biodegradable, which significantly reduces the negative environmental impact of

PLA waste, the material recycling and investigation of changes in the properties of PLA upon its multiple processing are important. The first reason for this is that the generation of industrial waste is inevitable in various industrial processes where an online grinding of the off cuts and blending with neat polymer is integrated [42]; the second reason is the possibility to reuse the post-production (used) PLA waste because it would be advisable to extend their service lives before finally discarding them to bio-disposal facilities such as composting plants [43, 44]. Finally, the low environmental impact of PLA would be compromised if the recycling of industrial wastes cannot be achieved [42]. A study [42] on the reprocessing of PLA (containing 92% L-lactide and 8% D-lactide) showed that only the tensile modulus remains constant with thermomechanical cycles of up to seven injection moldings. In contrast, stress and strain at break, modulus, hardness, and rheological factors generally show decreased values. The viscosity of PLA decreases greatly (from 3960 to 713 Pa s) after only one injection cycle. Zero viscosity (g0) of PLA as a function of injection number is presented in Figure 4.1. The mechanical properties of the recycled PLA had become too poor for any industrial application of the polymer.

This dramatic effect of repeated processing cycles was mainly attributed to the degradation of PLA by chain scission during processing, leading to a significant decrease in the molecular weight. To understand the degradation phenomena in PLA, two stabilizers, namely quinone (PLA-Q) and tropolone (PLA-T), were blended with PLA. Quinone was found to be an efficient stabilizer that traps free radicals and maintains the PLA chain length at the processing temperatures. This confirms that the main mechanism of the degradation of PLA is through formation of free radicals and not by the hydrolysis caused by moisture [42]. A recent accelerated hygrothermal aging test using three-times reprocessed

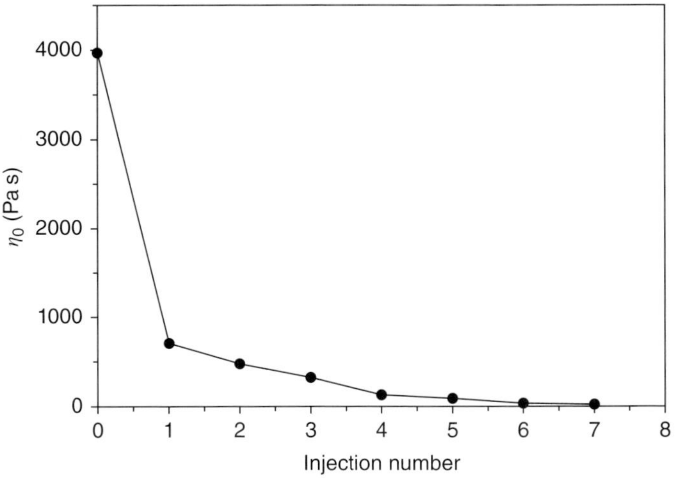

Figure 4.1 Zero viscosity of PLA as a function of injection number. (Pillin *et al.* [42] Reproduced with permission of Elsevier.)

and virgin PLA showed that the water diffusion rate decreased with increasing number of reprocessing cycles and increased with higher hygrothermal aging temperatures. The degradation was confirmed to be due to the chain scission reactions promoted by temperature; however, in presence of water in the polymer matrices, hydrolytic chain scission also occurred, which was especially noticeable for reprocessed PLA and higher temperatures. The effects of hygrothermal aging were particularly significant at higher temperatures and for PLA reprocessed more than once [45]. It shows that poly(L-lactic acid) (PLLA) and the aliphatic polyester (Mater-Bi TF01U/095R) can be mechanically recycled up to 5 and 10 times, respectively, with an acceptable loss in mechanical and thermal properties, whereas the used starch-based waste (Mater-Bi YI014U/C) should be destined to composting because its recyclability is very poor [46]. PLLA is semicrystalline and shows high mechanical strength, whereas PLA (poly(D,L-lactic acid)) is essentially amorphous, or has a low crystallinity, because of its intrinsic structure [47]. In a study by Helfenbein [48], the processing equipment and the processing parameters were modified to suit the processing of PLA sheets directly and economically into recycled PLA pellets. These recycled PLA pellets, which can be added in 20–50% blend with virgin PLA, not only reduced the material cost but also helped to protect the environment by sending less waste into landfill. Recently, a pilot project was started in PLA recycling by the yogurt maker Stonyfield Farm, Londonderry, NH, which has begun working with Nature Works on the mechanical recycling of PLA (Ingeo) trim scrap from its thermoforming/filling/sealing packaging lines [49]. Tsuneizumi *et al.* [50] conducted chemical recycling of PLA-based polymer blends PLA/PE and PLA/poly(butylene succinate) (PLA/PBS). Two routes associated with the chemical recycling of PLA/PE blend were employed:

Direct separation of PLA and PE first by their different solubilities in toluene, followed by the chemical recycling of PLA using montmorillonite;

The selective degradation of PLA in the PLA/PE blend by montmorillonite in a toluene solution at 100 °C forming the lactic acid (LA) oligomer with a small molecular weight. The PE remained unchanged and was quantitatively recovered by the reprecipitation method for material recycling (Figure 4.2).

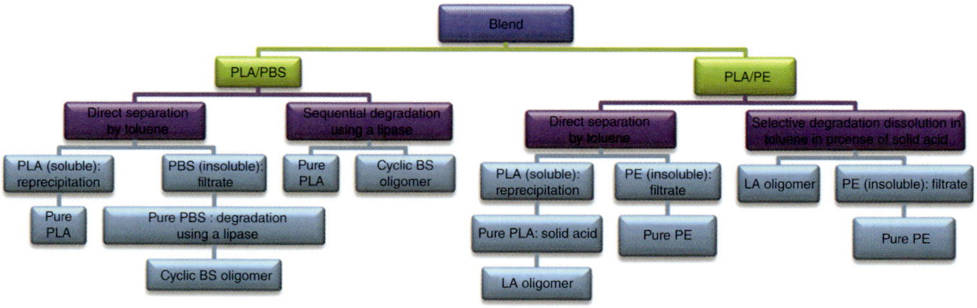

Figure 4.2 Chemical recycling of PLA/PE and PLA/PBS.

4.5 Poly(Vinyl Chloride)

The low cost and high performance of PVC products combined with the wide range of properties that can be obtained from different formulations has contributed to the widespread use of PVC in construction products. There is a long time lag between PVC consumption and the accumulation of PVC waste due to the long life of PVC products, which can be up to 50 years. It is obvious that all the PVC that is being produced will become waste some day. Various studies have reported the recycling of PVC and its systems since the beginning of the last decade.

For recycling PVC (from two-component mixed plastics such as PVC/PET, PVC/PP, PVC/PE, or PVC/PS) mechanically, triboelastic technology was most commonly used. Yarahmadi *et al.* reported mechanical recyclability of PVC sheets used in building-floor applications [51]. The results has shown that PVC floorings as plastic waste can be mechanically recycled without upgrading and without the addition of new plasticizer. For investigating the effect of wood fiber content on the recyclability, the recyclability of pure PVC and wood-fiber-reinforced PVC were compared through its effect on the mechanical properties of both PVC and the composites [52]. The results showed that the addition of the wood fiber to PVC improves its recyclability where it was found that up to five processing cycles, the composite properties remained stable. However, the flexural strength increased, whereas the other mechanical properties remained almost constant after 10 processing cycles, especially after 20 cycles. The small amount of PVC that is used in packaging poses a number of problems: PVC containers appear to be identical to bottles made from other resins, but even a small amount of PVC can render a large batch of recycled material useless. In addition, the manufacture of PVC products and their incineration are suspected of being highly polluting, although the producers contest this claim. There are no packaging applications presently using PVC that cannot be met by another resin. We assume in the scenario that PVC in packaging will be replaced by other resins, namely HDPE, LDPE, PP, and PET. This change will reduce PVC wastes and increase the demand for recycling of these other resins.

During recent years, the world has been facing a waste crisis due to the generation of huge amounts of plastic wastes, including PVC, by industries and households. Moreover, considering that about half of PVC is in use in long-life products and the lifespan of such PVC products is about 30–40 years, a significant increase in PVC waste generation is also expected in the near future. It has been suggested that material and/or chemical recycling of the PVC wastes may be a suitable solution to the problem of environmental pollution arising from landfilling or incineration of such MSWs. However, as was emphasized, a simple recycling to produce a secondary material may not by itself solve the problems of disposing of PVC wastes, which is due to the existence of many types of incompatible polymers in a typical PVC waste stream. In fact, for a successful and economical recycling in which a recyclate has an acceptable characteristic can be produced, the first step is to facilitate and improve the PVC separation methods. Therefore, the review also is focused on the conventional and recent PVC separation methodologies.

4.6 Polyethylene

PEs are one of the most widely used plastics characterized by a density in the range 0.918–0.965 g cm^{-3} resulting in a range of toughness and flexibility. Their major application is as packaging film, although their outstanding dielectric properties mean that they are also widely used as an electrical insulator. Other applications of PEs, including domestic ware, tubing, squeeze bottles, and cold water tanks, are also well-known (Table 4.1).

Jin *et al.* studied the effect of mechanical recycling on the rheological and thermal properties of low-density polyethylene (LDPE) [53]. LDPE samples were subjected to extensive extrusion cycles up to 100 cycles. The results showed that the complex viscosity of the samples at a low test frequency of 0.628 rad s^{-1} (Figure 4.3a) increased with increasing number of extrusion cycle, which was attributed to the crosslinking reactions that took place throughout LDPE chains during recycling process due to the presence of the reactive carbon radicals [54, 55]. The same tend was observed in the melt flow index (MFI) measurements, where it was noted that the MFI of the material decreased as the number of extrusion cycle increased (Figure 4.3b). The results of this work revealed that the processability of LDPE is affected only after the 40th extrusion cycle. Also, the results reported by Kartalis *et al.* [56] on the recycling of low-density polyethylene/medium density polyethylene (LDPE/MDPE) blend revealed that even after five successive extrusion cycles, the material shows significant processing stability.

Table 4.1 Summary of the main products of LDPE/HDPE blend degradation at reaction temperature of 360 °C over various catalysts.

Degradation results	Catalyst type				
	USY (zeolitic catalysts)	ZSM-5 (zeolitic catalysts)	MOR (zeolitic catalysts)	ASA (non-zeolitic catalysts)	MCM-41 (non-zeolitic catalysts)
Yield (wt% feed)					
Gaseous	87.5	93.1	90.2	85.6	87.3
Liquid	3.7	3.3	4.3	4.7	5.6
Residue	8.8	3.6	5.5	9.7	7.1
Involatile residue	4.7	2.4	2.9	7.4	5.2
Coke	4.1	1.2	2.6	2.3	1.9

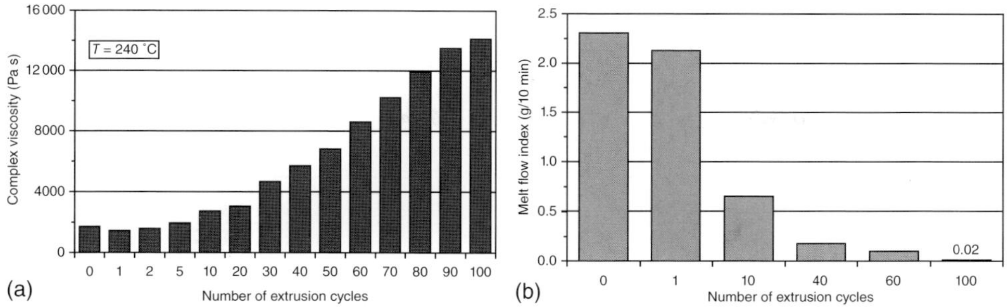

Figure 4.3 (a) Complex viscosity and (b) MFI of LDPE after extrusion cycles. (Jin *et al.* [53] Reproduced with permission of Elsevier.)

Puente *et al.* [57] studied the chemical recycling of LDPE using fluid catalytic cracking (FCC) at 500 °C in the presence of various commercial catalysts. The process was performed in a solution of LDPE using toluene as solvent. The results showed that the FCC products were qualitatively similar in all the catalysts and the contribution was centered mainly on the gasoline fraction, with high aromatic content, although the production of gases was also important, with a high proportion of valuable light olefins C_3–C_4. Isoparaffins C_4–C_5 were significant as well. Hajekova and Bajus [58] studied the chemical recycling of LDPE and PP waste using a two-step thermal cracking method; in the first step, the polymer waste was decomposed individually in a batch reactor at 450 °C and was converted to wax/oil products. In the second step, the wax/oil products were dissolved in heavy naphtha to obtain steam-cracking feedstock. The selectivity and kinetics of co-pyrolysis for 10 mass% solutions of wax/oil from LDPE or PP with naphtha in the temperature range 740–820 °C at residence times of 0.09–0.54 s using industrial ethylene units were studied. The results showed that it was possible to perform the recycling of polyalkenes via the co-pyrolysis of polyalkene oils and waxes with conventional liquid steam cracking feedstock on already existing industrial ethylene units.

Achilias *et al.* [59] studied the chemical recycling of PEs (LDPE and HDPE) and PP obtained from various applications including packaging film, bags, pipes, and food-retail outlets using two techniques:

Dissolution/reprecipitation using different solvents and nonsolvents;
Catalytic pyrolysis using FCC method.

In the first technique, the polymers were dissolved in xylene and reprecipitated using *n*-hexane, and the pure polymers were then dried. In the second one, the polymers were thermally cracked in the presence of an FCC catalyst. The pure polymers obtained from the first technique were mechanically tested and compared with those of the waste, whereas the products of the second technique were characterized using spectroscopic methods (Figure 4.4).

Waste polymer could be converted into lighter fractions at 430 °C during 45 min with yields of 40–50% depending on the catalysts. HZSM-5 increased the yields

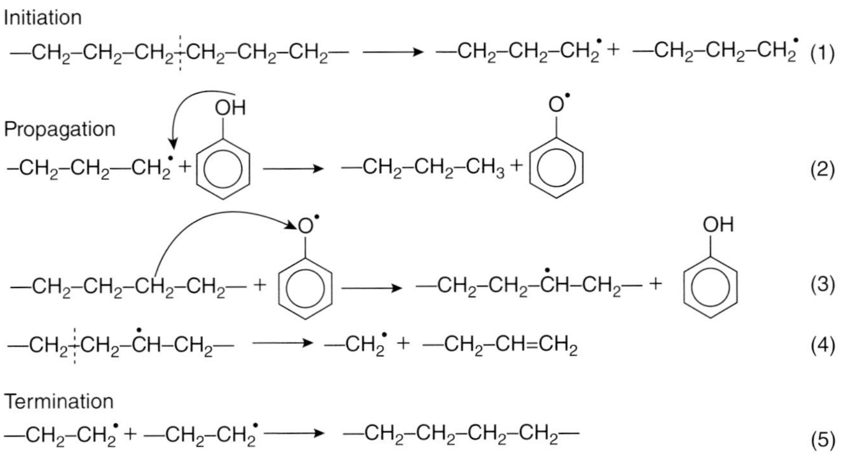

Figure 4.4 Role of phenol in the thermal cracking of HDPE. (Vicente *et al.* [33] Reproduced with permission of Elsevier.)

of both gaseous (C1 – C4) and liquid (C5 – C23) products, while FCC and clinoptilolite catalysts increased the yield of liquid. With catalysts of 2%, the structure of the products were significantly modified, because the hydrocarbons got rather branched and the position of olefinic double bonds also changed in case of thermocatalytic cracking. The degradation of HDPE was well followed via the decrease of the average molecular weight of heavier residue fractions (C24+). Liquid hydrocarbons and residues are of low sulfur and nitrogen content from additives of raw material; therefore they may be favorable for energy production or for raw materials of the petrochemical industry. Film, mainly HDPE and LDPE, accounts for the largest use of plastic and have been identified as the next major source of feedstock for recycling [60]. The source for recycled film is likely to be limited to commercial and institutional sectors, such as dry-cleaning bags, shrink wrap, and merchandise bags, rather than household film, which is often contaminated by food and other materials.

4.7 Polypropylene

Bourmaud *et al.* studied [30, 61] the effect of recycling on the rigidity of PP/vegetal fiber composites prepared by the extrusion process. The results showed that the rigidity of the composites nearly remained constant after the processing cycles due to the good stabilization of the fibers' aspect ratio after recycling. Rheological, mechanical, and structural properties of the composites were determined and compared after each extrusion cycle. The results showed that the melt viscosity of the composites decreased with the processing number in the same way as for pure

PP [31]; also the mechanical properties of the composites decreased with the processing number. All noted changes in the composites' properties were attributed to the changes in the structural properties of the composites during the processing cycle. The obtained results from this work were useful for optimizing the recycling process and for a better use of the recycled materials in components design [62].

4.8 Polystyrene

Brennan *et al.* [63] studied the effect of recycling on the properties of acrylonitrile/butadiene/styrene (ABS), high-impact polystyrene (HIPS), and the blend of ABS waste and HIPS waste. Mechanical and thermal properties of the virgin and recycled polymers (ABS and HIPS) were determined and compared. The results showed that in the two cases (ABS and HIPS) the effect of recycling on the tensile strength and tensile modulus of the blend, which had a small portion of one of the two components (ABS or HIPS), was negligible but the strains at break and impact strength of the blend reduced considerably compared to virgin and pure components (ABS and HIPS). These results revealed that the blending of small amounts of ABS or HIPS could result in improving the tensile strength and modulus of the blend after recycling. Elmaghor *et al.* [64] used poly(ethylene-*co*-vinyl acetate) and poly(styrene-*b*-ethylene/butylenes-*b*-styrene) as compatibilizers for a ternary blend prepared from waste polymers including PS, HDPE, and PVC using a single-screw extruder. For improving in the compatibility between the components of the prepared blend further, the extrudates were subjected to γ-irradiation. The results showed that both the compatibilizers and irradiation improved the mechanical properties of the blend, where impact strength and ductility of the blend were sharply enhanced and the tensile strength showed only moderate improvement. The chemical recycling of PS waste was reported first in the work of Lee *et al.* [11] by using clinoptilolites as catalysts; they found that clinoptilolites possess a good catalytic activity for the degradation of PS with very high selectivity to aromatic liquids. Arandes *et al.* [65] studied the thermal cracking of PS and polystyrene-butadiene (PSB) dissolved in light cycle oil (LCO) in the presence of mesoporous silica as a catalyst. The cracking of the PS/LCO blend produced high yields of styrene, whereas the cracking of the PSB/LCO blend resulted in a stream of products of petrochemical interest.

4.9 Poly(Ethylene Terephthalate) (PET)

PET is a semicrystalline thermoplastic polyester widely used in the manufacture of apparel fibers, disposable soft-drink bottles, photographic films, and so on. It has been in use largely in the production of textile fibers since the mid-1940s. The world production of PET in 2002 was 26 million tons, which was expected to rise to 55 million ton ins 2010 [66]. In 1980s, PET also began to be used popularly for the production of disposable soft-drink bottles with a high growth rate [67]. Although the nontoxicity, durability, and crystal clear transparency are the major advantages of PET, its nonbiodegradability is a serious cause of concern to the environmentalists. PET accounts for more than 8% by weight and 12% by volume of the world's solid waste [68]. Since landfilling of such nonbiodegradable waste has severe limitations, chemical recycling is the best choice. The major sources of PET fiber waste generation are the manufacturing waste and the post-consumer waste. One of the chemical recycling techniques comprises of glycolytic depolymerization of PET waste, conventionally carried out in boiling ethylene glycol (EG) under atmospheric pressure using metal acetate catalyst to obtain the virtual monomer bis(2-hydroxyethylene terephthalate) (BHET) [69–72]. PET has been used extensively in blow-molding applications to produce bottles. The blow-molding process starts with the injection of PET into a cold mold to produce an amorphous preform. With single-stage injection blow molding, the preform is transferred directly to an air-blowing unit where the preform is stretched and blown into a bottle mold. In two-stage injection blow molding, the preform is reheated to around 10 °C above T_g and then blown into a bottle mold in a different air blowing unit (Figure 4.5).

Navarro *et al.* [73] prepared a blend of virgin HDPE and PET waste using the extrusion process in an effort to improve the performance of PET waste. The prepared blend, in different ratios, was injected to obtain test samples for mechanical characterization. Thermal and rheological properties of the prepared blends were also evaluated. It was found that the presence of HDPE in the blend reduced the

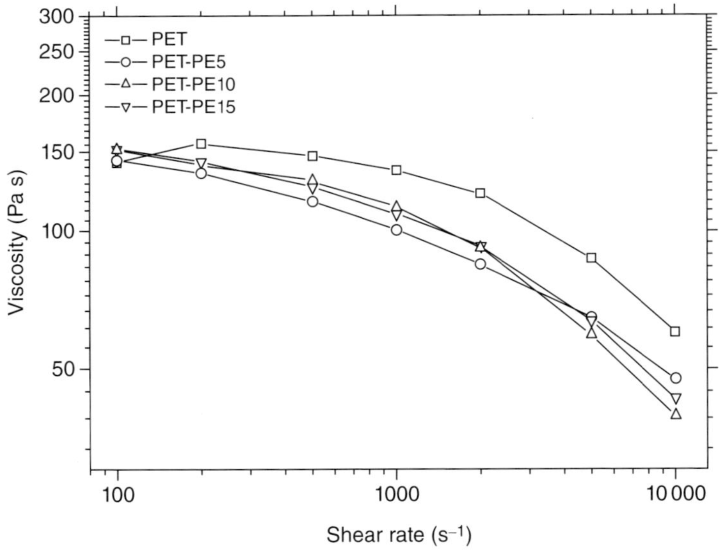

Figure 4.5 Effect of HDPE content on the viscosity of the recycled PET. (Navarro *et al.* [73] Reproduced with permission of Elsevier.)

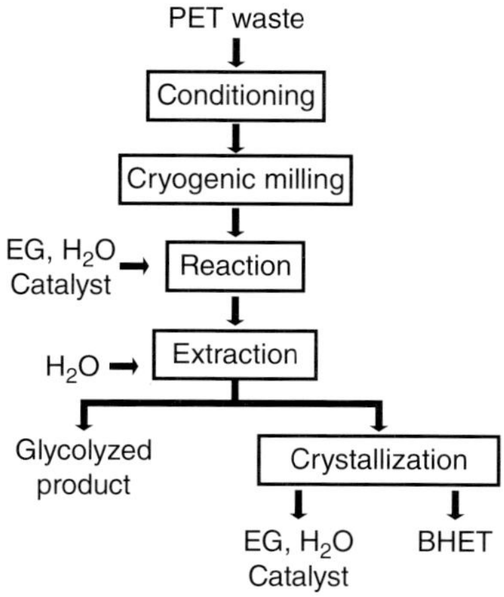

Figure 4.6 Reaction and analytical procedure for the glycolysis of PET wastes. (Fonseca *et al.* [74] Reproduced with permission of Elsevier.)

melt viscosity of the blend, indicating good flowability compared to PET waste (Figure 4.6). However, incompatibility between HDPE and PET in the blend was detected at a content of HDPE higher than 5%, resulting in poor mechanical properties compared to PET. The results of this work also indicated that it would be possible to modify PET waste using a small amount of virgin HDPE (<5%). The using of PET waste in the preparation of polymer systems was also reported by Shi *et al.* [75]; in their work, a blend of PE waste and ABS was made, and for reducing the interfacial tensions between PET and ABS in the blend, SiO_2 was incorporated. The results of this study showed that the mechanical properties of composites improved with increasing SiO_2 content.

For chemical recycling of PET waste, microwave irradiation in the presence of EG and zinc acetate as catalysts was used [76]. The yield of the main product (BHET) was nearly same as that obtained by conventional electric heating. However, the time taken for completion of reaction was reduced drastically from 8 h to 35 min, leading to substantial saving in energy. Shukla *et al.* [72] used EG and sodium sulfate for the chemical recycling of PET waste. The BHET obtained as the main product of the reaction was used to hydrophobically disperse dyes for synthetic textiles. Fonseca *et al.* [74, 77] investigated the effect of various metal salts (zinc acetate, sodium carbonate, sodium bicarbonate, sodium sulfate, and potassium sulfate) as depolymerization catalysts in the presence EG at different temperatures for PET recycling, as shown in Figure 4.7.

Also in another work by Achilias *et al.* [79], PET waste was depolymerized in the presence of ethanolamine (aminolytic) under microwave irradiation to enhance the waste degradation. The main product of the reaction was bis(2-hydroxyethyl) terephthalamide (BHETA). The results showed that the complete degradation of PET waste occurred at temperatures higher than 260 °C. By comparing with the previous work reported using of diethylene glycol [80], it could be said that the presence of diethylene glycol caused in complete degradation of PET at a lower

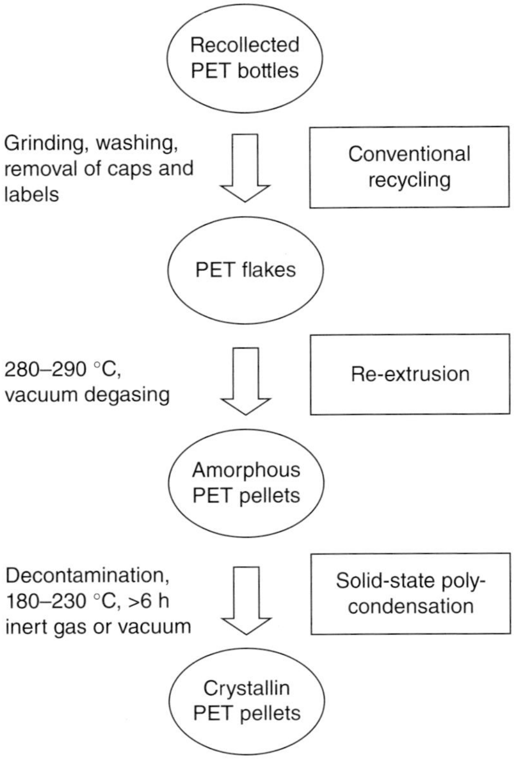

Figure 4.7 Scheme of PET super-clean recycling processes based on pellets. (Welle [78] Reproduced with permission of Elsevier.)

temperature compared to when using ethanolamine. PP packaging is being recycled into textiles, low-value items like fiberfills, and high-value outdoor specialty clothing. Large volume, high-value textiles also use recycled PET [60]. We assume that textiles will use recycled materials for 5% of their plastic requirements.

The recycling dataset used for PET recycling is based on a process from the Ecoinvent database for the primary production and on literature data for secondary production. The primary production is the process of Ecoinvent database: "Polyethylene terephthalate, granulate, amorphous, at plant/RER," which describes the primary production of amorphous PET.

The polymerization of virgin PET starts from the monomers EG and terephthalic acid or terephthalic acid methyl ester. In the first step, the monomers react to form a so-called PET pre-polymer. The viscosity of this pre-polymer, however, is too low for packaging, fibers, or any other application. Therefore, the viscosity of the PET polymer has to be increased by a further condensation reaction of the PET material in the melt phase at temperatures up to 280 °C. The reaction product of the esterification (water or methanol) and excessive monomer (EG) are removed under vacuum. At the end of the polymerization reaction, the melt is extruded into PET pellets. The viscosity of these pellets is still too low for PET packaging applications. Therefore, a further condensation reaction follows, which is typically done in a solid-state post-condensation (SSP) reaction. The end product of this reaction is crystalline pellets. On the other hand, due to the reversible character of the trans-esterification reaction, polyesters can be degraded during processing, for example, extrusion, bottle manufacturing, or recycling, especially in the presence of moisture. During this degradation, the polymer backbone is partly broken, and residual polymer chains with shorter chain length are generated. In principle, manufacturing in the presence of moisture (or alcohols) will lead to a decrease in the chain length and therefore in a decrease of the viscosity of the polymer. Because moisture cannot be completely excluded during PET processing, the hydrolysis reaction occurs during processing and recycling in any case. For bottle-to-bottle recycling, it is important that the viscosity of the PET polymer is not decreased too much; otherwise it is not possible to introduce the recyclate into the PET packaging market. Therefore, special recycling steps based on high-temperature and vacuum treatments have been introduced in order to increase the viscosity of the recyclate. On the other hand, such a degradation reaction (hydrolysis, glycolysis) can be also used for PET recycling.

The PET recycling processes based on pellets use as main decontamination process the last step of the PET virgin processes, the SSP. As input material, conventionally recycled PET flakes are used. In one of the first steps of the super-clean process, the washed PET flakes were re-extruded to pellets. The pellets were subsequently further deep-cleansed using the SSP technology. A scheme of PET super-clean recycling processes based on pellets is shown in Figure 4.7. The SSP reaction is carried out either in batch or continuous-working production lines. The process parameters, for example, residence time, temperature, vacuum, or inert gas stream, are very similar in both cases. Typically, the residence time in solid-state reactions is between 6 h and up to about 20 h, depending on the applied temperature and the desired increase in viscosity of the PET material. The temperatures are in the range of about 180 and 220 °C. The fact that the recycling facilities for the super-clean recycling are very similar to those of virgin processes, so that the same equipment could be used in both processes, was most probably one reason why the super-clean processes, based on pellets with the first mechanical or "secondary" recycling processes, entered the market.

But there are also some critical points. Because of the light weight of the PET bottles, the yield of PET flakes from the collected bottle fraction is small, because of the added weight of caps and labels. This gives economic pressure to the recycling companies. In addition, additives like the acetaldehyde scavenging substance anthranilamide or other functional additives like oxygen scavengers, which are also more and more introduced into PET bottles, might have a negative influence on the optical properties of the PET recyclates. These additives might result in a partial degradation during recycling, and the clear PET bottle flakes will become slight ly brownish. So the concentration of additives should be kept low. This example shows that for successful PET bottle-to-bottle recycling, all partners in the value-added chain have to be in close contact: the additives manufacturers, the PET virgin polymer suppliers, the preform and bottle manufacturers, the filling companies, as well as the recycling companies.

4.10 Applications

New composite materials have been produced from recycled five-layered PA-PE film, recycled PE, and fillers in form of fly ash (FA) with a particle size up to 41 μm. Based on the tested characteristics of those materials, the potential uses of the new materials for specific technical products were determined. So far, for the production of technical devices, virgin materials are regularly used. Recycled materials have shown similar characteristics as those of the original materials and meet the requirements to ensure trouble-free operation of the devices. Modification of recycled materials with FA increases the thermal resistance of the composite, which has a beneficial effect on the tribological and wear properties in long-term exploitation in the form of increased load and improved processing properties of the material (decreased shrinkage after injection molding). With loads up to 0.45 MPa in the range of a working temperature of up to 40 °C, the composites show a coefficient of friction of 0.26 (for PA 0.33). The filler content in the composite is significant [81]. PE has a relatively long time for the electric charge on the surface to decay. The phenomenon of contact electrification and its effects are described in the literature [82, 83]. Due to the static charge followed by the uncontrolled discharge (spark rise) and ignition, explosion of gas, fuel vapors, flammable chemicals, and steaming, and so on, may occur for PE products working in a hazardous environment. Therefore, it is necessary to discharge the charges from the surface and to obtain low voltage. Such opportunities were obtained for PA-PE recyclates and composites based on these recycled materials. The composites modified with PE have good tribological properties and may be a substitute for similar composites made from original materials. Particularly, the electrical properties of these materials are preferred, especially the resistance to low-power electric arc and tribological charge. The modification of composites with PE improves their electrical resistance. Modification with FA can ensure the dimensional stability of products, which is important especially for precision products. These composites may be processed with the methods used for thermoplastic materials (injection, extrusion, and pressing). The use of multilayer film waste reduces pollution and saves virgin materials because they can successfully replace them in technical articles. With increasing use of these materials, any use of waste has an ecological and economic importance [81]. Manufacturers have developed new recycling technologies since the implementation of the recycling law. Because of the limited number of appliances covered, manufacturers have developed appliance-based recycling processes. Processing lines are specially designed for each type of appliance.

Manufacturers, recycling plants, and academic researchers have implemented LCA to reduce the environmental impacts of home electrical appliances through their life cycles. Plant-level [84, 85] and national-level [86] impact analysis of carbon dioxide emissions suggest that material recycling of plastics from home electrical appliances is the best option for recycling in terms of reducing greenhouse gas (GHG) emission. Thus, manufacturers have aimed to develop material recycling technologies for plastics.

References

1. Christensen, T.H. and Damgaard, A. (2010) in *Solid Waste Technology and Management*, vol. **1 and 2** (ed. T.H. Christensen), John Wiley & Sons, Ltd., Chichester, pp. 203–210.
2. PlasticsEurope (2005) Plastics Europe Eco Profiles, http://www.plasticseurope .org/plastics-sustainability/eco-profiles .aspx (accessed 9 May 2016).
3. Moir, J.L., APRI Recycling Seminar, Melbourne, July 21, 1994, 1-4, 6-14.
4. Sanchez, E.M.S. (2007) Ageing of PC/PBT blend: mechanical properties and recycling possibility. *Polym. Test.*, **26** (3), 378–387.
5. Zaverl, M., Seydibeyoglu, M.O., Misra, M., and Mohanty, A. (2012) Studies on recyclability of polyhydroxybutyrate-co-valerate bioplastic: multiple melt processing and performance evaluations. *J. Appl. Polym. Sci.*, **125** (Suppl. 2), E324–E331.
6. Rigamonti, L., Grosso, M., Møller, J., Martinez Sanchez, V., Magnani, S., and Christensen, T.H. (2014) Environmental evaluation of plastic waste management scenarios. *Resour. Conserv. Recycl.*, **85**, 42–53.
7. Kiran, N., Ekinci, E., and Snape, C.E. (2000) Recycling of plasticwastes via pyrolysis. *Resour. Conserv. Recycl.*, **29** (4), 273–283.
8. Miller, S.J., Shah, N., and Huffman, G.P. (2005) Conversion of waste plastic to lubricating base oil. *Energy Fuels*, **19** (4), 1580–1586.
9. Chen, F.-Z. (2005) Reclaiming technology of Japanese used plastics. *World Plast.*, **23** (3), 50–62.
10. Scott, D.S., Majerski, P., Piskorz, J., Radlein, D., and Barnickel, M. (1999) Production of liquid fuels from waste plastics. *Can. J. Chem. Eng.*, **77** (5), 1021–1027.
11. Lee, S.-Y., Yoon, J.-H., Kim, J.-R., and Park, D.-W. (2002) Degradation of polystyrene using clinoptilolite catalysts. *J. Anal. Appl. Pyrolysis*, **64** (1), 71–83.
12. Kim, J.-S., Lee, W.-Y., Lee, S.-B., Kim, S.-B., and Choi, M.-J. (2003) Degradation of polystyrene waste over base promoted Fe catalysts. *Catal. Today*, **87**, 59–68.
13. Walendziewski, J. and Steininger, M. (2001) Thermal and catalytic conversion of waste polyolefines. *Catal. Today*, **65**, 323–330.
14. Lee, S.Y., Yoon, J.H., Kim, J.R., and Park, D.W. (2001) Catalytic degradation of polystyrene over natural clinoptilolite zeolite. *Polym. Degrad. Stab.*, **74**, 297–305.
15. Miskolczi, N., Bartha, L., Antal, F., and Cs, D. (2005) Determination of sulphur in liquids obtained by thermal cracking of waste polymers and commercial fuels with different analytical methods. *Talanta*, **66**, 1264–1271.
16. Sterling, W.J., Walline, K.S., and McCoy, B.J. (2001) Experimental study of polystyrene thermolysis to moderate conversion. *Polym. Degrad. Stab.*, **73**, 75–82.
17. Madras, G., Kumar, S., and Chattopadhyay, S. (2000) Continuous distribution kinetics for ultrasonic degradation of polymers. *Polym. Degrad. Stab.*, **69**, 73–78.
18. Westphal, C., Perrot, C., and Karlsson, S. (2001) Py-GC/MS as a means to predict degree of degradation by giving microstructural changes modelled on LDPE and PLA. *Polym. Degrad. Stab.*, **73**, 281–287.
19. Moriya, T. and Enomoto, H. (1999) Characteristics of polyethylene cracking

in supercritical water compared to thermal cracking. *Polym. Degrad. Stab.*, **65**, 373–386.

20. Miskolczi, N., Bartha, L., and Gy, D. (2005) Thermal degradation of polyethylene and polystyrene from packaging industry over different catalysts into fuel-like feed stocks. *Polym. Degrad. Stab.*, **66**, 235–245.
21. Hana, S.O., Lee, D.W., and Han, O.H. (1999) Thermal degradation of crosslinked high density polyethylene. *Polym. Degrad. Stab.*, **63**, 237–243.
22. Koshiduka, T., Ohkawa, T., and Takeda, K. (2003) Computer simulation of thermal degradation of poly(butylene terephthalate) and analytical problems of terephthalic acid in scission products. *Polym. Degrad. Stab.*, **79**, 1–11.
23. Park, J.W., Cheon, S.O., Lee, H.P., Kim, H.T., and Yoo, K.O. (2000) A kinetic analysis of thermal degradation of polymers using a dynamic method. *Polym. Degrad. Stab.*, **67**, 535–540.
24. Dawood, A. and Miura, K. (2001) Pyrolysis kinetics of g-irradiated polypropylene. *Polym. Degrad. Stab.*, **73**, 347–354.
25. Sakata, Y., Uddin, M.A., Muto, A., Kanada, Y., Koizumi, K., and Murata, K. (1999) Catalytic degradation of polyethylene into fuel oil over mesoporous silica (KFS-16) catalyst. *J. Anal. Appl. Pyrolysis*, **43**, 15–25.
26. Christensen, T.H. (Ed.). (2011). *Solid waste technology & management* (Vol. 1). Chichester, West Sussex, UK: Wiley. doi: 10.1002/9780470666883.ch6
27. Plastics Europe (2008) *An Analysis of Plastics Production, Demand and Recovery for 2007 in Europe*, Association of Plastics Manufacturers, Brussels.
28. Cui, J. and Forssberg, E. (2003) Mechanical recycling of waste electric and electronic equipment: a review. *J. Hazard. Mater. B*, **99**, 243–263.
29. Vallim, M., Araujo, J., Spinace, M., and Paoli, M. (2009) Polyamide-6/high-density polyethylene blend using recycled high-density polyethylene as compatibilizer: morphology, mechanical properties, and thermal stability. *Polym. Eng. Sci.*, **49**, 2005–2014.
30. Bahlouli, N., Pessey, D., Raveyre, C., Guillet, J., Ahzi, S., Dahoun, A. *et al.* (2012) Recycling effects on the rheological and thermomechanical properties of polypropylene based composites. *Mater. Des.*, **33**, 451–458.
31. Gao, Z., Kaneko, T., Amasaki, I., and Nakada, M. (2003) A kinetic study of thermal degradation of polypropylene. *Polym. Degrad. Stab.*, **80**, 269–274.
32. Sasse, F. and Emig, G. (1998) Chemical recycling of polymer materials. *Chem. Eng. Technol.*, **21**, 777–789.
33. Vicente, G., Aguado, J., Serrano, D.P., and Sanchez, N. (2009) HDPE chemical recycling promoted by phenol solvent. *J. Anal. Appl. Pyrolysis*, **85** (1), 366–371.
34. Achilias, D., Kanellopoulou, I., Megalokonomos, P., Antonakou, E., and Lappas, A. (2007) Chemical recycling of polystyrene by pyrolysis: potential use of the liquid product for the reproduction of polymer. *Macromol. Mater. Eng.*, **292**, 923–934.
35. Stoclet, G., Seguel, R., and Lefebvre, J. (2011) Morphology, thermal behavior and mechanical properties of binary blends of compatible biosourced polymers: polylactide/polyamide11. *Polymer*, **52**, 1417.
36. Hamad, K., Kaseem, M., and Deri, F. (2011) Rheological and mechanical characterization of poly (lactic acid)/polypropylene polymer blends. *J. Polym. Res.*, **18**, 1799–1806.
37. Huneault, M.A. and Li, H. (2011) Comparison of sorbitol and glycerol as plasticizers for thermoplastic starch in TPS/PLA blends. *J. Appl. Polym. Sci.*, **119**, 2439–2448.
38. Soroudi, A. and Jakubowicz, I. (2013) Recycling of bioplastics, their blends and biocomposites: a review. *Eur. Polym. J.*, **49** (10), 2839–2858. doi: 10.1016/j.eurpolymj.2013.07.025
39. Duigou, A., Pillin, I., Bourmaud, A., Davies, P., and Baley, C. (2008) Effect of recycling on mechanical behaviour of biocompostable flax/poly(l-lactide) composites. *Composites Part A*, **39**, 1471–1478.
40. Pillin, I., Montrelay, N., Bourmaud, A., and Grohens, Y. (2008) Effect of thermo-mechanical cycles on

the physico-chemical properties of poly(lactic acid). *Polym. Degrad. Stab.*, **93**, 321–328.
41. Piemonte, V. (2011) Bioplastic wastes: the best final disposition for energy saving. *J. Polym. Environ.*, **19** (4), 988–994.
42. Pillin, I., Montrelay, N., Bourmaud, A., and Grohens, Y. (2008) Effect of thermomechanical cycles on the physico-chemical properties of poly(lactic acid). *Polym. Degrad. Stab.*, **93** (2), 3218.
43. Żenkiewicz, M., Richert, J., Rytlewski, P., Moraczewski, K., Stepczyńska, M., and Karasiewicz, T. (2009) Characterisation of multi-extruded poly(lactic acid). *Polym. Test.*, **28** (4), 412–418.
44. Badia, J.D., Stromberg, E., Karlsson, S., and Ribes-Greus, A. (2012) Material valorisation of amorphous polylactide. Influence of thermomechanical degradation on the morphology, segmental dynamics, thermal and mechanical performance. *Polym. Degrad. Stab.*, **97** (4), 670–678.
45. Badia, J.D., Santonja-Blasco, L., Martinez-Felipe, A., and Ribes-Greus, A. (2012) Hygrothermal ageing of reprocessed polylactide. *Polym. Degrad. Stab.*, **97** (10), 1881–1890.
46. Lopez, J.P., Girones, J., Mendez, J.A., Puig, J., and Pelach, M.A. (2012) Recycling ability of biodegradable matrices and their cellulose-reinforced composites in a plastic recycling stream. *J. Polym. Environ.*, **20** (1), 96–103.
47. La Carrubba, V., Carfi Pavia, F., Brucato, V., and Piccarolo, S. (2008) PLLA/PLA scaffolds prepared via thermally induced phase separation (TIPS): tuning of properties and biodegradability. *Int. J. Mater. Form.*, **1** (Suppl. 1), 619–622.
48. Helfenbein, D. (2011) *Development of Recycling of PLA Sheet into PLA Pellets without Undergoing Chemical Process*, Atlanta, GA.
49. Naitove, M.H. (2012) Conference report: bioplastics are breaking out of their 'green' niche. *Plast. Technol.*, **58** (4), 13–17, 58–62.
50. Tsuneizumi, Y., Kuwahara, M., Okamoto, K., and Matsumura, S. (2010) Chemical recycling of poly(lactic acid)-based polymer blends using environmentally benign catalysts. *Polym. Degrad. Stab.*, **95**, 1387–1393.
51. Yarahmadi, N., Jakubowicz, I., and Martinsson, L. (2003) PVC floorings as post-consumer products for mechanical recycling and energy recovery. *Polym. Degrad. Stab.*, **79**, 439–448.
52. Augier, L., Sperone, G., Garcia, C., and Borredon, M. (2007) Influence of the wood fibre filler on the internal recycling of poly(vinyl chloride)-based composites. *Polym. Degrad. Stab.*, **92**, 1169–1176.
53. Jin, H., Gutierrez, J., Oblak, P., Zupančič, B., and Emri, I. (2012) The effect of extensive mechanical recycling on the properties of low density polyethylene. *Polym. Degrad. Stab.*, **97**, 2262–2272.
54. Waldman, W.R. and De Paoli, M.A. (1998) Thermo-mechanical degradation of polypropylene, low-density polyethylene and their 1:1 blend. *Polym. Degrad. Stab.*, **60**, 301–308.
55. Kabdi, S.A. and Belhaneche-Bensemra, N. (2008) Compatibilization of regenerated low density polyethylene/poly(vinyl chloride) blends. *J. Appl. Polym. Sci.*, **110**, 1750–1755.
56. Kartalis, C.N., Papaspyrides, C.D., and Pfaendner, R. (2000) Recycling of post-used PE packaging film using the restabilization technique. *Polym. Degrad. Stab.*, **70**, 189–197.
57. Puente, G., Klocker, C., and Sedran, U. (2002) Conversion of waste plastics into fuels: recycling polyethylene in FCC. *Appl. Catal., B*, **36**, 279–285.
58. Hajekova, E. and Bajus, M. (2005) Recycling of low-density polyethylene and polypropylene via copyrolysis of polyalkene oil/waxes with naphtha: product distribution and coke formation. *J. Anal. Appl. Pyrolysis*, **74**, 270–281.
59. Achilias, D.S., Roupakias, C., Megalokonomos, P., Lappas, A.A., and Antonakou, E.V. (2007) Chemical recycling of plastic wastes made from polyethylene (LDPE and HDPE) and polypropylene (PP). *J. Hazard. Mater.*, **149**, 536–542.
60. Leaversuch, R. (1993) Polyethylene film becoming next big source of reclaim feedstock. *Mod. Plast.*, 64–65.
61. Bourmaud, A. and Baley, C. (2009) Rigidity analysis of polypropylene/vegetal

fibre composites after recycling. *Polym. Degrad. Stab.*, **94** (3), 297–305.
62. Phuong, N. and Gilbert, V. (2008) Preparation of recycled polypropylene/organophilic modified layered silicates nanocomposites part I: the recycling process of polypropylene and the mechanical properties of recycled polypropylene/organoclay nanocomposites. *J. Reinf. Plast. Compos.*, **27**, 1983–2000.
63. Brennan, L.B., Isaac, D.H., and Arnold, J.C. (2002) Recycling of acrylonitrileebutadieneestyrene and high impact polystyrene from waste computer equipment. *J. Appl. Polym. Sci.*, **86**, 572–578.
64. Elmaghor, F., Zhang, L., and Li, H. (2003) Recycling of high density polyethylene/poly(vinyl chloride)/polystyrene ternary mixture with the aid of high energy radiation and compatibilizers. *J. Appl. Polym. Sci.*, **88**, 2756–2762.
65. Arandes, J., Erena, J., Azkoiti, M., Olazar, M., and Bilbao, J. (2003) Thermal recycling of polystyrene and polystyrene-butadiene dissolved in a light cycle oil. *J. Anal. Appl. Pyrolysis*, **70**, 747–760.
66. Fradet, A. and Tessier, M. (2003) in *Synthetic Methods in Step-Growth Polymers* (eds M.E. Rogers and T.E. Long), John Wiley & Sons, Inc., New Jersey, p. 17.
67. Gargiulo, C. and Belletti, G. (1997) Polyester fiber from 100% recycled PET bottles apparel. *Chem. Fibers Int.*, **47**, 16–20.
68. Sivaram, S. (1997) Proceedings of National Seminar on Recycling and Plastics Waste Management, September 24–26, 1997, pp. 283–288.
69. Baciu, D., Chiritescu, P., and Dudeanu, E. (1993) Method and installation for waste exploitation in polyester yarns. Patent RO103337.
70. Boos, F., Schnittker, N., and Seelig, J. (1998) Production of polyethylene terephthalate from waste PET. Patent DE19643479.
71. Pilati, F., Toselli, M., Stramigioli, C., Baldi, G., Capra, M., Osella, M. *et al.* (1996) Process to prepare bis (2-hydroxyethyl) terephthalate. Patent EP723951.
72. Shukla, S.R. and Kulkarni, K.S. (2002) Depolymerization of poly(ethylene terephthalate) waste. *J. Appl. Polym. Sci.*, **85**, 1765–1770.
73. Navarro, R., Ferrandiz, S., Lopez, J., and Segui, V.J. (2008) The influence of polyethylene in the mechanical recycling of polyethylene terephthalate. *J. Mater. Process. Technol.*, **195**, 110–116.
74. Fonseca, R., Ingunza, I., Rivas, B., Arnaiz, S., and Ortiz, J.I. (2010) Chemical recycling of postconsumer PET wastes by glycolysis in the presence of metal salts. *Polym. Degrad. Stab.*, **95**, 1022–1028.
75. Shi, G., He, L., Chen, C., Liu, J., Liu, Q., and Chen, H. (2011) A novel nanocomposite based on recycled Poly(ethylene terephthalate)/ABS blends and nano-SiO_2. Trans Tech Publications. *Adv. Mater. Res.*, **150–151**, 857–860.
76. Pingale, N.D. and Shukla, S.R. (2008) Microwave assisted ecofriendly recycling of poly (ethylene terephthalate) bottle waste. *Eur. Polym. J.*, **44**, 4151–4156.
77. Fonseca, R., Ingunza, I., Rivas, B., Giraldo, L., and Ortiz, J. (2011) Kinetics of catalytic glycolysis of PET wastes with sodium carbonate. *Chem. Eng. J.*, **168**, 312–320.
78. Welle, F. (2011) Twenty years of PET bottle to bottle recycling—An overview. *Resources, Conservation and Recycling*, **55** (11), 865–875, doi: 10.1016/j.resconrec.2011.04.009
79. Achilias, D., Tsintzou, G., Nikolaidis, A., Bikiaris, D., and Karayannidis, G. (2011) Aminolytic depolymerization of poly(ethylene terephthalate) waste in a microwave reactor. *Polym. Int.*, **60**, 500–506.
80. Achilias, D., Redhwi, H., Siddiqui, M., Nikolaidis, A., Bikiaris, D., and Karayannidis, G. (2010) Glycolytic depolymerization of PET waste in a microwave reactor. *J. Appl. Polym. Sci.*, **118**, 3066–3073.
81. Tartakowsk, Z. (2010) Recycling of packaging multilayer films: new materials for technical products. *Resour. Conserv. Recycl.*, **55**, 167–170.
82. Blythe, A.R. and Bloor, D. (2005) *Electrical Properties of Polymers*, Cambridge University Press, Cambridge.

83. Pionteck, J. and Wypych, G. (2007) *Handbook of Antistatics*, ChemTec Publishing Inc, Toronto.

84. Kawamura, Y. and Suzuki, H. (2005) LCA evaluation of waste home appliances recycling. *Resour. Process.*, **52** (4), 199–202 [in Japanese, with English abstract].

85. Sanou, M. and Fujisaki, K. (2001) Environmental assessment for treatment of used electric and electronic appliances. *J. Jpn. Soc. Des. Eng.*, **36** (11), 473–478 [in Japanese].

86. Ito, Y., Mori, S., and Dowaki, K. (2004) A model of the recycling of electric appliances in Kanto Area. Technical Meeting on Metabolism Society and Environmental Systems. Japan: Institute of Electrical Engineers, August 11, 2004, pp. 15–20 [in Japanese, with English abstract].

5
Recycling of Rubber

Valiya Parambath Swapna and Ranimol Stephen

5.1
Introduction

The single largest application of rubber is in the manufacture of tires. A survey through the rubber industry shows that the world consumption and waste production per year are in millions of tons. A statistical report on the percentage of world rubber production, consumption, and demand surplus is presented in Figure 5.1, which was prepared on the basis of reported data [1] in the year 2014. Both natural and synthetic rubbers (SRs) are used in various sectors; broadly their use falls into two distinct sectors: the tire sector and the general rubber goods (GRG) sector. Approximately 50% of rubber produced is used in the manufacture of tires and 50% for GRG. According to the figures presented in one report [2], rubber waste production is a global problem because around 1.7 billion new tires are produced per annum as well as over 1 billion waste tires. World natural rubber (NR) demand is expected to rise at a rate of 4.4% in the year 2015, and the amount of waste production will also increase globally. NR consumption is expected to increase to 17.0 million tons in 2023 from 12.4 million tons in 2015. Similarly, the demand for SR is expected to increase to 22.0 million tons in 2023 from 16.8 million tons in 2015 [3]. Assumptions about the growing rate of production and consumption of rubber in various sectors circuitously highlight the increase in rate of generation of million tons of rubber waste.

Waste tires and GRG create problems such as landfilling, health, and environmental challenges. Accumulation of waste tires, which are nonbiodegradable polymers due to the presence of fillers, steel cord, organic, and inorganic components, is a major environmental concern. Different agencies in the world are now recycling waste tires and other rubber goods into useful products instead of polluting the environment.

Recycling of Polymers: Methods, Characterization and Applications, First Edition. Edited by Raju Francis.

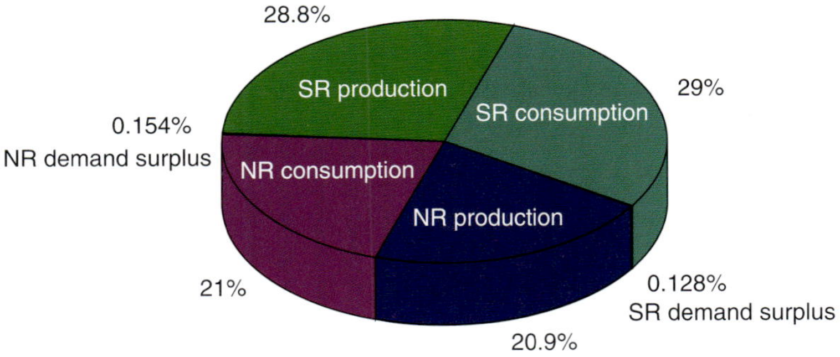

Figure 5.1 World production and consumption of natural rubber (NR) and synthetic rubber (SR).

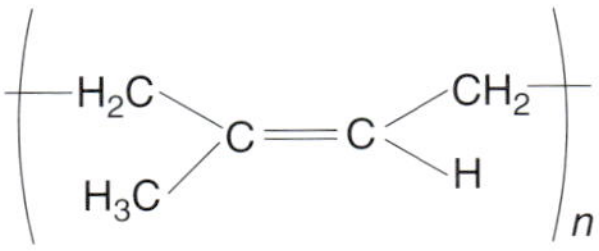

Figure 5.2 Chemical structure of natural rubber (NR).

5.2 Rubber

NR is obtained from the sap of a tree called *Hevea Brasiliensis*. NR consists of naturally occurring monomer units called isoprene. NR is *cis*-1,4-polyisoprene, whose structure is given in Figure 5.2. Initially, NR was used for the manufacture of sports goods. The great discovery of Charles Goodyear in 1839 of sulfur vulcanization of rubber is considered as a major breakthrough, resulting in the wide range of applications of rubber in different sectors. Upon sulfur vulcanization, the physical properties of rubber, such as strength, elasticity, resistance to heat and electricity, impermeability to gases, and abrasion resistance, are improved.

SRs are made by the polymerization of various organic molecules called *monomers*. Typical examples for SRs are styrene–butadiene rubber (SBR), ethylene–propylene diene monomer (EPDM), nitrile rubber (NBR), butyl rubber (IIR), and ethylene-*co*-vinyl acetate (EVA). There is remarkable variations in the physical properties of SRs when compared to NR.

NR has considerable tensile strength, good elasticity and resilience, and appreciable resistance to tearing, flexing, and abrasion. Upon stretching, crystallization can be produced in NR, which gives its high strength in the uncompounded state. During stretching, rearrangement and orientation of the chains occur to form crystallites. The main drawbacks of NR are its poor resistance to aliphatic and aromatic solvents and petrol as well as to oxidation and ozone cracking. A unique feature of NR is its low hysteresis and heat build-up. Moreover, rubber is an excellent insulator for heat and electricity. Therefore, it is extensively used in the manufacture of insulating cables and other electrical materials. Rubber can be made conducting by incorporating various conducting fillers such as carbon black, carbon nanotubes, metal particles, and certain ceramics. Some SRs show properties that closely resemble those of NR, whereas some SRs differ significantly from NR [4].

5.3 Recycling of Rubber Products

Rubber waste in landfills lasts for hundreds of years, and its accumulation results in the production of toxic gases, mosquito-borne diseases, and tire fires, which cause air and water pollution [5, 6]. The recycling of waste rubber materials is the best way to protect the environment and conserve petroleum resources from which the raw materials are derived. Waste rubber is used as fuel for the cement kilns and paper mills. The recycling of energy and materials from waste tires has high commercial application. The United States is the largest producer of waste tires in the world, producing 291.8 million tires in 2009. In the European Union, 3.4×10^6 tonnes of end-life tires are produced yearly. Recently, all developed countries have enacted laws for the disposal of waste rubber because of the risk associated with such waste. Recycling of rubber waste is an extremely challenging problem, since it is very essential to carry it out in an environmentally safe and economically productive manner for the development of the world. The main problem that arises during the recycling process is the rupture of sulfide bonds without the breakage of the main chain. The rubber material, which contains antioxidants, reinforcing fillers, curing agents, and carbon black, also makes the recycling process difficult. The composites of NR/ground tire rubber show weak mechanical properties due to the fewer reactive sites on the surface of ground tire rubber and the weak bonds between them. Increasing the number of reactive sites and improving the flexibility of the hydrocarbon chain of ground tire rubber can be achieved through the cleavage of sulfur bonds on the surface of ground tire rubber. Industries are interested in recycled rubber for the production of high-quality and at the same time cheap rubber products. The recycling of rubber products is not a trivial process because of their crosslinked structure, which restricts reprocessing [7–10]. Efficient methods of devulcanization include chemical, mechanical, biological, thermal, microwave, and ultrasonic techniques [11–13].

5.3.1 Chemical Process

Chemical reactions such as oxidation cleavage, ozonolysis, hydrogenation, catalytic reaction, metathesis reaction, and cyclization are attractive routes for the recycling of rubber. These reactions have great potential for the degradation of rubber to low-molecular weight (MW) products [14–19].

Through the chemical oxidation reaction of NR, it is possible to produce carbonyl group end-functionalized (telehelic) oligomers. It can easily be modified through functionalization into various versatile telehelic oligomers [20]. Periodic acid (H_5IO_6) is a best oxidant for the breakage of double bonds in NR via a single step. The H_5IO_6 cleavage reaction is mainly applied in the recycling of waste tire rubber [21, 22]. H_5IO_6 is a very cheap, nontoxic, and highly effective means for the chemical degradation of NR in a controlled manner. During NR degradation with H_5IO_6, the reaction starts with the epoxidation of carbon–carbon double bonds, followed by the cleavage of the oxirane units. The product is a telechelic polymer with the carbonyl group at the chain ends. This can act as a precursor for the synthesis of crosslinked, linear, and block copolymers [23].

N_2O is another most interesting oxidation reagent. It selectively oxidizes alkenes to carbonyl compounds (mostly ketones) via 1,3-dipolar cycloaddition of N_2O into C=C bond. Oxidation of *cis*-1,4-polyisoprene with N_2O leads to a reduction in the initial molecular weight of the polymer due to the breakage of some C=C bonds. N_2O reacts exclusively with the C=C bonds, remaining inert toward the other bonds [24–26]. N_2O destroys the vulcanite present in waste tire chips and rubber crumb and converts it into plastic reclaimable products (Figure 5.3). The extent of vulcanite destruction increases with the C=C bond change to ketone [27–31].

Oxidative degradation through the epoxide intermediate is an outstanding technique to produce telehelic liquid polymers with advanced properties [32, 33]. Conventionally used epoxidation systems, such as CH_3COOH/H_2O_2, posses very low epoxidation efficiency to NR. A new catalytic system has been produced recently by combining sodium tungstate (Na_2WO_4) with the conventional CH_3COOH/H_2O_2 system. This catalytic system sharply raises the epoxidation and degradation efficiency of NR. The most interesting fact regarding the $Na_2WO_4/CH_3COOH/H_2O_2$ system is that it induces hydrolytic degradation and produces a low-MW carbonyl telehelic NR [34–36].

Olefin metathesis degradation is an interesting chemical approach for the degradation of waste rubber. During metathesis, rearrangement of the double bonds between carbon atoms of olefins takes place through the cleavage and

Figure 5.3 Recycling of tire using nitrous oxide. (Dubkov *et al.* [27] Reproduced with permission of Elsevier.)

reassembly of C=C bonds [37]. Metathesis degradation in unsaturated polymers can be achieved by using a transition-metal–carbene complex as catalyst. The complexes of different metals, such as tungsten, osmium, ruthenium, and molybdenum, can act as metathesis catalysts [17, 38–40]. Specifically, Grubbs catalysts (ruthenium catalyst) and Schrock catalysts are more effective olefin metathesis reagents [37]. The recycling of waste rubber through metathesis produces end-functionalized (telehelic) oligomers. Degradation of SBR and synthetic polyisoprene can be carried out using second-generation Grubbs catalysts, and the product obtained is low-MW acetoxy telehelic oligomers. The MW of the resulting oligomers can be adjusted by changing the ratio of Ru to polyisoprene. Telehelic polymers produced from virgin rubber can be used as a precursor for the synthesis of linear crosslinked polymers or block copolymers [41].

Most recycling techniques are carried out with the help of a devulcanizing agent in order to enhance the efficiency. Organic disulfides, mercaptans, aliphatic amines, trialkyl phosphites, and triphenylphosphine are the commonly used devulcanizing agents [11, 42]. Diphenyl disulfide (DD) is the best devulcanizing reagent for NR and butyl rubber when using supercritical CO_2 as the reaction medium. Supercritical CO_2 acts as a good reaction medium for devulcanization because it is cheap, nontoxic, non-inflammable, and chemically inert, and it can easily be removed by releasing pressure [43–45].

The chemical triphenylphosphine has the ability to convert polysulfides into mono or disulfide. The effective removal of polysulfide bonds can be done by using propanethiol, hexane-1-thiol, or piperidine. Dithiothreitol can change a disulfide bond into a thiol bond. Methyl iodide is also used for efficient devulcanization.

5.3.2 Physical Methods

One of the major problems in the recycling of rubber-like materials is their crosslinked molecular structure. Therefore, several reclamation methods have been introduced based on two approaches: devulcanization of cured rubber, and the surface modification of the waste particles to get better properties during reuse. Mechanochemical, microwave, and ultrasonic techniques are efficient physical methods for devulcanization.

5.3.2.1 Mechanochemical Techniques

Generally, mechanochemical techniques are employed for the reclamation of waste rubber because of the presence of sulfur crosslinks. High-shear mechanical milling technology can be used to reclaim ground rubber, and partial decrosslinking can also be achieved by this technique [46]. This process does not require any chemicals for devulcanization and hence it does not generate byproducts. Das and coworkers [47, 48] carried out the devulcanization process at 110 °C for 10 min in an open two-roll cracker-cum-mixing mill. They found that the recovery of 70–80% properties of the original gum vulcanizate is possible for the reclaimed rubber with this technique. Devulcanization of rubber products based on stress-induced structural changes and chemical reactions is environmentally friendly, cost effective, and simple. Lu *et al.* [49, 50] carried out the devulcanization of NR in a specially designed pan-mill-type mechanochemical reactor in which the ground rubber was milled for nearly 30 cycles. As the number of milling cycles increased, the mechanical properties of the reclaimed rubber also increased. Another interesting mechanochemical method for devulcanization is chemical treatment along with two-roll milling. A chemical combination of benzoic acid (78%), zinc oxide (5%), stearic acid (5%), hydroquinone (8%), and rosin (4%) was used for the devulcanization process [51]. The function of benzoic acid is to donate proton that break the sulfur bonds and render the sulfur passive.

Cavalieri *et al.* [52] analyzed the effect of mechanical milling in the devulcanization process of ground rubber in the presence of a phenolic antioxidant. Characterization of the reclaimed samples at high-energy milling in the presence of the phenolic antioxidant suggested the breakage of the high-MW polymer chain (C–C breakage) and sulfur crosslinks.

5.3.2.2 Microwave Technique

The microwave technique is an efficient method for recycling rubber. The technique improves the degradation process by reducing processing time and energy; continuity and handling of a bulk amount of materials are the other advantages. Molecular interaction of rubber with the electromagnetic field results in the absorption of microwave radiation and raises the temperature of the material. This process uses microwave energy (300 MHz to 300 GHz) to rupture sulfur–sulfur and sulfur–carbon crosslink in the rubber, without rupturing the carbon–carbon chemical bond in the main polymeric chain. Microwave treatment remolds the three-dimensional network structure of thermosetting rubbers by a thermomechanical process into new recycled products [53–55]. Microwave heating can be improved by combining it with mechanical processing. Industrial wastes of SBR (a thermoplastic polymer) can be efficiently recycled by microwave devulcanization. During microwave treatment of SBR, mainly the polysulfidic bonds are broken, as well as the sulfoxide and sulfide groups, and which are released in the form SO_2 and volatile organic compounds with low molecular weights. Most monosulfidic crosslinks are not cleaved by this technique because those bonds possess higher dissociation energy (302 kJ mol^{-1}) than disulfidic (273 kJ mol^{-1}) or polysulfidic (256 kJ mol^{-1}) crosslinks. Microwave-devulcanized SBR industrial wastes can effectively used as fillers for the preparation of composites [56, 57].

5.3.2.3 Ultrasonic Technique

Ultrasound devulcanization of rubber is the most sophisticated method. The process uses high-frequency (generally above 20 kHz, which is beyond the range of human hearing) ultrasound waves at a very high rate (within a time of several seconds), resulting in the sudden breakage of the three-dimensional network of rubber. Industries are strongly attracted to this treatment because it does not

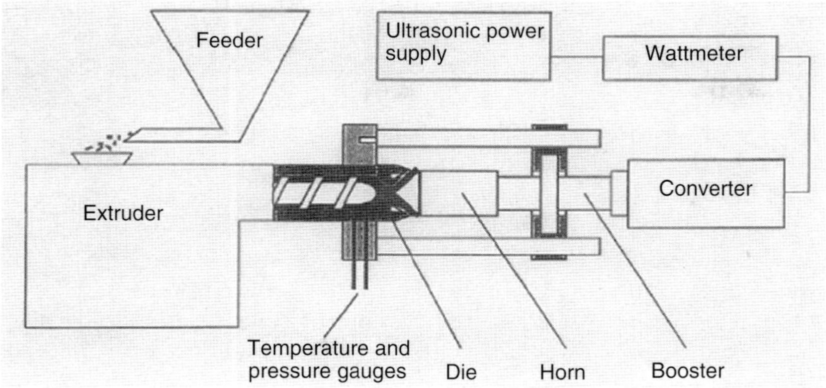

Figure 5.4 Ultrasound reactor with a conical chamber [58]. (Reproduced with permission from Wiley, permission received on December 2015.)

need any chemical agents. The rubber materials treated ultrasonically become soft and moldable, and can be reshaped and reprocessed similar to unvulcanized neat rubber. The processing conditions, such as reactor barrel temperature, ultrasonic amplitude, flow rate, and gap size, influence the devulcanization process. Ultrasonic degradation can be applied to NR, SR, ground tire rubber, butadiene rubber, SBR, polyurethane rubber, and ethylene–propylene–diene rubber.

The ultrasound equipment has a feed screw, an ultrasonic power supply, a converter, and a 1 : 1 booster providing longitudinal vibrations of 20 kHz frequency of the water-cooled horn with a selected amplitude. Through the feed screw, the rubber is transferred into the extruder, as shown in Figure 5.4. The rate of feed influences the flow rate of rubber. The rubber from the extruder passes into the gap between the horn and the die plate, where it undergoes longitudinal vibrations perpendicular to the flow direction. Water jackets are used to cool the reactor from getting hot by the ultrasound exposure. The degree of devulcanization is determined by the amplitude of ultrasound waves and the residence time [58]. The grooved barrel reactor enables continuous devulcanization with an output as high as $2.53\,g\,s^{-1}$ [13, 53, 59, 60]. During devulcanization, a two-phase material is generated by the partial breakup of crosslinks. One phase is called *sol*, which is a soluble, fully devulcanized portion, and other phase is an insoluble portion of considerably reduced crosslink density, called a *gel*. The devulcanization is a thermomechanical process that cleaves both the sulfur crosslinks and the C–C bonds of the main chain. Thus, the devulcanization of the rubber is accompanied by some degradation of the polymer. Hence, the final material exhibits properties different from those of the initial vulcanized rubber [61].

Ultrasound processing of butyl rubber (copolymer of isobutylene and isoprene) gum can result in change in the curing behavior and physical properties, and decrease in the molecular weight. All the properties greatly depend on the ultrasound amplitude [62, 63].

5.3.2.4 Twin-Screw Extruder

This method is widely used for industrial applications. The co-rotating twin-screw extruder is a very powerful, convenient, and promising method for devulcanization of NR and EPDM rubber. Temperature, screw speed, screw configuration, and the devulcanizing agent are the important parameters determining the efficiency of devulcanization and the quality of the product obtained. Continuous processing and high speed are the advantages of this method. The co-rotating twin-screw extruder can attain 1000–3000 rpm speed and apply a powerful shear stress on the extruded material. The large shear stress and the temperature from the extruder cause the breakage of the crosslink network in the ground rubber [64–67].

5.3.3 Biological Process

In recent years, many studies have been reported on the recycling of rubber by the biological method. Biological degradation occurs with the presence of bacteria, fungi, or some other microorganisms that can synthesize suitable enzymes needed for the degradation [68–70]. From the point of view of protecting energy resources and the environment, biological method is superior to physical or chemical processes. Physical recycling consumes huge energy, while chemical processes result in the release of poisonous side products such as SO_x and CO_2, which results in secondary pollution. There are reports that the sulfur content in the waste rubber is reduced to 30% through 30 days of microbial desulfurization (using the bacteria *Thiobacillus perometabolis*), whereas only 8% reduction is achieved by chemical processes (using the chemical di-(cobenzanidopheny)disulfide) [71]. The biological process presents many advantages, such as simplicity, minimum equipment requirement, non-release of any toxic chemicals, minimum use of energy, and cleavage of only the sulfur crosslinked bonds on the surface without any major damage to the main chain, that make the process most promising for rubber recycling. Microbial desulfurization is the most successful biological method used for rubber recycling. Microorganisms have the ability to selectively break the sulfur crosslinks in rubber through reduction or oxidation (sulfur to sulfate) and increasing the number of reactive sites on the surface of ground rubber. Microorganisms only break the sulfur–sulfur crosslinks on the surface of the rubber, since the contacting surfaces between rubber and the culture medium give rise to no attraction, and desulfonation takes place through two distinctive phases. The devulcanzation of waste rubber leads to effective recycling with technological and economic advantages. *T. ferrooxidans, T. thiooxidans, T. thioparus,* and *Sulfolobus acidocaldarius* are the most commonly used microorganisms for the desulfurization of the rubber. Anaerobic sulfur reduction of ground rubber occurs with thermophilic archaeon *Pyrococcus furiosus.* The additives in the rubber are highly toxic toward *P. furiosus.* Hence, ethanol-leached ground rubber is treated with the microorganism. Microbially treated crumb rubber shows enhanced mechanical properties [72]. However, microbial desulfurization is not an accepted method in the commercial field because of the difficulty in maintaining the environmental conditions for microorganisms. Moreover, antiaging agents such as *N*-(1,3-dimethylbutyl)-*N*-phenyl-*p*-phenylenediamine (Dusantox 6PPD) and curing agents such as tetramethyl thiuram disulfide (TMTD) and tetramethylthiuram monosulfide (TMTM) in the rubber materials can suppress the growth of microorganisms. Hence, cryogrinding followed by the killing of toxic microorganisms attached to rubber by extraction with ethanol is applied before the treatment with bacteria. *Resinicium bicolor* is a lignin-degrading wood-rotting fungus, which also has the ability to eliminate toxic additives from rubber products and allow the growth of desulfurizing bacteria in the medium containing rubber.

The typical dimensions of *T. ferrooxidans* is 1.0–1.5 μm in length and 0.5 μm in diameter, and it has a short rod-shaped structure (Figure 5.5). It has found application in biorefining of diesel fuel, metallurgy, and rubber desulfurization. *T. ferrooxidans* is found in the soil of an iron ore mine in Xinlong, Hbei province, China. *T. ferrooxidans* have extreme metabolic activity toward sulfur, so sulfur uses *T. ferrooxidans* as an energy source to convert into SO_4^{2-}; that is, *T. ferrooxidans* carries out the oxidation desulfurization reaction on the surface of ground rubber, as shown in Figure 5.6 [74]. The oxidation reaction may be catalyzed by the enzymes in the microorganism, which enhances the reprocessing ability and the surface reactivity of the ground rubber. If *T. ferrooxidans* is added to the medium, its biomass increases very quickly. But with the addition of ground rubber, the biomass of *T. ferrooxidans* reduces due to the presence of residual chemicals on the ground rubber, preventing the growth of *T. ferrooxidans*. The sulfur content of the rubber was reduced to 7.8% after 30 days, while sulfone group on the rubber surface increased during desulfurization. The SO_4^{2-} concentration in the medium increased from 4.56 to 4.74 g l^{-1} within 20 days. NR filled with

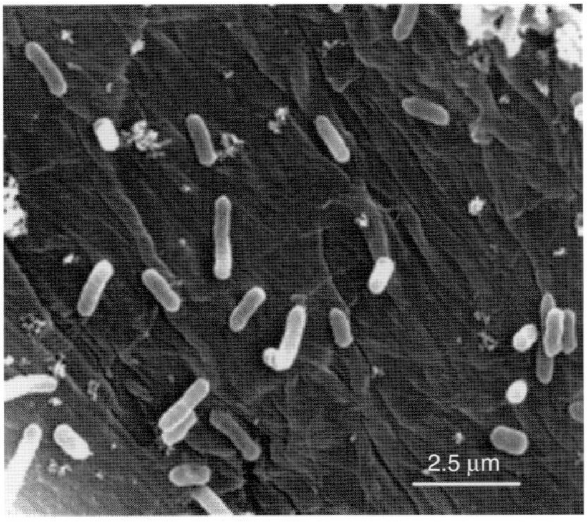

Figure 5.5 Environmental scanning electron microscope image of *T. ferrooxidans*. (Li *et al.* [73] Reproduced with permission of Elsevier.)

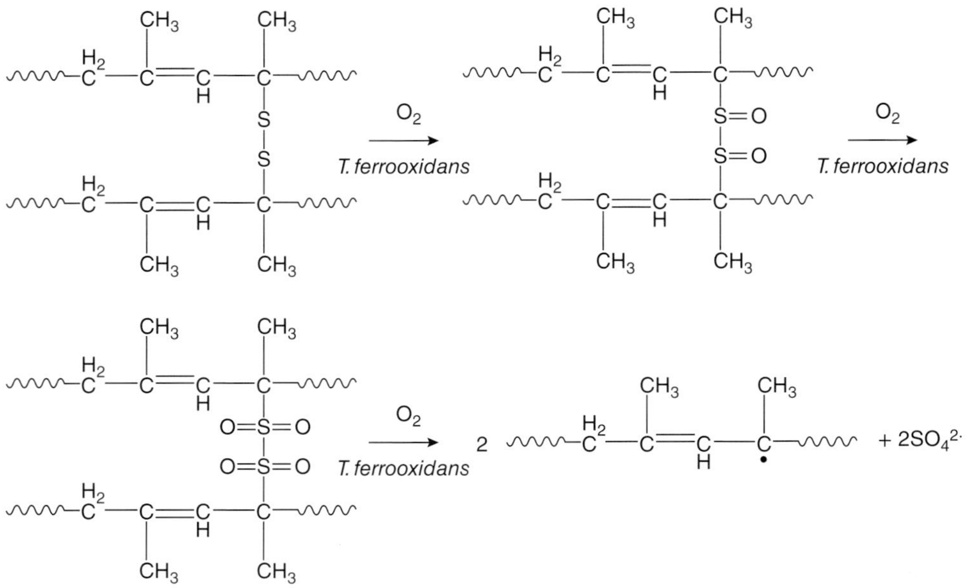

Figure 5.6 Schematic representation of desulfurization of the sulfur linkage by *T. ferrooxidans*. (Jiang *et al.* [74] Reproduced with permission of Wiley.)

desulfurized ground rubber showed improved mechanical properties compared to those of neat NR. The crosslink were ruptured during desulfurization of the ground rubber, which allowed strong permeability for intermolecular chains and formed strong interfacial forces and chemical bonds with NR. The main chains of the polymer were not oxidized by *T. ferrooxidans* [73].

Thiobacillus spp. have strong sulfur-oxidizing capacity, and so they are used in desulfurization on the surface of the ground rubber. The NR/desulfurized ground rubber composite shows superior mechanical properties due to enhanced interfacial interaction between NR and the desulfurized ground rubber. *Thiobacillus* spp. can increase the oxygen content on the surface of ground rubber by around 30% [75].

Sphingomonas sp. is a bacterium producing an enzyme that can degrade sulfur hetrocyclic compounds and polycyclic aromatic hydrocarbons. It is a strict aerobe and gram-negative. *Sphingomonas* sp. is road-shaped with a length of 1.5–2 μm and diameter 0.5 μm (Figure 5.7). *Sphingomonas* sp. is found in coal mine soils in Sichuan Province, China. The bacterium exhibits low biomass in the first 3 days. However, when detoxicated ground rubber and glucose are added to the medium on the third day, desulfonation was initiated. The growth of *Sphingomonas* sp. slightly decreased, but maintained good growth even during desulfonation of ground rubber. The biomass of *Sphingomonas* sp. increased rapidly when glucose was added without ground rubber. The metabolism of *Sphingomonas* sp. resulted in the rupture of the conjugated double bonds in the waste rubber and reduced the sulfur content on the surface up to 22.9% [76, 77].

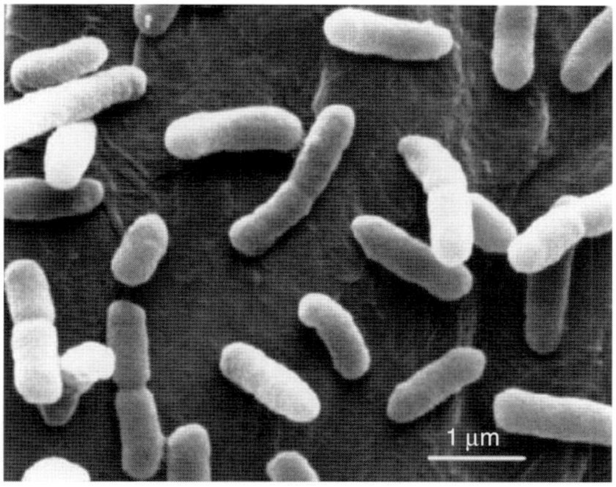

Figure 5.7 Environmental scanning electron microscope image of *Sphingomonas* sp. (Li *et al.* [76] Reproduced with permission of Springer.)

The binding energy of sulfur, sulfite, and sulfate are 164, 166–168, and 169–171 eV, respectively. In the X-ray photoelectron spectroscopy (XPS) spectra, the desulfurized ground rubber showed a large peak at 167–169 eV, corresponding to oxidized sulfur group or sulfite, but no peak was found at 164 eV [78]. The SBR composite with desulfurized ground rubber showed improved mechanical properties and lesser crosslinking density. The desulfurization of rubber can be checked by detecting (i) the decrease of sulfur on the surface of rubber, (ii) the increase of sulfur products in the medium, and (iii) changes in the properties of microorganisms in the treated rubber.

Alicyclobacillus sp. is a good desulfurizing microbe for waste rubber. The microbe oxidizes the sulfide bonds present on the rubber surface into sulfur oxide groups, and the oxidation process enhances the surface activity, processability, and hydrophilic property of the rubber. *Alicyclobacillus* sp. is very sensitive to some rubber additives and to surfactants such as Tween 80. So the proper caution was taken while introducing waste latex rubber and Tween 80 into the culture media. At 2% and 5% (w/v) of waste latex rubber added into the medium, there was no toxic effect on *Alicyclobacillus* sp. The surfactant Tween 80 improves the contact between the lipophilic latex rubber and hydrophilic *Alicyclobacillus* sp. However, the separate addition of the surfactant into the *Alicyclobacillus* sp. culture medium inhibits the growth of *Alicyclobacillus* sp. The toxic effect of Tween 80 on *Alicyclobacillus* sp. can be eliminated by premixing of waste latex rubber with Tween 80 and then adding to the medium having the microbe [79]. *Acidithiobacillus ferrooxidans YT-1* improves the rubber recycling efficiency when the microbe was cultured in a medium containing Fe^{2+}. High Fe^{2+} concentration was good for the growth of *A. ferrooxidans YT-1* [80].

Bacterial species such as *Actinomycets*, gram-negative and positive bacteria such as *Actinomadura*, *Bacillus*, *Xanthomons*, *Actinoplanes*, *Micromonospora*, *Streptomyces*, and *Gardonia*, and many other species are commonly used for the degradation of waste rubber [81, 82]. Moreover, the bacteria and fungi species such as *Fusarium solani* and *Penicillium variables* (which are the filamentous fungi) also have the ability to degrade the waste rubber [69]. The fungus belonging to the basidiomycetes *Ceriporiopsis subvermispora* (a lignin-degrading wood-rotting basidiomycete) has the ability to directly devulcanize rubber. The fungus oxidatively decomposed the sulfide bonds in the waste rubber and reduced the sulfur content of the rubber by 29% within 200 days. *C. subvermispora* cleaves the S–C bond preferentially to the S–S bonds in the polysulfide bonds, and hence reduces the crosslinking density. The white rot fungus *Dichomitus squalens* exhibits a behavior quite contrary to that of *C. subvermispora*, in that it did not decrease the weight or crosslinking density of vulcanized rubber. Both *C. subvermispora* and *D. squalens* are powerful lignin degraders and secrete maganese peroxide (MnP) and laccase (Lac) as the lignin-degrading enzymes [83].

5.4 Applications of Recycled Rubber

In recent years, waste rubber has found many applications. It can be used as a supplementary fuel for cement kilns, paper mills, and the steel industry, as road paving, building, and construction materials, in marine application, as a refuse-derived fuel and oil absorbent, in the manufacture of rubber-modified asphalt and rubber mats, in energy generating processes, and in the reproduction of rubber materials.

5.4.1 Sound-Insulation Materials

With the rapid development of modern technologies, noise reduction and waste disposal are of significant concern to human health and the ecosystem. Continuous population rise and improved quality of life are the major factors responsible for uncontrolled pollution. Developed societies have initiated legal remedies against waste and noise pollution [7, 84, 85]. It is essential to develop new ecofriendly, cost-effective, light-weight, high-quality sound insulation materials for a wide range of frequencies. The existing techniques are unable to manufacture materials that produce less or no sound. Recently developed insulation materials from recycled rubber show excellent sound absorbing property. This is encouraging, as it is both financially effective and ecofriendly compared to any other commercially available product. Various industrial rubber wastes can be converted into good sound absorbents, thus solving the existing problem of sound and environmental pollution. Such materials can be used for flooring of buildings for better sound insulation. These new underlayers show better noise reduction performance than existing commercially available underlayers. Low-cost floors manufactured from recycled rubber show good sound insulation compared to conventional floors. Even a small thickness of the layer shows better sound insulation property [9, 86–88, 116].

Using materials with good insulation property is the best choice for the construction of buildings. These materials potentially reduce energy usage, and help in bringing down sound and environmental pollution and global warming. A concrete wall has lower insulation properties compared to a clay brick wall because of the high density of the concrete wall. The porosity of crumb rubber concrete is lower than that of normal concrete. Because of the presence of air bubbles in cement paste or porous aggregates in concrete, sound absorption in the normal concrete increases, which in turn results in the dissipation of sound energy in the pores. In rubber-filled concrete, the rubber decreases the permeable voids in the concrete [89].

Crumb rubber mixed concrete has a lower density than plain concrete. The bulk density of plain concrete is around 2530 kg m^{-3}, whereas that of crumb rubber-mixed concrete in the range 1800–2100 kg m^{-3}. Crumb rubber-mixed concrete exhibits low heat transfer, high heat resistivity, better sound absorption, lower noise, and lower thermal conductivity when compared to normal concrete. The sound absorption coefficient (α) or the noise reduction coefficient (NRC) represents the power of a material to absorb sound. Waste rubber-mixed concrete shows improved sound absorption even a higher range of frequencies than normal concrete. Rubber-mixed concrete exhibits lower thermal conductivity (k), poorer heat transfer rate, and superior heat resistivity compared to normal concrete [90]. Noise pollution from the traffic frequently causes severe environmental problem in worldwide. Nowadays, sound insulation materials are gaining much importance to control noise. Wood/waste tire rubber composite gives good sound insulator because wood and recycled rubber are viscoelastic materials. Viscoelastic substance have superior sound resistance compared to non-viscoelastic materials [91].

5.4.2 Civil Engineering Applications

Waste rubber has found many applications in civil engineering, for example, as an additive to Portland cement concrete, as modifiers in asphalt paving mixtures, as light weight fillers, in tires, as crash barriers, and in artificial reefs. Crumb rubber shows good compatibility and interaction with asphalt binder, resulting in enhancement in the properties and performance of rubberized asphalt pavements. Compared to its use in asphalt mixtures, the use of recycled rubber in Portland cement concrete has not flourished because of the incompatibility in chemical composition between crumb rubber and cement paste and substantial difference in the stiffness [92–96]. Waste rubber is the natural candidate for the construction of buildings and roadways. The use of crumb rubber from waste tires in road construction has increased in many countries, especially in Spain. Bitumen added to crumb rubber, incorporated via a wet process, shows enhanced physical and chemical performance of asphalt pavements; moreover, it helps to reduce the thickness of the pavement overlays, which is very economical. Waste crumb rubber-modified binders reduce the sound produced by vehicle tire–pavement interaction [97–99].

5.4.3 Oil Absorbent

The growth of petroleum products and crude oil in the marine environment and marine oil transportation are leading to dangerous oil pollution. Oil spill prevention is very essential for the protection of the ecosystem [100]. The high cost of the commercially available oil-absorbing materials limits their usage. Cheaply available waste rubber can act as an oil absorbent and will prevent oil pollution in the marine environment, which is also advantageous to the rubber recycling industry. Recycled rubber is an excellent hydrocarbon absorbent or an oil absorbent because of its hydrophobic or oil-philic and flexible characteristics [101–103].

Ecosol is a commercially available oil absorbent. It possesses high porosity and absorb oil very quickly: 103.3 g of motor oil per gram of sorbent. The difficulty regarding Ecosol is that it can be used only once for oil absorption and cannot be reused. Moreover, it is highly expensive: the cost is around US\$ 65 kg^{-1}. In contrast, waste tire rubber is very competitive to commercially available sorbents, because is cheap and absorbs 2.2 g motor oil per gram of sorbent. The most attractive thing about waste tire rubber is that 1 g of waste tire rubber can be reused for more than 100 times without reducing its oil absorption efficiency. The sorption efficiency enhances when the particle size of the rubber and the temperature of the environment decrease [104].

The oil-absorption ability of waste rubber is, however, significantly lower than presently available commercial absorbent. Waste rubber can be modified through graft copolymerization/blending. The grafting of 4-*tert*-butylstyrene (tBS) onto waste rubber enhances the oil-absorption properties of waste rubber up to 24.0 gg^{-1}. The chemical crosslinker also influences the oil absorption property of the rubber material [105, 106].

Petroleum derivatives are the major sources of water pollution. Compounds such as toluene, xylene, and ethylbenzene in oil can make the water pollution more serious. The compounds are very dangerous to human health. Toluene and xylene compounds can cause damage to kidney, liver, and even the central nervous system in humans. Recycled rubber from waste tires is a highly efficient and cost-effective oil absorbent. It can effectively remove organic compounds from water: 5 $g\,l^{-1}$ of tire crumb rubber can remove 81% xylene and 60% toluene from 5 ppm aqueous solution within 30 min [107].

5.4.4 Energy Production

Recently, waste tire has found another significant and interesting use in the field of energy production. Waste rubber is a very good fuel, with a high caloric value of about 3.3×10^4 kJ kg^{-1}, which is the highest for an industrial waste and can substitute for coal. Tire rubber produces oil, char, and gas during pyrolysis. The design of the reactor, temperature, and raw materials influence the resulting product. Tire derived fuel reduces the consumption of diesel fuel and also finds application in concrete industry and in the generation of electricity. Pyrolysis of tire rubber at 550 °C and atmospheric pressure produces oil, char, and gas, but the major component is liquid. The tire-derived oil fraction is composed of aliphatic (C_5–C_{24} organic compounds) and aromatic compounds. The oil exhibit properties similar to those of commercially available heating fuel. The gas derived from the tire is composed of methane and *n*-butane hydrocarbons, CO, CO_2, and H_2. The gas can be converted into electric power using a co-generation turbine [108–110].

5.4.5 Zinc Fertilizer

Zinc is an essential element for plant life. The deficiency of Zn is frequent in wheat-growing areas of the earth. A commercial Zn fertilizer such as $ZnSO_4$ contains, as by product, high levels of cadmium, which will pollute the soil, plants, and even the food-chain [111, 112]. Waste rubber is a good source for Zn and some other nutrients for crops. Ground rubber may be available at a much cheaper price than the commercially available $ZnSO_4$ fertilizer. Ground rubber is a harmless fertilizer source for Zn. No hazard from lead or cadmium contamination can occur in plants growing on soils fertilized with ground rubber because purified Zn is used in rubber products during manufacture. For most grown-up plants, 4–200 mg kg^{-1} zinc is sufficient. But the plants growing in acidic media of soil fertilized with ground rubber contain more than 200 mg kg^{-1} of zinc, because the acidic media enhance the Zn uptake [113–116].

5.5 Concluding Remarks

With the fast development of the modern world, the use of rubber products has increased tremendously. Used rubber materials are discarded in the environment after their limited lifetime. It is expected that the world's total rubber consumption will grow to 29.2 million tons in 2015, and the production of rubber waste will rise to 4 million tons. It represents a serious environmental, social, and economic issue for the sustainable development of the world. Natural degradation of rubber is very difficult owing to the formation of the crosslinked network by sulfur in the polymeric chain, which results in a stable three-dimensional structure in vulcanized rubber and the presence of stabilizers and other additives. Although much progress has been made in recycling of ground rubber, it is still a challenge to develop an effective method to reclaim vulcanized waste rubber.

References

1. Rubber Statistical Bulletin, International Rubber Study Group (IRSG), Rubber Statistical Bulletin, **69**, 10–12, April–June 2015.
2. Sekhar, G.B. (2014) Proceedings of the Tire Technology Expo, Tire Technology Expo, Cologne, Germany.
3. IRSG (2014) Report from International Rubber Study Group (IRSG), 08 July 2014.
4. Blow, C.M. (1982) *Rubber Technology and Manufacture*, 2nd edn, Butterworth Scientific.
5. Isayev, A.I. (2005) *Science and Ttechnology of Rubber*, 3rd edn (eds J.E. Mark, B. Erman, and F.R. Eirich), Elsevier Inc, pp. 663–701. ISBN: 978-0-12-464786-2.
6. Fukumori, K., Matsushita, M., and Mouri, M. (2006) Dynamic devulcanization and dynamic vulcanization for recycling of crosslinked rubber. *Kautsch. Gummi Kunstst.*, **59**, 405–411.
7. Morillas, J.M.B., Escobar, V.G., Gozalo, G.R., and Gómez, R.V. (2010) Possible relation of noise levels in streets to the population of the municipalities in which they are located. *J. Acoust. Soc. Am.*, **128**, 86–92.
8. Singh, S., Nimmo, W., Gibbs, B.M., and Williams, P.T. (2009) Waste tyre rubber as a secondary fuel for power plants. *Fuel*, **88**, 2473–2480.
9. Santini, A., Morselli, L., Passarini, F., Vassura, I., Carlo, S.D., and Bonino, F. (2011) End-of-life vehicles management: Italian material and energy recovery efficiency. *Waste Manage.*, **31**, 489–494.
10. Abraham, E., Cherian, B.M., Elbi, P.A., Pothen, L.A., and Thomas, S. (2011) Recent advances in the recycling of rubber waste in *Recent Developments in Polymer Recycling* (eds A. Fainleib and O. Grigoryeva), 47–100. ISBN: 978-81-7895-524-7.
11. De, D., Das, A., De, D., Dey, B., Debnath, S.C., and Roy, B.C. (2006) Reclaiming of ground rubber tyre (GRT) by a novel reclaiming agent. *Eur. Polym. J.*, **42** (4), 917–927.
12. Sombatsompop, N. and Kumnuantip, C. (2006) Comparison of physical and mechanical properties of NR/carbon black/reclaimed rubber blends vulcanized by conventional thermal and microwave irradiation methods. *J. Appl. Polym. Sci.*, **100** (6), 5039–5048.
13. Sun, X.M. and Isayev, A.I. (2007) Ultrasound devulcanization: comparison of synthetic isoprene and natural rubbers. *J. Mater. Sci.*, **42** (17), 7520–7529.
14. Fan, P. and Lu, C. (2011) Surface graft copolymerization of poly(methyl methacrylate) onto waste tire rubber powder through ozonization. *J. Appl. Polym. Sci.*, **122**, 2262–2270.
15. Mahittikul, A., Prasassarakich, P., and Rempel, G.L. (2009) Hydrogenation of natural rubber latex in the presence of [Ir(cod)(PCy3)(py)]PF6. *J. Mol. Catal. A: Chem.*, **29**, 135–141.
16. Klinklai, W., Kawahara, S., Mizumo, T., Yoshizawa, M., Sakadapipanich, J.T., Isono, Y. *et al.* (2003) Depolymerization and ionic conductivity of enzymatically deproteinized natural rubber having epoxy group. *Eur. Polym. J.*, **39** (8), 1707–1712.
17. Solanky, S.S., Campistron, I., Laguerre, A., and Pilard, J.F. (2005) Metathetic selective degradation of polyisoprene: low-molecular-weight telechelic oligomer obtained from both synthetic and natural rubber. *Macromol. Chem. Phys.*, **206**, 1057–1063.
18. Lena, J.I.C., Sesenoglu, Ö., Birlirakis, N., and Arseniyadis, S. (2001) Oxidative cleavage of unsaturated 1,2-diols using chiral lead-tetracarboxylates obtained by in situ metathesis. *Tetrahedron Lett.*, **42**, 21–24.
19. Riyajan, S.A., Tanaka, Y., and Sakdapipanich, J.T. (2007) Preparation of cyclized deproteinized natural rubber in latex state via a combination of benzotrichloride and sulfuric acid system, and its properties. *Rubber Chem. Technol.*, **80**, 365–377.
20. Pilard, J.F., Campistron, I., Sadaka, F., and Laguerre, A. (2015) Method for preparing telechelic oligomers from tire waste. US Patent 8962765 B2.

21. Phinyocheep, P., Phetphaisit, C.W., Derouet, D., Campistron, I., and Brosse, J.C. (2005) Chemical degradation of epoxidized natural rubber using periodic acid: preparation of epoxidized liquid natural rubber. *J. Appl. Polym. Sci.*, **95**, 6–15.
22. Ritoit, S.G., Reyx, D., Campistron, I., Laguerre, A., and Singh, R.P. (2003) Telechelic *cis*-1,4-oligoisoprenes through the selective oxidolysis of epoxidized monomer units and polyisoprenic monomer units in *cis*-1,4-polyisoprenes. *J. Appl. Polym. Sci.*, **87**, 42–46.
23. Sadaka, F., Campistron, I., Laguerre, A., and Pilard, J.F. (2012) Controlled chemical degradation of natural rubber using periodic acid: application for recycling waste tyre rubber. *Polym. Degrad. Stab.*, **97**, 816–828.
24. Panov, G.I., Dubkov, K.A., and Kharitonov, A.S. (2009) in *Modern Heterogeneous Oxidation Catalysis* (ed. N. Mizuno), Wiley-VCH Verlag GmbH & Co. KGaA, Weinheim, pp. 217–252.
25. Hermans, I., Moens, B., Peeters, J., Jacobs, P., and Sels, B. (2007) Diazo chemistry controlling the selectivity of olefin ketonisation by nitrous oxide. *Phys. Chem. Chem. Phys.*, **9**, 4269–4274.
26. Starokon, E.V., Dubkov, K.A., Babushkin, D.E., Parmon, V.N., and Panov, G.I. (2004) Liquid phase oxidation of alkenes with nitrous oxide to carbonyl compounds. *Adv. Synth. Catal.*, **346**, 268–274.
27. Dubkov, K.A., Semikolenov, S.V., Ivanov, D.P., Babushkin, D.E., Panov, G.I., and Parmon, V.N. (2012) Reclamation of waste tyre rubber with nitrous oxide. *Polym. Degrad. Stab.*, **97**, 1123–1130.
28. Avdeev, V.I., Ruzankin, S.P., and Zhidomirov, G.M. (2003) Mechanism of direct oxidation of cyclohexene to cyclohexanone with nitrous oxide. Theoretical analysis by DFT method. *Chem. Commun.*, **1**, 42–43.
29. Semikolenov, S.V., Dubkov, K.A., Ivanov, D.P., Babushkin, D.E., Matsko, M.A., and Panov, G.I. (2009) Ketonisation of a nitrile-butadiene rubber by nitrous oxide. Comparison with the ketonisation of other type diene rubbers. *Eur. Polym. J.*, **45**, 3355–3362.
30. Dubkov, K.A., Semikolenov, S.V., Babushkin, D.E., Echevskaya, L.G., Matsko, M.A., Ivanov, D.P. *et al.* (2006) New reaction for the preparation of liquid rubber. *J. Polym. Sci., Part A*, **44**, 2510–2520.
31. Ivanov, D.P., Dubkov, K.A., Babushkin, D.E., Semikolenov, S.V., and Panov, G.I. (2009) Ketonisation of 1,5 cyclooctadiene by nitrous oxide. *Adv. Synth. Catal.*, **351** (11-12), 1905–1911.
32. Wang, Q.F., Zhang, X.W., Wang, L., and Mi, Z.T. (2009) Kinetics of epoxidation of hydroxylterminated polybutadiene with hydrogen peroxide under phase transfer catalysis. *Ind. Eng. Chem. Res.*, **48** (3), 1364–1371.
33. Mahha, Y., Salles, L., Piquemal, J.Y., Briot, E., Atlamsani, A., and Bregeault, J.M. (2007) Environmentally friendly epoxidation of olefins under phase-transfer catalysis conditions with hydrogen peroxide. *J. Catal.*, **249** (2), 338–348.
34. Zhang, J., Zhou, Q., Jiang, X.H., Du, A.K., Zhao, T., Kasteren, J., and Wang, Y.Z. (2010) Oxidation of natural rubber using a sodium tungstate/acetic acid/hydrogen peroxide catalytic system. *Polym. Degrad. Stab.*, **95**, 1077–1082.
35. Shi, H., Wang, X., Hua, R.M., Zhang, Z., and Tang, H. (2005) Epoxidation of alpha, beta unsaturated acids catalyzed by tungstate (VI) or molybdate (VI) in aqueous solvents: a specific direct oxygen transfer mechanism. *Tetrahedron*, **61** (5), 1297–1307.
36. Gregorio, J.R., Gerbase, A.E., Mendes, A.N.F., Marcico, P.D., and Rigo, A. (2005) Pre-determination of the epoxidation degree of polydienes using methyltrioxorhenium-CH_2Cl_2/H_2O_2 biphasic catalytic system. *React. Funct. Polym.*, **64** (2), 83–91.
37. Grubbs, R.H. (2004) Olefin metathesis. *Tetrahedron*, **60**, 7117–7140.
38. Schrock, R.R. and Hoveyda, A.H. (2003) Molybdenum and tungsten imido alkylidene complexes as efficient

olefin-metathesis catalysts. *Angew. Chem. Int. Ed.*, **42**, 4592–4633.
39. Castarlenas, R., Esteruelas, M.A., and Onate, E. (2005) N-heterocyclic carbene–osmium complexes for olefin metathesis reactions. *Organometallics*, **24**, 4343–4346.
40. Lapinte, V., Fontaine, L., Montembault, V., Campistron, I., and Reyx, D. (2002) Ring-opening metathesis polymerization (ROMP) of isomerically pure functional monomers and acyclic diene metathesis depolymerization (retro-ADMET) of functionalized polyalkenamers. *J. Mol. Catal. A: Chem.*, **190**, 117–129.
41. Sadaka, F., Campistron, I., Laguerre, A., and Pilard, J.F. (2013) Telechelic oligomers obtained by metathetic degradation of both polyisoprene and styrene butadiene rubbers. Applications for recycling waste tyre rubber. *Polym. Degrad. Stab.*, **98**, 736–742.
42. Rajan, V.V., Dierkes, W.K., Joseph, R.J., and Noordermeer, W.M. (2006) Recycling of NR based cured latex material reclaimed with 2,2′-dibenzamidodiphenyldisulphide in a truck tire tread compound. *J. Appl. Polym. Sci.*, **102**, 4194.
43. Kojima, M., Tosaka, M., and Ikeda, Y. (2004) Chemical recycling of sulfur-cured natural rubber using super critical carbon dioxide. *Green Chem.*, **6**, 84–89.
44. Jiang, K., Shi, J., Ge, Y., Zou, R., Yao, P., Li, X., and Zhang, L. (2012) Complete devulcanization of sulfur-cured butyl rubber by using supercritical carbon dioxide. *J. Appl. Polym. Sci.*, **127** (4), 2397–2406. doi: 10.1002/APP.37542
45. Kojima, M., Tosaka, M., Ikeda, Y., and Kohjiya, S. (2005) Devulcanization of carbon black filled natural rubber using supercritical carbon dioxide. *J. Appl. Polym. Sci.*, **95**, 137–143.
46. Zhang, X., Lu, C., Zheng, Q., and Liang, M. (2011) An environmental friendly approach for recycling of post-vulcanized fluoroelastomer scraps through high- shear mechanical milling. *Polym. Adv. Technol.*, **22**, 2104–2109.
47. Jana, G.K. and Das, C.K. (2005) Recycling natural rubber vulcanizates through mechanochemical devulcanization. *Macromol. Res.*, **13** (1), 30–38.
48. Jana, G.K., Mahaling, R.N., Rath, T., Kozlowska, A., Kozlowski, M., and Das, C.K. (2007) Mechanochemical recycling of sulfur cured natural rubber. *Polymery*, **52**, 131–136.
49. Zhang, X.X., Lu, C.H., and Liang, M. (2007) Devulcanisation of natural rubber vulcanisate through solid state mechanochemical milling at ambient temperature. *Plast. Rubber Compos.*, **36**, 370–376.
50. Zhang, X.X., Lu, C.-H., and Liang, M. (2007) Preparation of rubber composites from ground tire rubber reinforced with waste- tire fiber through mechanical milling. *J. Appl. Polym. Sci.*, **103**, 4087–4094.
51. Hassan, M.M., Aly, R.O., Abdel Aal, S.E., El- Masry, A.M., and Fathy, E.S. (2013) Mechanochemical devulcanization and gamma irradiation of devulcanized waste rubber/high density polyethylene thermoplastic elastomer. *J. Ind. Eng. Chem.*, **19**, 1722–1729.
52. Cavalieri, F., Padella, F., and Cataldo, F. (2003) Mechanochemical surface activation of ground tire rubber by solid-state devulcanization and grafting. *J. Appl. Polym. Sci.*, **90**, 1631–1638.
53. Leonellia, C. and Masonb, T.J. (2010) Microwave and ultrasonic processing: now a realistic option for industry. *Chem. Eng. Process.*, **49**, 885–900.
54. Bani, A., Polacco, G., and Gallone, G. (2011) Microwave-induced devulcanization for poly(ethylene-propylene-diene) recycling. *J. Appl. Polym. Sci.*, **120**, 2904.
55. Scuracchio, C.H., Waki, D.A., and da Silva, M.L.C.P. (2007) Thermal analysis of ground tire rubber devulcanized by microwaves. *J. Therm. Anal. Calorim.*, **87**, 893–897.
56. Zanchet, A., Carli, L.N., Giovanela, M., Brandalise, R.N., and Crespo, J.S. (2012) Use of styrene butadiene rubber industrial waste devulcanized by microwave in rubber composites for automotive application. *Mater. Des.*, **39**, 437–443.
57. Hirayama, D. and Saron, C. (2012) Chemical modifications in

styrene–butadiene rubber after microwave devulcanization. *Ind. Eng. Chem. Res.*, **51**, 3975–3980.
58. Massey, J.L., Parr, J.C., Wagler, T.A., Meerwall, E., Hong, C.K., and Isayev, A.I. (2007) Ultrasound devulcanization of unfilled natural rubber networks, studied via component molecular mobility. *Polym. Int.*, **56**, 860–869.
59. Feng, W. and Isayev, A.I. (2006) Recycling of tire-curing bladder by ultrasonic devulcanization. *Polym. Eng. Sci.*, **46** (1), 8–18. doi: 10.1002/pen.20449
60. Yun, J. and Isayev, A.I. (2003) Recycling of roofing membrane rubber by ultrasonic devulcanization. *Polym. Eng. Sci.*, **43**, 809.
61. Yun, J., Yashin, V.V., and Isayev, A.I. (2004) Ultrasonic devulcanization of carbon black–filled ethylene propylene diene monomer rubber. *J. Appl. Polym. Sci.*, **91**, 1646–1656.
62. Scuracchio, C.H. and Bretas, R.E.S. (2004) Blends of PS with SBR devulcanized by ultrasound: rheology and morphology. *J. Elastomers Plast.*, **36**, 145–175.
63. Feng, W. and Isayev, A.I. (2005) High-Power Ultrasonic Treatment of Butyl Rubber Gum: Structure and Properties. *J. Polym. Sci., Part B: Polym. Phys.*, **43** (3), 334–344.
64. Si, H., Chen, T., and Zhang, Y. (2012) Effects of high shear stress on the devulcanization of ground tire rubber in a twin-screw extruder. *J. Appl. Polym. Sci.*, **128** (4), 2307–2318. doi: 10.1002/APP.38170
65. Yazdani, H., Ghasemi, I., Karrabi, M., Azizi, H., and Bakhshandeh, G.R. (2013) Continuous devulcanization of waste tires by using a co-rotating twin screw extruder: effects of screw configuration, temperature profile, and devulcanization agent concentration. *J. Vinyl Add. Tech.*, **19** (1), 65–72. doi: 10.1002/vnl.20322
66. Balasubramanian, M. (2009) Cure modeling and mechanical properties of counter rotating twin screw extruder devulcanized ground rubber tire—natural rubber blends. *J. Polym. Res.*, **16**, 133–141.
67. Maridass, B. and Gupta, B.R. (2004) Performance optimization of a counter rotating twin screw extruder for recycling natural rubber vulcanizates using response surface methodology. *Polym. Test.*, **23**, 377–385.
68. Ignatyev, I.A., Thielemans, W., and Beke, B.V. (2014) Recycling of polymers: a review. *ChemSusChem*, **7** (6), 1–16.
69. Stevenson, K., Stallwood, B., and Hart, A.G. (2008) Tire rubber recycling and bioremediation: a review. *Biorem. J.*, **12** (1), 1–11.
70. Tsuchii, A. and Tokiwa, Y. (2001) Microbial degradation of tyre rubber particles. *Biotechnol. Lett*, **23**, 963–969.
71. Kim, J.K. and Park, J.W. (1999) The biological and chemical desulfurization of crumb rubber for the rubber compounding. *J. Appl. Polym. Sci.*, **72**, 1543–1549.
72. Bredberg, K., Persson, J., Christiansson, M., Stenberg, B., and Holst, O. (2001) Anaerobic desulfurization of ground rubber with the thermophilic archaeon *Pyrococcus furiosus* – a new method of rubber recycling. *Appl. Microbiol. Biotechnol.*, **55**, 43–48.
73. Li, Y., Zhao, S., and Wang, Y. (2011) Microbial desulfurization of ground tire rubber by *Thiobacillus ferrooxidans*. *Polym. Degrad. Stab.*, **96**, 1662–1668.
74. Jiang, G., Zhao, S., Luo, J., Wang, Y., Yu, W., and Zhang, C. (2010) Microbial desulfurization for NR ground rubber by *Thiobacillus ferrooxidans*. *J. Appl. Polym. Sci.*, **116**, 2768–2774.
75. Li, Y., Zhao, S., and Wang, Y. (2012) Improvement of the properties of natural rubber/ground tire rubber composites through biological desulfurization of GTR. *J. Polym. Res.*, **19**, 9864.
76. Li, Y., Zhao, S., and Wang, Y. (2012) Microbial desulfurization of ground tire rubber by *Sphingomonas* sp.: a novel technology for crumb rubber composites. *J. Polym. Environ.*, **20**, 372–380.
77. Hu, M., Zhao, S., Li, C., Wang, B., Yao, C., and Wang, Y. (2014) The influence of different Tween surfactants on biodesulfurization of ground tire

rubber by *Sphingomonas* sp. *Polym. Degrad. Stab.*, **107**, 91–97. doi: 10.1016/j.polymdegradstab.2014.04.025

78. Moulder, J.F., Stickle, W.F., Sobol, P.E., and Bomben, K.D. (1992) *Handbook of X-Ray Photoelectron Spectroscopy*, Perkin Elmer Corporation Physical Electronics Division, Eden Prairie, MN.
79. Yao, C., Zhao, S., Wang, Y., Wang, B., Wei, M., and Hu, M. (2013) Microbial desulfurization of waste latex rubber with *Alicyclobacillus* sp. *Polym. Degrad. Stab.*, **98**, 1724–1730.
80. Li, Y., Zhao, S., Zhang, L., Wang, Y., and Yu, W. (2013) The effect of different Fe2+ concentrations in culture media on the recycling of ground tyre rubber by *Acidithiobacillus ferrooxidans* YT-1. *Ann. Microbiol.*, **63**, 315–321.
81. Linos, A. and Steinbuche, A. (2001) in *Biopolymers* (eds T. Koyama and A. Steinbüchel), Wiley-VCH Verlag GmbH, pp. 321–359.
82. Yikmis, M. and Steinbüchel, A. (2012) Historical and recent achievements in the field of microbial degradation of natural and synthetic rubber. *Appl. Environ. Microbiol.*, **78**, 4543–4551.
83. Sato, S., Honda, Y., Kuwahara, M., Kishimoto, H., Yagi, N., Muraoka, K., and Watanabe, T. (2004) Microbial scission of sulfide linkages in vulcanized natural rubber by a white rot basidiomycete, *Ceriporiopsis subvermispora*. *Biomacromolecules*, **5**, 511–515.
84. Zhang, X., Lu, Z., Tian, D., Li, H., and Lu, C. (2012) Mechanochemical devulcanization of ground tire rubber and its application in acoustic absorbent polyurethane foamed composites. *J. Appl. Polym. Sci.*, **127** (5), 4006–4014 doi: 10.1002/APP.37721.
85. Mahjoob, M.J., Mohammadi, N., and Malakooti, S. (2009) An investigation into the acoustic insulation of triple-layered panels containing Newtonian fluids: theory and experiment. *Appl. Acoust.*, **70** (1), 165–171.
86. Hyun, J., Sung, C., Yong, S., and Woo, K. (2007) Status of recycling end-of-life vehicles and efforts to reduce automobile shredder residues in Korea. *J. Mater. Cycles Waste Manage.*, **9** (2), 159–166.
87. Benkreira, H., Khan, A., and Horoshenkov, K.V. (2011) Sustainable acoustic and thermal insulation materials from elastomeric waste residues. *Chem. Eng. Sci.*, **66**, 4157.
88. Sanz, R.M., Gisbert, A.V.N., Amorós, J.E.C., and García, F.P. (2012) A novel sound absorber with recycled fibers coming from end of life tires (ELTs). *Appl. Acoust.*, **73**, 402–408.
89. Sukontasukkul, P. (2009) Use of crumb rubber to improve thermal and sound properties of pre-cast concrete panel. *Constr. Build. Mater.*, **23**, 1084–1092.
90. Zhao, J., Wang, X.-M., Chang, J.M., Yao, Y., and Cui, Q. (2010) Sound insulation property of wood–waste tire rubber composite. *Compos. Sci. Technol.*, **70**, 2033–2038.
91. Shu, X. and Huang, B. (2014) Recycling of waste tire rubber in asphalt and portland cement concrete: an overview. *Constr. Build. Mater.*, **67**, 217–224.
92. Batayneh, M.K., Marie, I., and Asi, I. (2008) Promoting the use of crumb rubber concrete in developing countries. *Waste Manage.*, **28**, 2171–2176.
93. Bignozzi, M.C. and Sandrolini, F. (2006) Tyre rubber waste recycling in self-compacting concrete. *Cem. Concr. Res.*, **36**, 735–739.
94. Paje, S.E., Bueno, M., Terán, F., Miró, R., Jiménez, F.P., and Martínez, A.H. (2010) Acoustic field evaluation of asphalt mixtures with crumb rubber. *Appl. Acoust.*, **71**, 578–582.
95. Paje, S.E., Luong, J., Vázquez, V.F., Bueno, M., and Miró, R. (2013) Road pavement rehabilitation using a binder with a high content of crumb rubber: influence on noise reduction. *Constr. Build. Mater.*, **47**, 789–798.
96. Ongel, A. and Harvey, J. (2010) Pavement characteristics affecting the frequency content of tire/pavement noise. *Noise Control Eng. J.*, **58**, 563–571.
97. Sanz, R.M., Castizo, M.M., and Gómez, R.V. (2011) The performance of resilient layers made from recycled rubber fluff for impact noise reduction. *Appl. Acoust.*, **72**, 823–828.
98. Sandberg, U. (2010) Asphalt rubber pavement in Sweden–noise and rolling

resistance properties. Internoise 2010, Lisbon, Portugal.

99. Atta, A.M. and Arndt, K.F. (2005) Swelling and network parameters of high oil absorptive network based on 1-octene and isodecyl acrylate copolymers. *J. Appl. Polym. Sci.*, **97**, 80–91.

100. Lin, C., Hong, Y.J., and Huc, A.H. (2010) Using a composite material containing waste tire powder and polypropylene fiber cut end to recover spilled oil. *Waste Manage.*, **30**, 263–267.

101. Aisien, F.A., Ebewele, R.O., and Hymore, F.K. (2011) Mathematical model of sorption kinetics of crude oil by rubber particles from scrap tires. *Leonardo J. Sci.*, **18**, 85–96.

102. Ceylan, D., Dogu, S., Karacik, B., Yakan, S.D., Okay, O.S., and Okay, O. (2009) Evaluation of butyl rubber as sorbent material for the removal of oil and polycyclic aromatic hydrocarbons from seawater. *Environ. Sci. Technol.*, **43**, 3846–3852.

103. Lin, C., Huang, C.-L., and Shern, C.-C. (2008) Recycling waste tire powder for the recovery of oil spills. *Resour. Conserv. Recycl.*, **52**, 1162–1166.

104. Wu, B. and Zhou, M.H. (2009) Recycling of waste tyre rubber into oil absorbent. *Waste Manage.*, **29**, 355–359.

105. Aisien, F.A., Hymore, F.K., and Ebewele, R.O. (2003) Potential application of recycle rubber in oil pollution control. *Environ. Monit. Assess.*, **85**, 175–190.

106. Nolea, L.A.A., Pereza, O.P., and Velazqueza, F.R.R. (2011) Sorption study of toluene and xylene in aqueous solutions by recycled tires crumb rubber. *J. Hazard. Mater.*, **185**, 107–111.

107. Raouf, M.E.A., Maysour, N.E., Azim, A., Azim, A.A., and Amin, M.S. (2010) Thermochemical recycling of mixture of scrap tyres and waste lubricating oil into high caloric value products. *Energy Convers. Manage.*, **51**, 1304–1310.

108. Lopeza, A.F., Centeno, T.A., Alguacila, F.J., and Lobato, B. (2011) Distillation of granulated scrap tires in a pilot plant. *J. Hazard. Mater.*, **190**, 285–292.

109. Vermeulena, I., Caneghem, J.V., Block, C., Baeyens, J., and Vandecasteele, C. (2011) Automotive shredder residue (ASR): reviewing its production from end-of-life vehicles (ELVs) and its recycling, energy or chemicals' valorisation. *J. Hazard. Mater.*, **190**, 8–27.

110. Chaney, R.L. (2007) Effect of ground rubber vs. ZnSO4 on spinach accumulation of Cd from Cd- mineralized California soil. Abstract, WEFTECH Residuals Conference, Denver, CO, April, 2007.

111. Chaney, R.L. (2010) in *Trace Elements in Soils*, Chapter 17 (ed. P. Hooda), Blackwell, Oxford, pp. 409–439.

112. Afyuni, M., Khoshgoftarmanesh, A.H., Dorostkar, V., and Moshiri, R. (2007) Zinc and cadmium content in fertilizer commonly used in Iran. International Conference of Zinc Crops, Istanbul, Turkey.

113. Newman, S.E. and Meneley, J.C. (2006) Adaptation of waste tire rubber for green house media and zinc fertilizer. Final Report, (Waste tire report). Colorado School of Mines, Golden, CO. http://ghex.colostate.edu/pdf_files/Waste_Tire_Report.pdf.

114. Khoshgoftarmanesh, A.H., Shariatmadari, H., Kalbasi, M., Karimian, N., and Khajehpour, M. (2004) Zinc efficiency of wheat cultivars grown on a saline calcareous soil. *J. Plant Nutr.*, **27**, 1953–1962.

115. Taheri, S., Khoshgoftarmanesh, A.H., Shariatmadari, H., and Chaney, R.L. (2011) Kinetics of zinc release from ground tire rubber and rubber ash in a calcareous soil as alternatives to Zn fertilizers. *Plant Soil*, **341**, 89–97.

116. Hong, Z., Bo, L., Guangsu, H., and Jia, H. (2007) A novel composite sound absorber with recycled rubber particles. *J. Sound Vib.*, **304**, 400–406.

6
Fibers

Raju Francis, Nidhin Joy, Anjaly Sivadas, and Geethy P. Gopalan

6.1
Introduction

Fiber-reinforced polymer materials are composites consisting of high-strength fibers (reinforcement) embedded in polymeric matrices. Fibers in these materials are the load-carrying elements and provide strength and rigidity, while the polymer matrices maintain the fibers alignment (position and orientation) and protect them against the environment and possible damage. A brief classification of various fibers as shown in Scheme 6.1. A pure polymer does not usually have the requisite mechanical strength for application in various fields. The reinforcement by high-strength fibers provides the polymer substantially enhanced mechanical properties and makes the fiber-reinforced polymer composites (FRPCs) suitable for a large number of diverse applications ranging from aerospace to sports equipment. One possibility for recycled fibers is for them to be employed as reinforcement in composite materials, which in turn could be used in the development of design and fashion products. Composite materials have at least two components or two stages wherein the physical and chemical properties are clearly different. Separately, these constituents maintain their characteristics, but when mixed, they form a composite whose properties are impossible to obtain with just one of them. The properties of the composites can be considered as a combination of the properties of the matrix and fibers in addition to the properties of interfaces between the fibers and the matrix [1, 2]. The objective of recycling is to employ residues and reuse them in order to save raw materials and decrease the amount of accumulated waste [3]. In addition, according to Scarlato and Pontin [4], recycling should not be the only way to obtain the raw material; however, it could be an alternative to reduce the accumulation of waste and also as a proposition for environmental education [4]. Recycling of used/degraded polymers is an important process to practice industrial economy and to maintain environmental safety. There are several disadvantages associated with the current landfilling of fibrous waste. First, a tipping fee is to be paid. Second, due to environmental concerns, there is increasing demand to ban polymers from landfills. Third, landfilling of polymers is a

Recycling of Polymers: Methods, Characterization and Applications, First Edition. Edited by Raju Francis.

waste of energy and materials. A variety of technologies have been developed in response to customer demands for recycled products and as alternatives to landfilling. Except in case of direct reuse, which is a common form of utilization for discarded textiles, some processing is involved to convert the waste into a product.

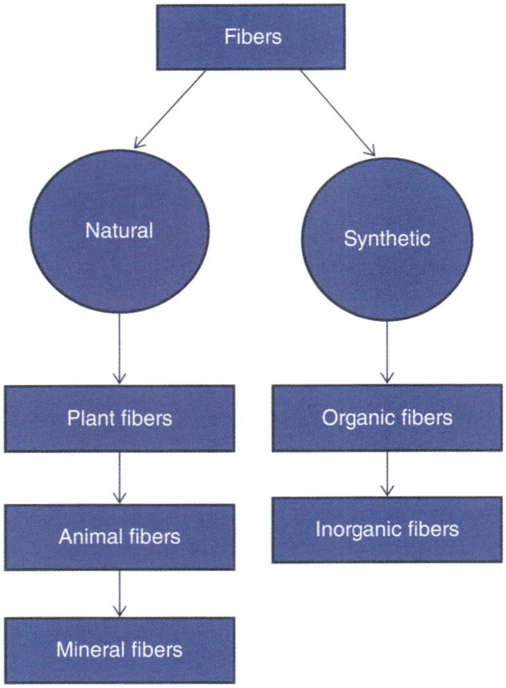

Scheme 6.1 Brief classification of fibers.

6.2
Natural Fibers

Natural fibers are divided on the basis of their origins, as coming from plants, animals, or minerals. Generally, plant or vegetable fibers are used to reinforce plastics. Natural fiber-reinforced composites with thermoplastic matrices have successfully proven their qualities in various fields of application. In polymer-based composites natural plant fibers are used as reinforcements due to their availability from renewable natural resources. They have high specific strength, modulus, light weight, low cost, and biodegradability. As a result of biodegradability natural plant fibers may present a healthy ecosystem. Low cost and good performance of these natural fibers are capable to fulfill the economic interest of industry. Composite manufacturing industries have to look for plant-based natural fiber reinforcements, such as flax, hemp, jute, sisal, kenaf, or banana as an alternative material, likely to replace solid wood. There is growing interest in using natural fibers in industrial products [5]. According to John and Thomas [6], who reviewed the various aspects of cellulosic biofibers and biocomposites, the main advantages of the use of natural fibers are their economical production with few requirements for equipment, low specific weight, and nonabrasiveness to molding equipment, smaller environmental impact, carbon dioxide neutrality, need for very little energy for production, worldwide availability, and possible energy recovery at the end of life. In short, natural organic polymers such as wood flour, starch, coconut fibers, and natural cellulose fibers are abundant, inexpensive, renewable, and fully biodegradable. They can be blended with plastics to produce materials that are more biodegradable while retaining the more desirable features of conventional plastics [7]. This section of recycling of natural fibers includes the most recycled fibers only.

6.2.1 Kenaf

Kenaf is well known as a cellulosic source with both economic and ecological advantages; it becomes available in 3 months after sowing the seeds [8]. Cellulose can be obtained from kenaf using classical and ultrasound-assisted extraction (UAE) and purification [9]. The fiber's major constituents are cellulose, lignin, hemicelluloses, and pectins. Cellulose represents a vast potential feedstock for a number of industries and has created a great deal of research interest. This is used for the production of liquid fuels (alcohol), pharmaceuticals, food, and chemical feedstock. The use of ultrasound decreases the total time of treatment; in addition, the purity of the obtained cellulose is very high.

A series of microcrystalline cellulose (MCC) materials were prepared from bleached kraft pulp of kenaf [10]. Hydrolysis of pulp was carried out using hydrochloric acid. All the MCC powders fitted log-normal distribution and MCC of kenaf core possessed the homogeneous dispersion of particles. Cellulose nanocrystals (CNCs) were first isolated from kenaf bast fibers by Kargarzadeh *et al.* using alkali and bleaching treatments [11]. CNCs were successfully extracted from this purified material using sulfuric acid while keeping the acid concentration and reaction temperature constant and varying the hydrolysis time, as shown in Figure 6.1. The chemical structure of the nanoparticles did not change during the acid hydrolysis process. Individually released CNCs showed an increased crystallinity when the reaction time was increased to 40 min. Neutralization of the acid sulfated groups induced an increase in the thermal stability of the nanoparticles.

The lignin from kenaf is, however, strikingly different from any other lignin that has been characterized. Lignins have been characterized from such a variety of plant materials that it seems unlikely to find a new type of lignin [12]. It actually represents a new lignin type, implicating new monomers (the 4-hydroxycinnamyl acetates) in its synthesis. Its uniqueness is revealed both in terms of its high syringyl content and in the extensive acetylation of the side-chain hydroxyls.

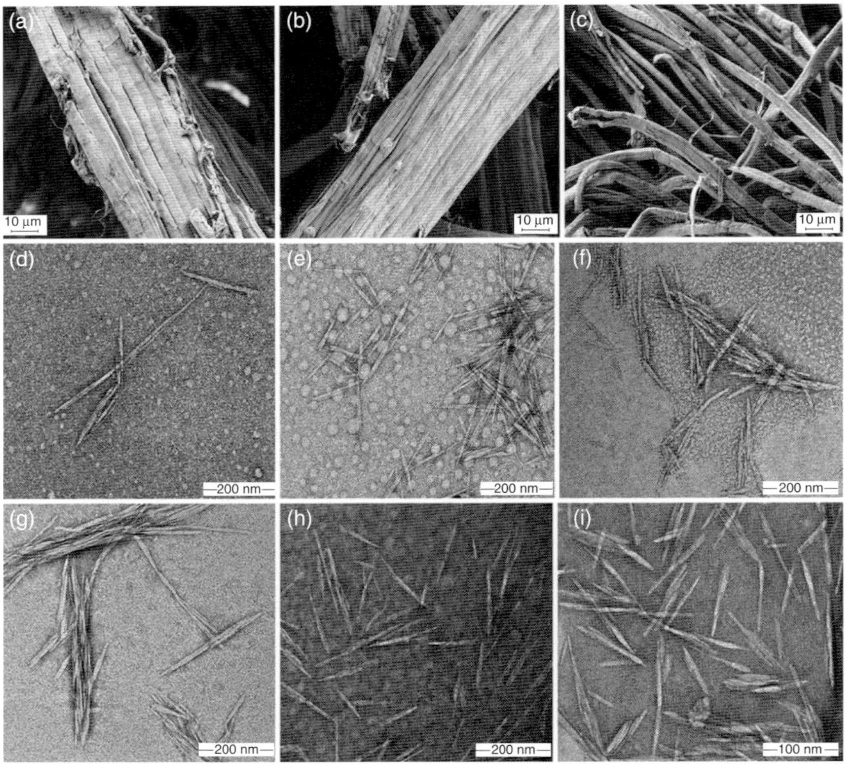

Figure 6.1 Morphological investigation. Field emission scanning micrographs of (a) a raw kenaf fiber, (b) alkali-treated fibers, and (c) bleached fibers. Transmission electron micrographs of cellulose nanocrystals produced after treatment for different hydrolysis times: (d) CNC20, (e) CNC30, (f) CNC40, (g) CNC60, (h) CNC90, and (i) CNC120. (Kargarzadeh *et al.* [11] reproduced with permission of Springer.)

Using steam injection pressing, Xu *et al.* successfully synthesized low-density, binderless particleboards from kenaf [13]. The synthesized particleboards had good mechanical properties and dimensional stability relative to their low board densities. The thermal conductivity of the particleboards showed values similar to those of insulation material (i.e., rock wool), and the sound absorption coefficient was high. In addition, the boards were free from formaldehyde emission. Kenaf core appears to be a potential raw material for low-density binderless panels suitable for sound absorption and thermally resistant interior products. Particleboard manufacturing from kenaf were determined by the parameters press temperature, press time, pressure, density, and shelling ratio [14]. These parameters had effect on the physical (thickness swelling) and mechanical (modulus of rupture and internal bond strength) properties but pressure is insignificant.

Lightweight laminate composites made from kenaf and polypropylene (PP) fibers were fabricated by press-forming [15]. The effects of the number of kenaf layers, heating time, and kenaf weight fraction on the flexural modulus of the composite specimen were investigated. The flexural modulus increased with increasing number of kenaf layers and heating time. The increase of the number of kenaf layers contributed to homogeneous PP dispersion in the composite board. This is because more kenaf layers resulted in better contact between kenaf and PP and prevented PP fibers from shrinking while heating.

6.2.2
Cotton

Cotton (Gossypium) is the most famous natural fiber and is native to tropical and subtropical regions around the world. Today, China, India, and the United States are the top three cotton producers in the world [16, 17]. Cotton fibers are grown in a protective capsule (boll) in the cotton plant and harvested on maturity [18]. Cotton is predominantly used as a textile fiber, and the growth and the demand for the cotton has been increasing. The cotton fiber is composed of cellulose (82.7%) and hemicellulose (5.7%). The cotton plant produces one of the purest forms of cellulose [19]. The hydrophilicity of the cotton fibers is high because of a large amount of cellulose (OH groups). The fibers have tensile strength between 287 and 800 MPa, modulus between 5.5 and 12.6 GPa, and elongation between 7% and 8%. The performance of cotton-based composites was studied by Mwaikambo *et al.* [20, 21]. Because of this high production, the amount of cotton plant waste and cotton gin waste is also high. It is estimated that over 2.25 million tons of cotton waste is generated every year across the cotton belt of the United States [22]. Various studies have focused on the use of cotton waste as livestock feed [23] as well as in composting, [24, 25] paper production [26], and energy production [27, 28]. In spite of the new methods for its use, today much of the available cotton waste in the world is still disposed of by returning it to the originating cropland. The field operations, on the other hand, are energy-intensive and tend to destroy the soil structure and increase the potential for erosion. Therefore, cotton waste must be utilized if sustainable agriculture is targeted. Nonwoven cotton-based composites were investigated for automotive applications by Kamath *et al.* [29]. The increasing number of fashion seasons in the retail market has led to shorter and shorter "lifetimes" for textiles garments and an increase in discarded clothing associated with the changing fashion. Accordingly, the percent contribution of textiles in municipal waste is increasing [30, 31], leading to an increase in landfill tax [32]. Therefore there is need for exploring alternative solutions that are more sustainable and mitigate the environmental impact of waste textiles. Some of the alternatives for the recycling of cotton waste garments involve conversion of the cotton-based waste garments by various methods into alternative renewable energy resources (Figure 6.2) [33–36]. While the approach of converting waste garments into renewable energy resources helps in reducing the environmental impact of waste garments, unfortunately there is no reduction in the pressure on water and land requirements for the production of cotton and synthetic fibers. Cotton fibers are used in many value-added products, and the demand for the fibers has increased, which makes the fibers expensive.

Figure 6.2 The cotton recycling symbol. (https://en.wikipedia.org/wiki/Cotton_recycling created under (CC BY-SA 3.0), http://creativecommons.org/licenses/by-sa/3.0/.)

Recycled cotton blend fabrics were used as reinforcement to reduce the cost [37, 38]. Cotton fiber composites were studied by Mueller and Krobjilowski. Ahmed *et al.* [39] studied filament-wound cotton fiber for reinforcing high-density polyethylene (HDPE) resin. Khalid *et al.* [40] also studied the use of cotton-fiber-reinforced epoxy composites along with glass-fiber-reinforced polymers. Cellulosic wastes are generated in significant amounts from the textile and garment factories, causing a further serious disposal problem. Cotton waste recycling to higher value-added products is one of the measures for both environmental and economic benefits. Therefore, potential methods of recycling cotton fabric waste have been pursued. Cotton waste is generated in large volumes from the textile and garment factories causing a further serious disposal problem. Hence, we have synthesized [41] MCC by acid hydrolysis of cotton fabric waste and used it as filler for producing biodegradable PVC films. The results show that it is possible to achieve PVC/MCC blend films with acceptable biodegradation and mechanical properties. Moreover, the thermal stability of PVC film was improved by the incorporation of the obtained MCC. However, considering the mechanical and thermal properties, the biodegradability, and the ease of processing, the amount of MCC in PVC film should not exceed 25 phr.

In India recycling of textiles is a domestic craft but now a days small scale industries and textile clusters are working on the second hand imported clothing. Recycled yarns, doormats, prayer rugs, blankets, and bed linen etc are created from second hand imported clothes. Local as well as international buyers in Japan and Australia using this remaining garments as industrial wipers for paints, chemicals, and construction industries. Mattresses and pillows are stuffed with the waste garments. Recycled cotton yarn is used to make bathroom mats [42]. Cotton textile fibers only account for one-third of the total raw cotton yield. The rest consists of byproducts such as cotton seeds, lints, and waste. Cotton seeds can be used to extract oil or used as cattle feed, and lint is used as raw material for viscose fibers [43].

Pre-consumer or post-consumer waste is coming under the classification of textile waste. The remaining or byproducts from textile, fiber or cotton industries are included in the pre consumer textile waste. Textiles in landfill biodegrade to form methane gas, which is released into the air; which is one of the most serious issues addressed by diverting textile waste from landfills. The volume of pre-consumer textiles used for recycling is 75%; this means that every year 750 000 tons of this waste is reused as a raw material. It is used in automotives, furniture, home furnishings, mattress, coarse yarn, paper, and other industries. Post-consumer textile waste is the waste of fleece, flannel, corduroy, cotton, nylon, denim, wool, and linen, which have already passed through the consumer market and are recycled and reconstituted into products for the consumer market once again. According to the Council for Textile Recycling, the textile recycling industry annually prevents 2.5 billion pounds of post-consumer textile product waste from entering the solid waste stream. This represents 10 pounds of post-consumer textile waste for every person in the United States. The quality of the mechanically recycled textile material cannot be the same as that of the first life cycle of the same material, hence there is need to promote chemical recycling of the waste garments into fibers for the second life cycle. Chemical conversion of the waste garments into new fibers must consider the separation of fiber blends and removal of finishes such as dyes and other functional finishes that may hinder the conversion process. In order to overcome the environmental and economic impact of the waste textiles, a closed-loop recycling technology is now being considered.

The first attempt to recycle cotton-based waste garments by regeneration into fibers was patented by Firgo *et al.* [44], where the process involved dissolution of the waste garments in *N*-methylmorpholine *N*-oxide (NMMO) solution, spinning, and regeneration of the cellulose fibers. The physical properties of the fibers were higher than those of fibers regenerated by other means and cotton fibers. However, in Firgo *et al.'s* patent, no consideration was given to the effects of finishes such as dyes and easy-care finishes on the dissolution of the waste garments. Subsequently Haule *et al.* [45] demonstrated that a typical easy-care finish applied to cotton garments was durable almost through the entire first life cycle of the garment. The choice of NMMO solution as a solvent for the waste cotton garments is due to the fact that the solvent can dissolve completely cellulose without any degradation, is 99% recyclable, and is safe to work with and in the environment in case of any spillages [46]. The dissolution of cellulose in NMMO solution is achieved by constant mixing at increasing temperature and reduced pressure so as to dehydrate the tertiary mixture of NMMO, water, and cellulose into a cellulose–NMMO mixture [47–49]. The process of dissolution of cellulose in NMMO solution and regeneration of fibers is known commercially as a *lyocell process* and the resulting fibers are generically known as *lyocell fibers*. The lyocell fibers are characterized by a high degree of orientation of the fibrils and weak intra-fibrillar hydrogen bonds, resulting in the fibers being susceptible to fibrillation under mechanical action and wet conditions [50, 51]. Although this tendency of wet fibrillation can make dyeing of lyocell fibers difficult, if controlled, fibrillation can make garments made from the lyocell fibers attractive in appearance and appealing to handle [51–54]. Additional beneficial features of the lyocell fibers are the relatively high elasticity and regain, which provide shapability and comfort to the garments. Overall, the lyocell process is considered an environmentally benign process, and the lyocell fibers have attractive mechanical and comfort properties.

Farahbakhsh *et al.* [55] examined the effect of cotton-derived cellulose in the micro and nano scales and the application of these materials as reinforcing fillers in low-density polyethylene (LPDE). The SEM (shown in Figure 6.3) and TEM images of the samples show fibrillation of cotton fibers; increasing the number of microgrinding passes resulted in smaller and more uniform nanofibrils. The microgrinding process increased the surface area with a marginal decrease in the

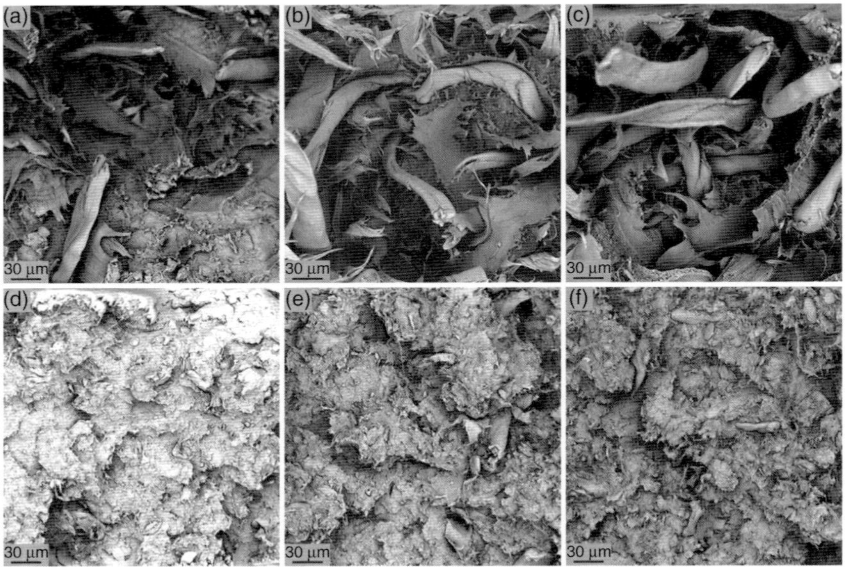

Figure 6.3 SEM images of the fracture point of composite films of pellet LDPE with a pCot filler content of (a) 2.5%, (b) 5%, and (c) 10% and of films fabricated with powdered LDPE with an MFC–pCot filler content of (d) 2.5%, (e) 5%, and (f) 10%. (Farahbakhsh *et al.* [55] reproduced with permission of Springer.)

crystallinity of cellulose. Nanofibrillated cotton was incorporated into LDPE by compounding in a twin-screw extruder and melt extrusion. Despite the absence of any surface compatiblizer, good interaction with the polymer was observed, resulting in improved mechanical and thermal properties of the nanocomposite with up to 10 wt% MFC–pCot (microfibrillated cellulose–pulverized cotton) filler. This work demonstrates the application potential of recycled cotton as promising sustainable filler choice that can be produced without any intensive purification process and applied to synthetic-based polymer nanocomposites.

Recycled textile cotton fiber is used for the production of ethanol [56]. The recycling of cotton can bring great benefits to the environment regarding sustainability in the textile and fashion industry. The production and consumption of a simple cotton shirt weighing 250 g (0.55 lb) translates into the consumption of 1.7 kg (3.75 lb) of fossil fuels, 450 g (1.0 lb) of waste in landfills, and the release of 4 kg (8.82 lb) of CO_2 into the atmosphere. Therefore, the reusing and recycling of fibers provide environmental and economic benefits [57].

6.2.3 Sisal

Sisal fiber is one of the forthcoming reinforcing materials; however its utilization has been more experiential than specialized until now. The utilization of 0.2% volume fraction of 25 mm sisal fibers leads to the lessening of free plastic shrinkage. This fiber composed of cellulose, hemicelluloses and lignin. The failure strength and the modulus of elasticity, besides the lengthening at rupture depend on the measure of orientation of the microfibers and the amount of the cellulose (Figure 6.4) [58]. Depending on the plant of origin sisal fiber shows some variation in their characteristics.

1. Advantages
   a. They are highly resistant to moisture.
   b. These fibers have good tensile strength and tensional resistance.
   c. They are highly resistant to heat.
   d. Sisal short fibers delay restrained plastic shrinkage by controlling crack development at early ages.
   e. Sisal fibers conditioned in a sodium hydroxide solution retain, respectively, 72.7% and 60.9% of their initial strength.
2. Disadvantages
   a. They decompose in alkaline environments or under biological attack.

Sisal fiber is mainly used for ropes, mats, carpets, and cement reinforcement. In developing countries, sisal fibers are used as reinforcement in house building. Sisal products are widely used in marine industries and agriculture because of their strength, durability, stretchability, affinity for certain dyes, and resilience in saltwater. Some potential innovations include the use of the material as an organic fertilizer, a supplement in ruminant feed, and a raw material in the production of some medicines [59]. The incorporation of sisal fiber in thermoplastics is an interesting subject because of many different factors. Sisal fiber combines good mechanical properties with a low specific mass. Moreover, their biodegradability can contribute to a healthy ecosystem, and their low production cost makes them very attractive for industries. They also contribute to less pollution because of their natural character and combustibility without the production of noxious gases or residues. Several studies have shown that vegetal fibers have only a very slight impact on the environment when compared to glass fibers. Wambua *et al.* have shown that hemp or sisal/PP composites have the potential to replace glass in many applications that do not require very high load-bearing capabilities [60]. Fung *et al.* studied the samples by injection molding after preimpregnation. This pre-impregnation was done with a special die assembly: sisal fibers were fed into the die, which was connected to a single-screw extruder. The major advantage of this technique is a relatively low barrel temperature, which avoids the thermal degradation of sisal fibers [61].

The biomass (sisal) left after fibers have been removed is now flushed away as waste. This sisal waste material is composed of water (85%), parenchymal tissue, short fibers, primary and secondary metabolites, and inorganic compounds [62]. Currently, this waste is mainly use in landfills or dumped in nearby rivers, where microorganisms degrade it. As a result, the untreated waste causes consumption of oxygen in the recipient water courses, which leads to oxygen-deficient water zones with a negative effect on fish and other living organisms. Thus, recycling of sisal waste is crucial because it reduces environmental pollution and may also provide other highly useful and commercializable materials [63].

Reports show the better recycling properties of sisal fibers. For an example, Bourmaud and Baley carried out nanoindentation and tensile tests on sisal fibers to obtain the transverse and longitudinal Young's moduli. The experimental values were then introduced into micromechanical models, taking the aspect ratio changes into account, to estimate the stiffness of the PP/sisal fiber injected composites before and after recycling [64]. The tensile modulus of PP is 1659 ± 29 MPa, whereas it is 3514 ± 72 MPa for the PP/sisal composite. This increase in tensile modulus values is interesting. This study of the evolution of PP/vegetal fiber rigidity after recycling evidences the good stability of the young modulus with injection cycles, thanks to an interesting stabilization of the fiber aspect ratio.

Microscopic, mechanical, rheological, and thermal investigations on the recycling of hemp- and sisal-fiber-reinforced PP composites were carried out by Bourmaud and Baley [65]. They concluded that a significant decrease of the fiber length takes place during reprocessing induced by the injection and mechanical grinding processes. The aspect ratio of vegetal fibers shows a significant decrease with reprocessing, and the initial value is low implying poor mechanical properties. Thermal analysis shows that vegetal fibers induce an increase in the degree of crystallinity and T_c. These effects are due to the nucleating ability of vegetal fibers for the crystallization of PP, sisal, or hemp fibers, accelerating the crystallization process. Rheological experiments evidence an important decrease of the composite viscosity. This decrease is certainly due to chain scissions induced by reprocessing and grinding, but also to the fiber length decrease with the number of cycles.

The sisal–PP composite sheets obtained by setting the factors predicted by the Taguchi analysis were mechanically recycled by pelletizing the sheets and feeding through the extruder [66]. The effects of recycling on the crystallinity, fiber length, mechanical properties, and stress relaxation were evaluated. By selecting a fiber mass of 30%, a polymer MFI of 1.3 g/10 min and 1% of lubricant, ~12%, ~20%, and >100% increases in tensile strength, impact strength, and tensile modulus, respectively, were observed. The fiber length dropped from 7 mm before extrusion to 6 mm and below after extrusion, and below 5 mm after recycling. A marginal change (±5%) in modulus values was found in both directions after recycling, and the ultimate tensile strength of the recycled sisal–PP specimens dropped from 41.4 ± 2.5 to 36.4 ± 0.75 MPa along the machine direction and increased from 19.2 ± 0.5 to 21.4 ± 0.1 MPa transverse (to the machine) direction mainly due to the decrease in the reinforcing fiber observed lengths. The recycled composites exhibited greater relaxation compared to the sisal–PP composites; this is because at higher temperatures the polymer matrix is in a softened state and the bonding between the fiber and matrix is expected to be weaker and the short fibers behave like polymer- rich areas and fail to share the imposed load, thus exhibiting greater relaxation at elevated temperatures (Figure 6.4).

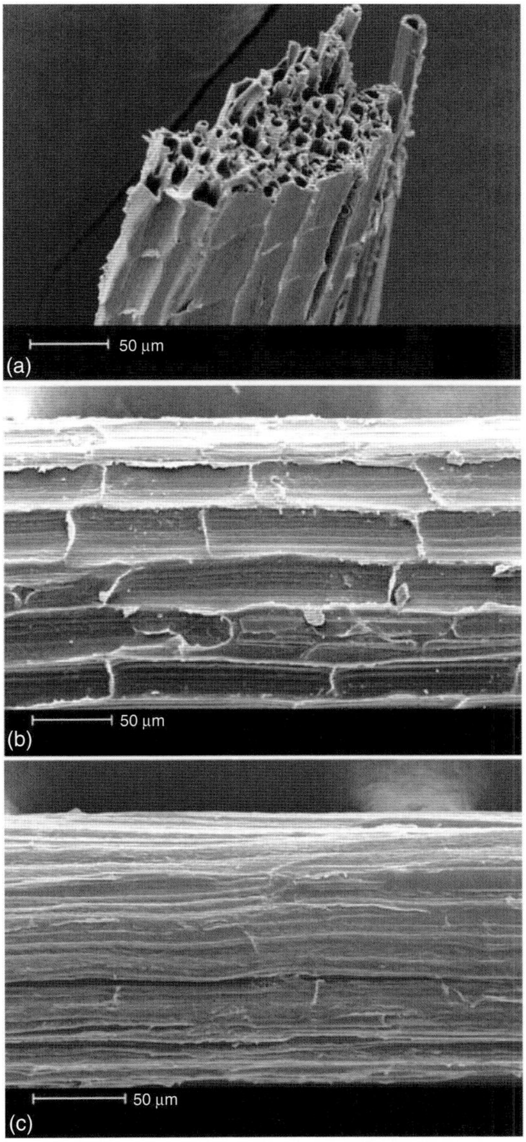

Figure 6.4 Sisal fiber SEM micrographs: fibrillar structure in fracture image (a), unmodified (b), and mercerized (c) fiber surface images. (Fávaro *et al.* [67] reproduced with permission of Express polymers letters.)

An efficient UAE of pectin from sisal waste was investigated and optimized by Maran and Priya [68]. Response surface methodology (RSM) based on a three-level four-factor Box–Behnken response surface design (BBD) was employed to optimize the extraction conditions (ultrasonic power, extraction temperature, extraction time, and solid–liquid ratio). Analysis of variance showed that the contribution of a quadratic model was significant for the pectin extraction yield. The experimental yield (29.32%)obtained under the optimal condition (ultrasonic power of 61 W, temperature of 50 °C, time of 26 min and SL ratio of 1 : 28 g ml^{-1}) was in good agreement with the predicted value. Therefore, UAE could be used as an alternative method to extract pectin from sisal waste with the advantages of lower extraction temperatures, shorter extraction time, and reduced energy consumption.

6.2.4 Asbestos

Asbestos is the name of group highly fibrous minerals with separable, long, and thin fibers (Figure 6.5). Separated asbestos fibers are strong and flexible enough to be spun and woven. Asbestos was being used as a building insulation material before the 1970s. There are three types of asbestos – white, blue, and brown. Asbestos was classified as a hazardous waste in 2005, as it is made of millions of tiny fibers, and when handled these fibers can be disturbed, potentially causing harm if breathed in. Facemasks and gloves should be used when handling asbestos. When asbestos fibers are inhaled, they can cause serious diseases, accounting for nearly 4500 deaths a year. Asbestos is regarded as the greatest single cause of work-related deaths. If the materials are in good condition they are safe to handle, but when they are damaged asbestos fibers become airborne, causing harm to the handlers [69].

Asbestos fibers are heat-resistant, making them useful for many industrial purposes. Because of their durability, asbestos fibers that get into lung tissue will remain for long periods. Asbestos has a long history. It was first used in Finland about 2500 BC to strengthen clay pots. In classical times, the indestructible

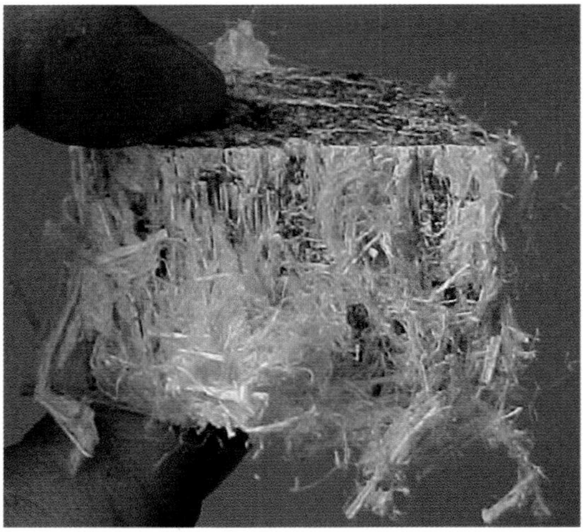

Figure 6.5 Asbestos fibers.

shrouds in which the ashes of eminent persons were preserved were woven from asbestos. The word "asbestos" comes from Greek, meaning "inextinguishable" or "indestructible." Its use as lamp wicks has continued through the ages up to the present day.

ACMs (asbestos-containing materials) causing radionuclide contamination. Low handling temperatures, low fumes emanation volumes, and the capacity of the silicate frameworks stuck ACM to synthetically spot radionuclide sources are imperative contemplations driving toward completion of pending agreements to utilize the procedure for radiation-contaminated ACM in atomic force plants and defence facilities. Asbestos abatement creates substantial volumes of conceivably unsafe material, which require safe transfer. Copper, lead, zinc and chromium are the most well-known dangerous metallic constituents. Recently, the only accepted disposal methods comprise of land filling or destroying the fiber mineral in the ACM by changing over it to another physical structure which is not typically fibrous. Pipes coated with asphaltic materials often use felt outer wraps that contain asbestos in high quantities. Such pipe wraps are banned under regulated asbestos coating materials according to the EPA (Environmental Protection Agency). Asbestos itself is a very toxic material that can cause various fatal lung diseases and cancers.

In high temperature applications asbestos fibers will degrade slowly and for the same reason asbestos found many uses in high temperature applications. Scientific examinations have been done to discover a technique to drastically reduce the mineral conversion temperature and time with a specific end goal to make the procedure economically feasible.

Numerous industrial and experimental facilities have been set up, particularly in the last 10 years, as a result of studies and researches on treating asbestos-containing waste (ACW) to stabilize it and to enable its reuse. Some of the stabilization processes reduce the hazards of ACW by immobilizing it in a cement or resonoid matrix. Other processes modify the fibrous structure of asbestos and transform it into an inert substance. One such inactivation process is mechanochemical transformation [70]. Such transformation of ACW can be likened to a cold "vitrification" process. Different types of reactions can be take place during a mechanochemical process: part of the mechanical energy transferred to solid systems is converted into heat, and part is utilized to cause fractures, compression, and slips at macro, meso, and microscopic levels, which affect the crystalline structure of solids.

Asbestos Recycling Inc.'s (ARI's) patented technology offers a thermochemical process for the mineralogical conversion of ACM to an inert, stable, nonhazardous, nonregulated, "clinker-like" end product with potential recycle value [71]. Polychlorinated biphenyls (PCBs), associated with ACM as a secondary waste, are totally destroyed.

Today, there are technologies to recycle asbestos into harmless silicate glass using thermal decomposition at very high temperatures, because both asbestos and glass fibers are silicate fibers [72]. This can then be turned into stoneware and ceramic products of various types. To overcome variable ACM chemistries, the vitrification approach usually leads to the addition of large amounts of glass formers and fluxes. Vitrification in conventional fuel-fired or all-electric glass furnaces also have the obvious disadvantage of not dealing with the metal contamination problem. Metal sorting is not practical.

A new asbestos removal system, called *Enviro System*, uses high-pressure water jets that strike the coating surface at pressures up to 20 000 psi. At these pressures, the water jet has the ability to scrape off the layer of asbestos in addition to corrosion products and soluble salts. The materials are collected without emission coming from the shroud, either as a mist or water dropping onto the ground. The collected material is transferred to a cyclone separator and then to a disposable container. The collected water is pumped as slurry into filtration units. The process also uses a 3800-gallon tank that provides water for the high-pressure pump [73].

Very high operating temperature needs for the conversion of asbestos fibers. It limits the equipment operating temperature, require high energy input and heavy maintenance. The choice of the demineralizing agent (e.g., boron additives) has proven to be very important in causing conversion at the lowest temperatures and in shortest times. In this process, asbestos fiber minerals could be converted to harmless nonfibrous, crystalline minerals due to thermodynamic and chemical demineralization mechanisms [74].

Asbestos is widely used in the earlier years due to its resistance to heat and chemical damage and this makes them challenging for other waste treatment options. Tetronics' plasma technology is a new technology for asbestos recycling [75]. Plasma system use high temperature to resist the asbestos hazard, in which other technologies are failed. The individual asbestos fibers are melted in plasma heat which eliminates its ability to cause detrimental harm to human health. Due to the ultrahigh temperature of the plasma the toxic asbestos will melt and vitrify into an inert reusable product called *plasmarok*, which can be used in building applications.

A variety of modern technologies are available for the disposal or recycling of asbestos fibers. They destroy hundreds of thousands of tons of asbestos fibers through thermal or chemical processes, producing an inert and nontoxic product that can be recycled for use in many construction applications.

6.3 Synthetic Fibers

Synthetic fiber is a chain of small units joined together in which each unit is actually a chemical substance that is produced synthetically (artificially), usually from petroleum products. Synthetic fibers have become very attractive in recent years as reinforcements for cementitious materials. They can provide effective and relatively inexpensive reinforcement for concrete and are alternatives to asbestos, steel, and glass fibers. If the fibers are further optimized, the same fiber-reinforced concrete property improvement could be achieved at a lower expense, or greater improvements can be obtained without increasing the reinforcement cost. Fiber types that have been incorporated into cement matrices include the following: polyethylene (PE), PP, acrylics (PAN), poly(vinyl alcohol) (PVA), polyamides (PAs), aramid, polyester (PES), and carbon. The properties of synthetic fibers vary widely, in particular with respect to the modulus of elasticity, an important characteristic when fibers are used for producing composites. Most commonly recycled synthetic fibers are given in the following section.

6.3.1 Nylon

Nylons come in many types, and the two most common used for textile and plastics industries are nylon 6 and nylon 66. Nylon generally performs the best among all synthetic fibers as carpet face yarn, but it is also the most expensive. Because of the higher value of nylon resin in comparison with other polymers used in carpets, nylon carpets have been looked at as a resource for making virgin nylon via depolymerization. The majority of PAs used commercially are nylon 6,6 or nylon 6, and the largest supply of waste for recycling of nylons is obtained from used carpets. The waste carpets are collected, sorted, and then subjected to a mechanical shredding process before depolymerization. The depolymerization of nylon 6 is a first-order endothermic reaction, which takes place in initiation and depropagation steps [76–78]. Water is an initiator for the depolymerization process. This process is endothermic and requires high temperatures, achieved by the use of superheated steam. Steam not only acts as a source of heat and water but also provides a better agitation for the reaction system. Temperatures above the boiling point of caprolactam (around 267 °C) are typically used. Under such conditions, a heterogeneous system is formed (a liquid polymer melt and a gas-phase caprolactam product). The caprolactam monomer is removed from the reactor along with the steam, resulting in a shift of the equilibrium toward further monomer formation. Nylon 6 solid wastes cannot be easily recycled to deliver fiber yarn since nylon 6 gets oxidized effectively during capacity and the trapped air in the voluminous fibrous wastes at the raised temperatures of melting results in gels. For reuse, the solid waste is usually depolymerized into the monomer. For granulate the fibrous waste an appropriate mill is to be utilized. The processed fiber waste is passed over to the forced feeding system with concurrent drying using the method of a pneumatic conveyor. The feeding system supplies the waste to the extruder with exact warming of the extruder screw and focal cooling arrangements. The melt is then separated, degassed, and released to give chips of standard measurements, which can be mixed with virgin crushes without running the danger of separation. During the extrusion process, additives can be introduced to get injection molding-grade products. The melt might be added persistently to the depolymerization unit to get caprolactam. Such items as tough heels for ladies shoes or machine parts can be made by using remelting strategies. The clean waste can be changed into soluble nylon by responding it with formaldehyde in a solvent. The soluble nylon have film forming properties, which enhances with the higher content of fiber viscosity and for that reason tire cord polymer and waste give a better product. During the extrusion process, additives can be introduced to get injection molding-grade products. The melt might be added persistently to the depolymerization unit to get caprolactam. Tough heels for ladies shoes or machine parts can be made by using remelting strategies. The clean waste can be changed into soluble nylon by responding it with formaldehyde in a solvent. The soluble nylon have film forming properties, which enhances with the higher content of fiber viscosity and for that reason tire cord polymer and waste give a better product. Nylon 6 waste is soluble in caprolactam oligomer solvent. This is precipitated after filtering the titanium oxide in the waste to get powdered nylon 6 without much loss of degree of polymerization. Such powders are utilized for covering metals by the fluidized-bed technique. High molecular weight nylon 6 powders with low COOH end group concentration and fine particle size have been created from the nylon 6 waste. This utilizes a mixture of solvent and nonsolvent to dissolve and precipitate the nylon 6. The suitable pair of solvents is formic acid-acetic acid. For the chromatographic separation of natural products, these powders were used. Nylon waste is treated with water to get an item with a low degree of polymerization. This procedure is once in a while utilized, as the item is not clean and pure enough to give a polymer of acceptable properties [80]. For the most part, the waste is depolymerised to give the monomer caprolactam or amino caproic acid. Nylon 6 strong wastes is changed over into caprolactam by warming with superheated steam to 200–400 °C if needed under pressure, within the sight of non-volatile organic or inorganic acids and alkalis ideally in 3–15 : 1 weight ratio of waste with acid/alkali hydroxide. Caprolactam is recouped from nylon 6 waste in great yield by going superheated steam through the liquid mass at 225–350 °C. The depolymerization can likewise be done within the sight of phosphoric acid alone and caprolactam distilled out. Oligomers are depolymerized into pure caprolactam by passing the oligomer through a fluidized bed of Al_2O_3 catalyst with steam at 290–400 °C. Aluminum oxide is calcined at 800 °C and has an average particle size of 0.2–0.8 mm. It is heated to 320 °C and fluidized with nitrogen at 360 °C. For over 1 h, the oligomer–caprolactam melt is passed through the fluidized bed in a nitrogen stream with 600 g steam added hourly for 4.5 kg oligomer to get 4.32 kg of pure caprolactam. As the water vapor content of the catalyst increases, the yield of caprolactam increases from 78% to 98% [80]. The oligomer can be converted into caprolactam by acid hydrolysis using phosphoric acid and superheated steam. The conversion is quantitative, and the recovered caprolactam is of very high purity. However, hydrochloric acid or sulfuric acid failed to give caprolactam from the oligomers. Thus it is possible to recover caprolactam from the total extractible quantitatively by recovering free caprolactam by steam distillation and hydrolyzing the oligomer. Caprolactam is recovered by heating the distillation residue in the presence of high-boiling hydrocarbons and NaOH at 250–500 °C and continuously removing the caprolactam from the reaction mixture off-gases. The off-gases are condensed producing caprolactam. The recovered caprolactam must meet the specifications of permanganate number (Pm No), volatile bases (VB), hazen color, UV transmittance, solidification point, turbidity, and so on, so that it can be used for further polymerization [81]. Disassembling the face fibers off the used carpet material using different mechanical separation methods provides nylon 6 fibers that can be used in several applications.

Nylon 66 (NY66) is an important semicrystalline thermoplastic polymer with good mechanical, thermal, and chemical properties. In the automobile industry, NY66 is widely used in high-performance molded applications such as in radiator end tank, engine cover, intake manifold, dipstick caps, thrust washer, liners for sheathed cables, and so on [82]. According to a recent industry estimate, an average car uses nearly 30 times NY66 in under-hood applications alone compared to that used in 1960s. Some of the routes followed for recycling the degraded NY66 products are extraction [83, 84], short-fiber reinforcement [85, 86], depolymerization [87, 88], melt blending [89], and recovery of energy by incineration [90, 91]. These conventional recycling processes were mainly used to either recover raw materials or manufacture second-class articles which have lower end-use properties than first-hand polymeric products. However, not sufficient attempts have been made to enhance the properties of recycled NY66 as much as first-hand NY66 products. Many researchers have used the concept of chain extension as a suitable recycling process for the degraded polymers, and some of the notable chain extenders used previously include bis-caprolactams [92–94], bis-oxazolines [95–97], organic phosphite esters [98], diisocyanates/diacid chlorides [99], and epoxy resins [100, 101].

Hong *et al.* reported [102] that the lab-scale glycolyzed NY66 (g-NY66) could be effectively recycled through melt-compounding using novel chain extenders. Overall, the results obtained from their study suggested that the recycling of g-NY66 by melt-compounding using chain extenders results in the linear NY66 (recycled) polymer, which can be formed into new products using the existing manufacturing lines employed for virgin NY66. Moreover, they can also be recycled many times using the above melt-compounding process, with properties almost comparable to those of virgin NY66 and hence have significant industrial applications. Most carpet fiber yarns in production today are either nylon 6 or 6.6, although a small percentage is made of PES, acrylics, PP, and other olefin fibers, wool, or cotton [103, 104]. Carpet recycling has emerged over the last few years because of the large amounts of nonbiodegradable carpet material accumulated in landfills. Recycling of the entire carpet materials poses a significant challenge because of the inhomogeneous nature of the material [72]. Moreover, it is even more difficult to recycle post-consumer carpet waste because of the dirt, cleaning chemicals, and other materials accumulated in the carpets over the years of use. Studies carried out in Europe reveal that post-consumer carpet is approximately 30% heavier than the new carpet owing to the dirt accumulated in the piles. It also contains a significant amount of contaminants, mainly the chemicals used for cleaning purposes. It is this inhomogeneous nature of carpets that makes effective recycling a very difficult and costly process.

Figure 6.6 shows the four main layers of carpet [105]. The top layer, or face yarn, is composed of nylon fibers tufted through a primary backing, which is usually

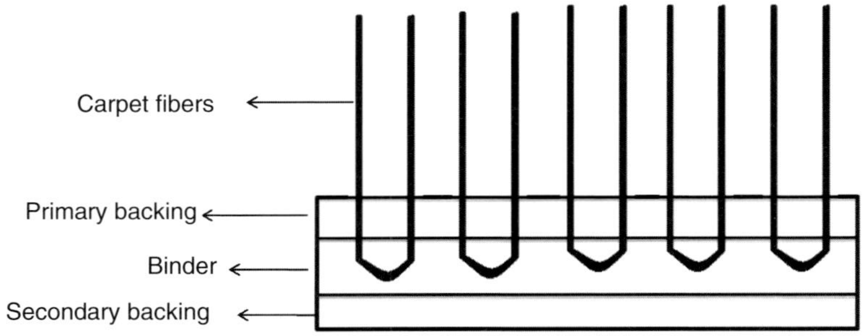

Figure 6.6 Carpet construction.

made of PP. Other fibers such as jute, PE, PESs, and rayon may also be used. Latex adhesive is applied under the primary backing in order to secure the face fiber. The adhesive is usually made of a styrene–butadiene copolymer (SBR), which is filled with inorganic materials such as $CaCO_3$ or $BaSO_4$ [106]. The fillers are added for soundproofing purposes. Finally, a secondary backing (same material as the primary backing) may optionally be added to the primary backing and bonded to it by the same SBR adhesive. The nylon face fibers, containing dyes, soil repellents (to improve the resistance to stains), and possibly other additives to improve the quality of the carpet, usually account for about half of the total carpet weight.

Depolymerization is the preferred route of carpet recycling, since it breaks down the carpet fibers (nylon 6 and/or nylon 6,6) into the corresponding monomers. This allows the recovery of the monomers, which can be repolymerized into new nylon products of high quality. Alternatively, the nylon component can be separated from the carpet waste by extraction. In this process, the entire waste carpet material is dissolved in a solvent at elevated temperatures. During extraction, the nylon from the fibers is recovered in its polymeric form and can be reused in injection-molding applications. The main problem associated with this approach is the selection of a suitable solvent that selectively dissolves the nylon fibers and does not react with or dissolve any of the other carpet components.

In 1994, Monsanto patented a process to recycle all the components of a post-consumer nylon 6,6 carpet, without separation, into a filled thermoplastic product suitable for injection molding [107] (Figure 6.7).They used a twin-screw extruder to accomplish high-intensity mixing of the thermoplastic from carpet samples. The recycled material contained 35–67 wt% nylon, 8–21 wt% PP,

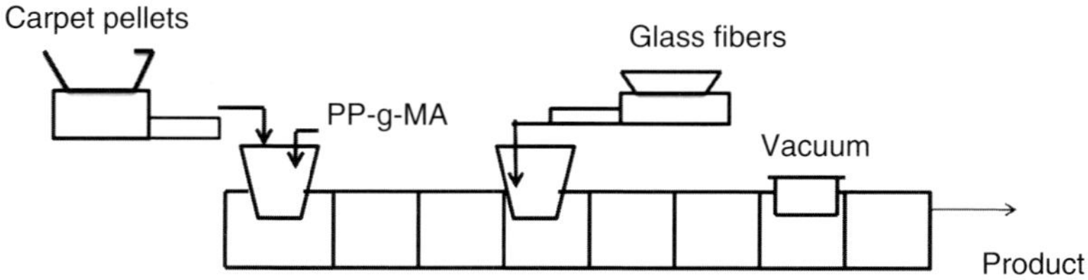

Figure 6.7 Extrusion process developed by Monsanto for recycling carpet waste.

5–29 wt% SBR, and 10–40 wt% inorganic filler. Hagberg and Dickerson [108] reported that a maleic anhydride grafted PP, PolyBond 3150, was added to compatibilize the nylon and PP in the carpet waste. Addition of 3 wt% PolyBond 3150 resulted in a tensile strength of about 40 MPa, tensile modulus of 3.1 GPa, and Izod impact strength of 35 $J\,m^{-1}$. The elongation at break was low, about 2.3%. These properties appeared promising enough for the product to compete with some virgin polymers.

Melt-blending of the entire carpet scrap generates a thermoplastic mixture that can be used for the manufacture of a lower quality plastic material. Such materials can be utilized in less "demanding" products. The method consists of melting the entire carpet waste, without previously separating into its components, to obtain a blend of different polymeric and inorganic materials. The low quality and lack of homogeneity of the resulting mixture are the main drawbacks of this method, restricting the number of applications in which its product can be used. Extraction methods attempt to separate and recover the PAs from the other carpet components without converting them back to the original monomers. Usually, the separated nylon obtained from such processes is injection molded into other products [109–113]. Tertiary recycling of carpet materials consists of reusing the carpet as a whole, by melting or extruding it to form a blended mixture, which is subsequently used in the production of injection-molded polymers and thermoplastics. An advantage of this process is the low cost, since no expensive separation and depolymerization procedures are required. However, depending on the chemical compatibility of the waste carpet components, melt blending usually results to low-quality melt-blended plastics with limited uses [89, 114, 115].

Lemieux *et al.* [116] reported that their pilot-scale rotary kiln experiments demonstrated the potential for use of post-consumer waste carpet as a fuel in cement kilns. Carpet has a high heating value, similar to that of a low-grade coal. The continuous feeding of shredded carpet fiber and ground carpet backing resulted in combustion without transient puffs and with almost no increase in CO and other products of incomplete combustion. Incomplete combustion products, including CO, THCs, VOCs, and PAHs, changed very little when waste carpet was co-fired with natural gas.

Zhang *et al.* [117] presented two techniques for the recycling of carpet. In the injection-molding process, the carpet waste is recycled "as is," without any additive, and the baseline properties of the molded samples are evaluated. In the compression-molding process, the approach followed is to reinforce the carpet waste with a fiber glass mat to form glass-mat-reinforced thermoplastics (GMTs). The market for GMTs using virgin materials is growing, particularly for use in the automotive sector. Polypropylene is the most common matrix used in GMTs. Therefore, carpet waste streams rich in PP were emphasized in this study, and it was found that the GMT from carpet waste has mechanical properties comparable those of GMT from virgin PP. Significant improvements in the impact strength of shear lint samples are achieved by adding just 20 wt% glass mat.

Bockhorn *et al.* [118] reported the kinetics of the thermal degradation of PA 6 under noncatalyzed and catalyzed conditions with dynamic and isothermal methods. The base-catalyzed degradation of PA 6 is best suited for an application in a procedure for e-caprolactam recovery from PA 6 waste. The applicability of this process has been demonstrated by investigations with a cycled spherical reactor in the laboratory scale. The measured degree of conversion was in good accordance with the calculated degree of conversion based on kinetic data from micro-scale experiments. Using samples of PA6: PP fibers from carpets, the basic catalytic separation of e-caprolactam from PP at 290 °C was performed. In case of samples of PA 6 from mechanically recycled carpets, ε-caprolactam with purity above 99% was obtained. It is also worth noting that the addition of synthetic carpet fibers to soil and concrete enhances their structural load-bearing properties [119]. For example, studies [120, 121] show that the additions of carpet waste fibers (nylon and PP) to concrete increased its toughness and ductility. In addition, studies [122] have shown that recycled carpet waste fibers reinforced in soil significantly increased its compression strength. Studies [123] have focused mainly on carpets having synthetic/man-made face fibers, but the isolation of these waste streams requires input of labor in the sorting stage of carpet waste processing. Some of the reported manufacturing processes also comprise several stages, involving mechanical separation of the carpets, fiber reprocessing procedures, and finally energy-intensive processing. However, energy-intensive processing reduces the environmental advantage on account of increased greenhouse gas emissions. Furthermore, some studies have utilized only a specific fraction of the carpet waste, which may not be an effective recycling approach if the remaining carpet constituents remain unrecycled. The addition of glass fibers to composites based on carpet waste offers improvements in the mechanical properties for load-bearing applications. However, glass fiber addition increases raw material costs and adds a further processing step during manufacture.

6.3.2
Polyester

Poly(ethylene terephthalate) (PET), commonly referred to as *polyester* in the textile industry, is considered to be one of the most important thermoplastic PESs [124]. PET is used for a broad range of applications, especially bottles for soft drinks but also for flexible packaging electronics, for solar panels, or as textile fibers and fabrics among many other applications. PET is a nondegradable plastic under normal conditions, as there is no known organism that can consume its relatively large molecules. Complicated and expensive procedures need to be operated in order for PET to degrade biologically [125].

The first recycling effort of post-consumer PET bottles in the world was in 1977 [126]. Since the price of virgin PET remains high, new and cheaper technologies for recycling PET are beneficial to the PET recycling industry by providing processors with relatively cheap PET. So the physical recycling of PET by melt reprocessing is very important, since it is relatively simple, requires low investments, utilizes established equipment, is flexible in terms of feedstock volume, and has little adverse environmental impact [127]. Mechanical recycling of PET includes the separation of waste, washing to remove dirt and contaminants, crushing and grinding to decrease PET particle size, re-extrusion, and reprocessing for the production of new PET goods [128]. It is possible to recycle thermoplastic PET mechanically, while it is impossible to recycle thermoset PET because it cannot be remolded by heat. Above all, the major disadvantage of mechanical recycling is the reduction of the molecular weight or intrinsic viscosity, by thermal and hydrolytic degradation. Chemical recycling is accompanied by transformation of the PET chain. By means of polymer chain scission, the PET polymer can be broken down into either its monomers or its oligomers and other chemicals [129–133].The commercially available chemical recycling technologies involve glycolysis [134], hydrolysis [135], methanolysis [136], and aminolysis [137] reactions. The classification of chemical recycling with each reaction product is shown in Figure 6.8.

However, during the melt reprocessing of any PET, the polymer undergoes chemical, mechanical, thermal, and oxidative degradation, which reduces its molar mass – and consequently its viscosity, melt strength, and mechanical properties – and limit its usefulness for many applications [139–141]. To compensate for polymer degradation, the use of chain extenders had been proposed in many studies as an effective method for increasing the molecular weight [142–144].

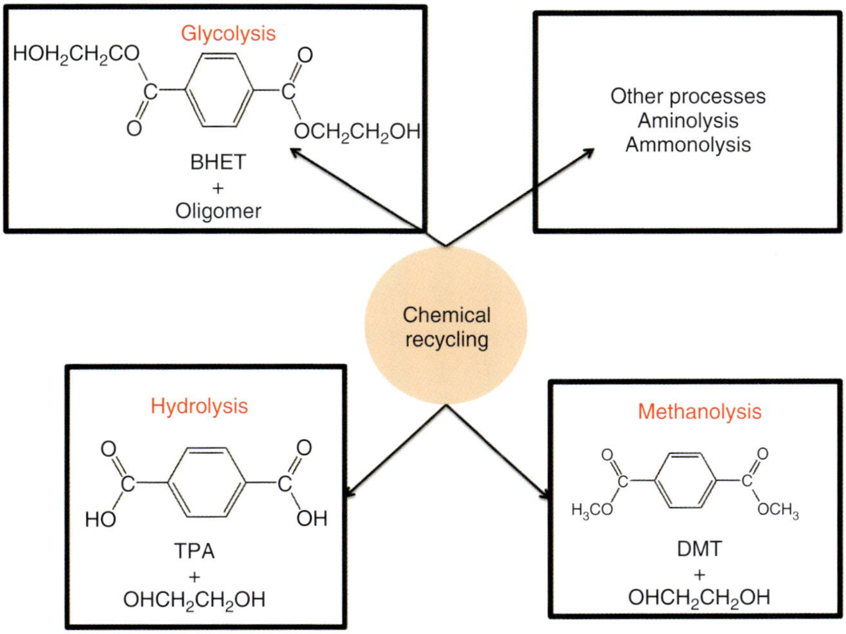

Figure 6.8 Classification of the chemical recycling methods with the products of each reaction. (Park and Kim [138].)

Duarte *et al.* [145] reported the use of a commercial Joncryl blend (POLYAD PR-002) with condensation polymers such as PET; it was compounded along with virgin and recycled post-consumer resin in a laboratory internal mixer. For virgin PET, only 0.5% additive was sufficient to "stabilize" the polymer (i.e., maintain its molar mass) during processing. For recycled post-consumer PET, 1.0% additive was required to accomplish the same result. Chemical modification of recycled post-consumer PET with 1.5% of Joncryl resulted in a molar mass increase, which was close to that of virgin PET added with 1.0% of Joncryl (Scheme 6.2).

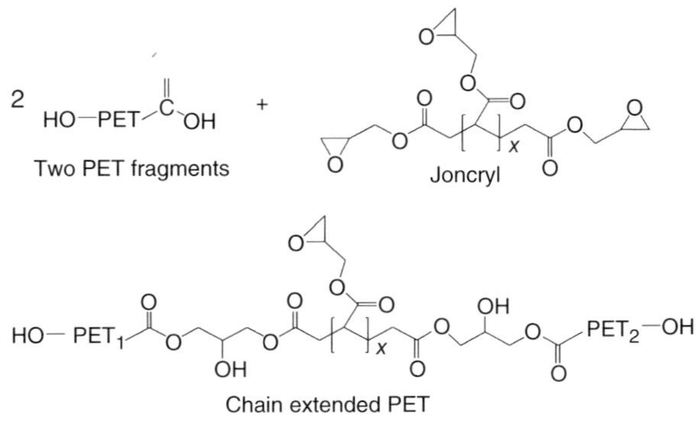

Scheme 6.2 Chain-extending reaction of PET by a multifunctional epoxidic oligomeric additive (Joncryl).

The Joncryl epoxidic chain extender additive was found to compensate virgin and recycled post-consumer PET degradation during processing and reprocessing and increase the molar mass of both. The proposed method appears to be a fast and simple one-step procedure to check the effectiveness chain extender additives and their dependence on the additive concentration and processing conditions, applicable to both virgin and recycled post-consumer PET. For a virgin product, this includes all steps from the extraction and transportation of raw materials and fuels, followed by all conversion steps until the product, that is, fiber – is delivered at the factory gate. The production of the end product (e.g., a shirt) and the use phase and the post-consumer waste management are excluded. For open-loop recycling, it is typically a problem to define the "cradle" stage of the recycled product. As default case, we choose the conventional "cut-off" approach to define the system boundary. Figure 6.9 illustrates the concept of the "cut-off" approach: the first life and second life are cut into two independent product systems. Based on the cut-off principle, the used bottles from the first life are considered to be waste; waste does not bear any environmental burden from the first life. In this study [146], we follow this rule and define the "cradle" of the second life as the collection and transportation of used PET bottles.

Used PET bottles are collected on a local scale, for example, they are from consumers and brought to a waste separation center where bottles are sorted out, baled, and compacted. The energy consumption related to sorting, baling, and

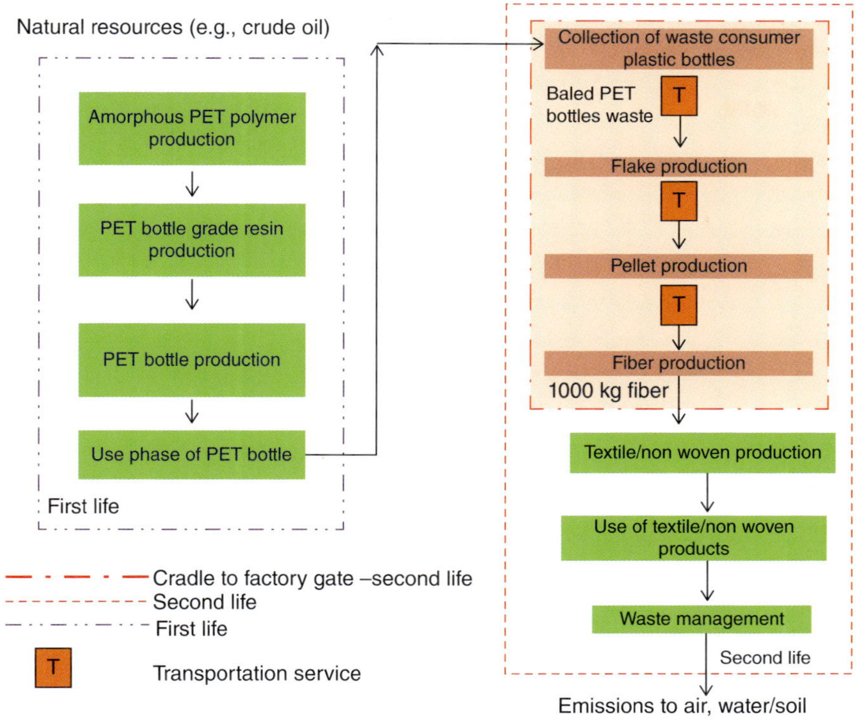

Figure 6.9 Cradle-to-factory gate system boundary of recycling PET fibers from waste PET bottles, splitting the first life and the second life based on the "cut-off" approach.

compacting is very small compared to the energy requirements of the recycling process [147, 148]. Figure 6.10 shows the flow sheet of the production of recycled PET flakes. After the baled bottles are opened, loose bottles are sorted by color and material type. Transparent (uncolored) bottles have a higher economic value than blue and green ones. The unwanted color fractions and unwanted materials (e.g., paper and metal) are either sold as byproducts or disposed of in local municipal solid waste (MSW) management facilities or landfilled, depending on the available local infrastructure. MSW can be incinerated with or without energy recovery.

Some producers wash the bottles with hot water to remove the labels before the sorting process. The plastics labels are either sold as byproducts (mainly consisting of LDPE and/or PVC) or sent to local MSW management. The bottles are then chopped into flakes, and put through a float separation step to separate PET from other plastics (e.g., HDPE caps) based on density differences. PE obtained from this step is sold as a byproduct. The PET flakes are then washed in a cleaning solution, rinsed, and dried. In some production lines, a second chopping step (also called *fine crushing*) is required to ensure that the PET flakes meet the quality requirements. Finally, the dried PET flakes are ready to be transported to pellet plant or a fiber plant.

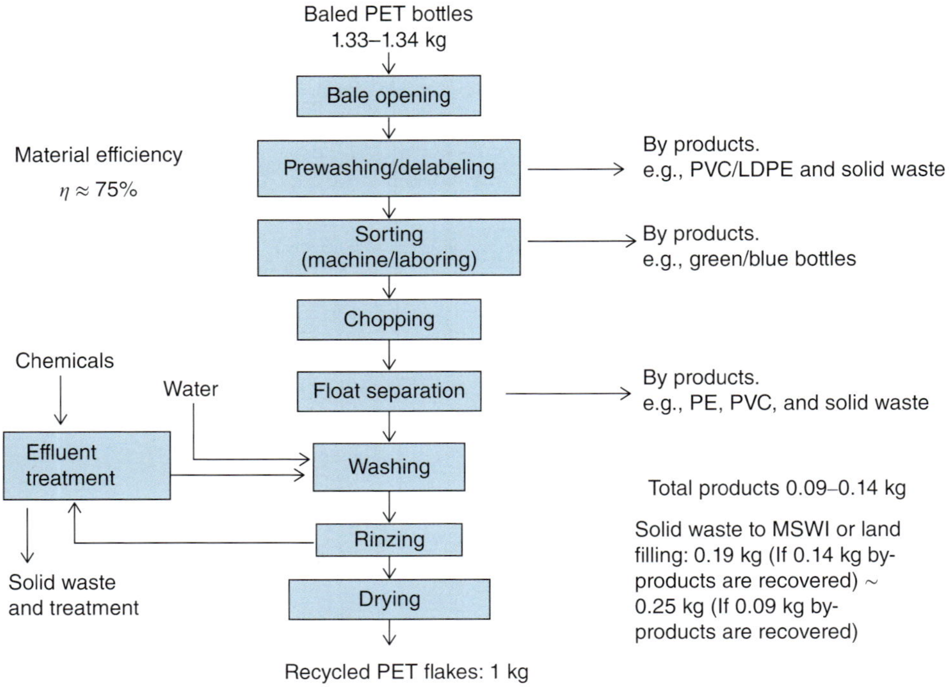

Figure 6.10 Flow sheet of the production of recycled PET flakes.

Dimitrov *et al.* [149] recycled the PET, which was studied by pyrolysis-gas chromatography/mass spectrometry (Py-GC/MS) to establish the potential remaining contamination and degradation products after material recycling process. Moreover, reactive pyrolysis-gas chromatography (Py-GC) in the presence of an organic alkali (thermally assisted hydrolysis and methylation (THM) such as tetramethyl ammonium hydroxide (TMAH) often provides useful information for the condensation type of polymer materials containing ester and/or carbonate linkages. It has been well accepted that the first step in the degradation of PET during pyrolysis is the classical ester scission reaction producing carboxylic acid and olefin end groups. This scission may then yield many different gaseous substances (primary products), which further react among themselves or with oxygen (if the decomposition has taken place in air atmosphere) (secondary products). The main products obtained by this organic alkali-assisted pyrolysis are trimethylamine (m/z 58), dimethyl terephthalate (m/z 163), and methyl (m/z 31) and ethyl (m/z 45) alcohol (Scheme 6.3). From the results, it is seen that their finally obtained concentration depends on the type of the studied PET samples. Alkali-assisted pyrolysis performed with TMAH gives us insight into thermal decomposition at lower temperatures.

Chemically recycled PET can be used for applications in the manufacture of polyurethane and unsaturated PES resin [150]. It is possible to use mechanically

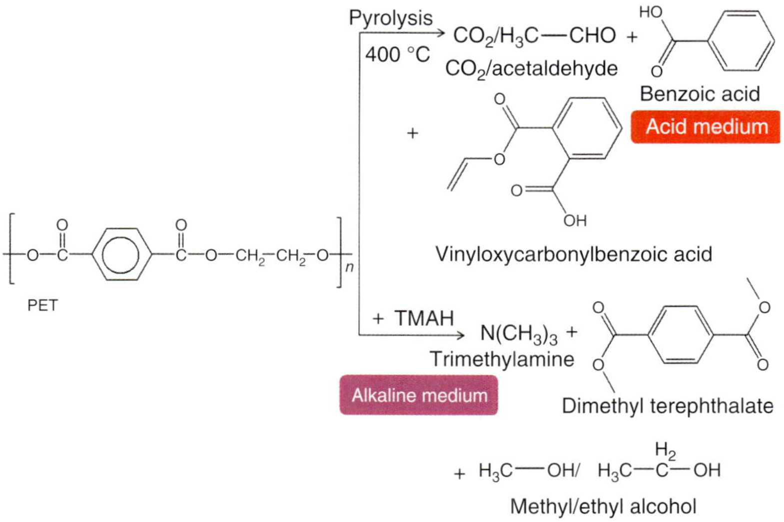

Scheme 6.3 Products of PET pyrolysis.

recycled PET for various applications in the textile industry, and the products obtained from chemical recycling of PET can be also used as feedstock for polyurethane and unsaturated PES resins. With these efforts to increase the use and specification of recycled PET as a replacement for virgin PET, the recycling of PET waste is an effective method for enhancing the environmental performance of the PET fashion and textile industries [119].

6.3.3 Glass Fiber

Glass fiber is a material made from extremely fine fibers of glass. It is also called *fiber glass*. A fiber is spun from an inorganic product of fusion that has cooled to a rigid condition without crystallizing. Glass melts are made by fusing (co-melting) silica with minerals, which contain the oxides needed to form a given composition. The molten mass is rapidly cooled to prevent crystallization and formed into glass fibers by a process also known as *fiberization*.

Fiber glass is a lightweight, extremely strong, and robust material. Although its strength is somewhat lower than that of carbon fiber (CF) and it is less stiff, the material is typically far less brittle, and the raw materials are much less expensive. It is used in the manufacture of structural composites, printed circuit boards, and a wide range of special-purpose products [151]. Glass fibers are useful because of their high ratio of surface area to weight. However, the increased surface area makes them much more susceptible to chemical attack. By trapping air within them, blocks of glass fiber make good thermal insulation, with a thermal conductivity of the order of 0.05 $W\,m^{-1}\,K^{-1}$.

6.3.3.1 Glass Fiber-Reinforced Plastics

Glass fibers reinforced plastics (GBRP) are obtained by reinforcing the glass fibers into to the plastic matrix. GFRP is a composite material manufactured by laminating unsaturated PES resin with glass fibers and fillers. It is a well-known composite material that has a high market value because of its superior characteristics in terms of weight and durability. The matrix comprises organic, PES, thermostable, vinylester, phenolic, and epoxy resins. PES resins are classified into bisphenolic and ortho or isophtalic [152]. The mechanical behavior of a fiber-reinforced composite basically depends on the fiber strength and modulus, the chemical stability, matrix strength, and the interface bonding between the fiber and matrix to enable stress transfer [153]. They have many desirable characteristics such as low cost, high tensile strength, high chemical resistance, and excellent insulating properties [154, 155]. GFRPs are strong and lightweight materials that find a wide range of uses in the industry.

Glass-fiber-reinforced composites are widely used in the automotive and aerospace industries due to their high strength and low weight. The other products range from wind turbines and sports equipment to automotive and construction components. However, GFRPs are difficult to recycle. This type of composite material is often produced with thermosetting polymers that cannot be remelted. In the past, several methods have been evaluated and tested, but these have not proven economically feasible or become established. Therefore, glass-fiber-reinforced composites often end up in landfill sites. The cost of landfilling is increasing and legislation requires that waste containing more than 10% organic material should not be used for landfilling. There is consequently a clear incentive to develop methods to recycle this material.

The techniques for recycling FRP waste fall into two main groups: (i) mechanical methods, involving comminution techniques to reduce the size of the FRP waste, and (ii) thermal methods that use heat to break down the waste into raw materials and, eventually, energy.

6.3.3.2 Mechanical Process

Mechanical recycling is presently the only process used commercially for thermosetting polymers. This method involves a series of operations, including cutting, crushing, and grinding, which successively reduce the recycled materials' size. The first processing stage consists of using slow-speed cutting or crushing mills to break the original material into 50–100 mm sized pieces.

6.3.3.3 Thermal Process

Thermolysis involves the volatilization of the resin, releasing the fibers and, eventually, fillers from the composite. But in the case of glass-fiber-reinforced composites, the fibers are strongly degraded.

6.3.3.3.1 Pyrolysis Process A diagram of a typical pyrolysis process is shown in Figure 6.3, in which the gases evolved are used as fuel to provide heat for the process.

Pyrolysis process involves heating of a combustible material in the absence of oxygen. In this process the material breaks down into lower molecular weight organic substances (liquids and gases) along with a solid carbon char product. It offers a method for recovering the material from the in a scrap composite. The plant designed for the batch processing of a tire pyrolysis the temperature is maintained to 700–1000 °C. This process produces a fuel gas, some liquid oil products, a solid residue comprising inorganic fibers and fillers, and a char residue. The use of this solid residue as filler in new sheet molding compound (SMC) were carried out. Incorporation of 30% ground solid residue into an SMC without adversely affecting the processing or the mechanical properties of the molded parts. Low process hydrolysis process was investigated [156] and there is a potential value in the solid products if the glass fiber could be recovered in a good condition. This process use a temperature of 400 °C. It was found that the presence of steam increases the rate of polymer degradation and enables the fibers to be separated more easily from the solid pyrolysis products. Cleaning of the resulting solid products is done by acid digestion to remove the calcium carbonate filler for recovery as calcium chloride flakes. Then the fibers were then separated and cleaned. Specific fracture energy retention of the fibers was measured. Fibers were retained about 50% of their specific fracture energy at a pyrolysis temperature of 400 °C and it is necessary to breakdown the polymer.

There has also been interest in the application of the technology of pyrolysis to recycle composite plastics [157–159]. Pyrolysis thermally degrades the plastic component to produce an oil and a gas product. The oil may be used as a liquid fuel, or returned for processing at the refinery, whilst the gas could be burnt to provide the energy requirements for the pyrolysis system. To fully recycle composite waste, viable applications are required for the pyrolysis solid residues, which consist of fiber, filler, and char. Possible reuse routes may include the replacement of virgin fiber and filler materials in the production of bulk molding compounds, or as filler in concrete [160, 161].

6.3.3.3.2 **Fluidized Bed Process** The fluidized-bed process involves thermal decomposition of the polymer matrix followed by release and collection of discrete glass fibers and filler particles. Oxygen is required to minimize char formation. Potentially, the fluidized bed is a convenient method of heating the feed material rapidly in an air stream and provides the attrition necessary to release monofilaments from the reinforcing phase. The fluidizing air is sufficient to elutriate the released fibers and fillers while retaining undegraded material. Organic contaminants such as mineral oils and facing paints are volatilized with the polymer matrix, while inorganic solids such as metal inserts sink in the fluidized bed and may be removed in a bed-regrading process. The separation of high-value fibers from the relatively low-value fillers can be done conveniently while each exists as discrete airborne particles. This is achieved in the present study using a novel rotating sieve separator situated in the freeboard. Its action removes long fibers from the fluidizing gas stream and allows fillers and short fibers to pass. These are removed subsequently from the gas stream using a cyclone. As the fluidized bed operates at a relatively low temperature (typically 450 °C), complete combustion of the organic constituents in the feed cannot be achieved. A secondary combustion chamber operating at a temperature in the region of 1000 °C is, therefore, necessary to fully oxidize the organics. This would produce a clean flue gas, subject to the removal of any acid gases resulting from the processing of halogenated resins in the scrap feed and allowing the recovery of energy. Figure 6.11 shows the idealized process.

6.3.3.4 Chemical Recycling

Chemical recycling operates at lower temperatures and uses a solvent to depolymerize the resin and release the fibers and, eventually, fillers. The interest is that valuable product can be recovered from the resin, and the fibers can be reused.

6.3.3.4.1 **Supercritical Fluids** Chemical recycling in supercritical conditions of water ($T > 374\,°C$ and $P > 221$ bar) and alcohols is more applied to carbon composite materials because of the value of carbon fibers (CFs) but without any interest in the products issued from the resin (which is more often epoxy). It has also been applied to glass-fiber-reinforced composite materials such as SMCs [162] or to PES resin model without fibers [163, 164].

Oliveux *et al.* reported the recycling of glass fibers from glass fiber–PES composite materials by subcritical hydrolysis, which depends on quality of recovered fibers, nature of recovered products, and process efficiency, using a commercial product. The study allowed an evaluation of the potential of the hydrolysis process to recycle glass-fiber-reinforced composites [165].

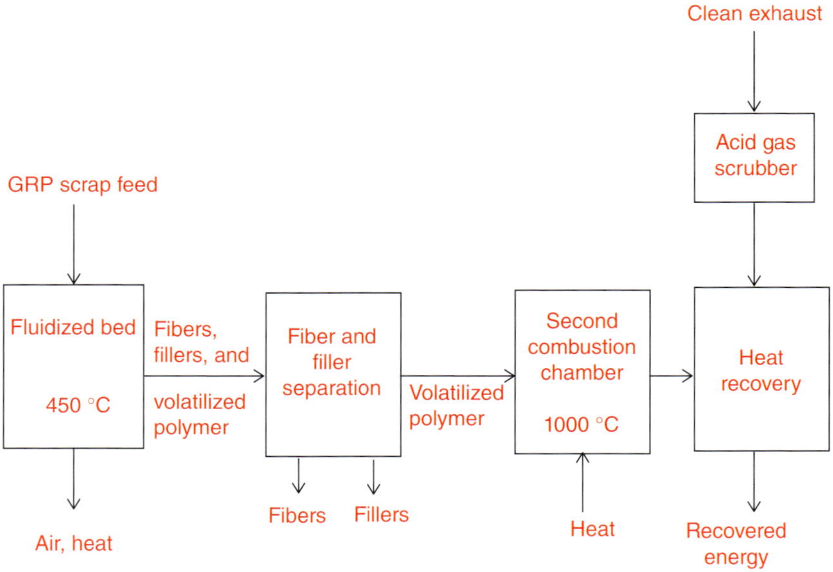

Figure 6.11 The fluidized-bed process [3].

Asokan *et al.* [166] assessed the recycling potential of ground GRP waste in concrete and explored the possibilities to improve its performance with the addition of superplasticizer for use in construction products for different applications (Figure 6.12). The quality of the concrete composite products depends on the characteristics of GRP waste power, particle size distribution, grinding process and the degree of grinding, aggregate mix proportion, and, finally, the additives' concentration. The concrete composites developed with ground GRP waste sample (5–15%) using 2% superplasticizer significantly improved the compressive strength, split tensile strength, shrinkage, initial surface absorption, total water absorption, and density, and complied with British standards for use in normal and structural concrete products for construction purposes. Potential applications of GRP waste in concrete include precast paving slabs, roof tiles, precast concrete wall elements, lightweight concrete, concrete paving blocks, and architectural cladding materials.

Glass fibers recovered after incineration or thermal decomposition of FRP waste are sometimes coated with char when the resin has not been completely decomposed. For the glass fibers to be recycled as reinforcement, they must first be cleaned to remove the char. If the fibers are to be used as insulation material, the char need not be removed, but the fiber can only be used as low-grade insulation. The insulation performance of glass fibers partly mixed with calcium carbonate (filler often used) is not known.

The end use of FRP recyclates should be based on the properties of the recyclates, particularly where these give added value, such as chemical and physical properties (e.g., phenolic composites for greater fire resistance), where the recyclates can give special surface effects and designs, for noise absorption (wall panels or in roads), use of fluffy material for high thermal insulation or for nonwoven materials, for viscosity modification of polymer mixes, and low-cost core material (in boats or in wood substitute products) as reinforcement [167].

(a) (b) (c)

Figure 6.12 Photographs (a) GRP waste ground sample, (b) separated glass fiber and polyethylene, and (c) GRP waste powder. (Asokan *et al.* [166] reproduced with permission of Elsevier.)

6.3.4 Carbon Fiber

CFs are one-dimensional filamentary form of carbon with an aspect ratio (length/diameter) greater than 100. The filaments are 7–15 pm in diameter, and they consist of small crystallites of "turbostratic" graphite, which is one of the allotropic forms of carbon. In the graphite crystal, carbon atoms are arranged in a hexagonal array in a plane, and the planes are stacked together, with covalent bonds acting within the planes and weaker van der Waals forces holding the planes together. CFs are unaffected by aggressive environments, abrasion resistant, stable at high temperatures, medically safe, as strong as steel fibers, and more chemically stable than glass fibers in alkaline environments. Moreover, CFs are of low density, especially when compared to steel fibers. Their strength to density ratio is one of the highest among all fiber types. The main drawback of CFs has been their high cost, because low cost is essential for most applications concerning cements. CFs possess an additional advantage of having a high electrical conductivity. Since cement itself is a poor electrical conductor, the presence of CFs greatly increases the electrical conductivity of cement. The high electrical conductivity makes the cement useful as a material for antistatic flooring, for walls of electromagnetic shield rooms, and so on [168–170].

Current uses of CF–cement composites include cladding, free-access floor panels, lightweight decorating frames, shell structures, and protective coatings for structural elements exposed to harsh environments [171].

Carbon-fiber-reinforced resin composites (CFRPs) have been widely applied in many fields, such as transport tools, sporting goods, and so on, which is made possible by their unique mechanical properties, corrosion resistance, and so on [172]. However, the production of CF involves high cost is energy-intensive. If CF with superior properties can be recycled in an economical way, the recovered CF can displace pristine counterparts in some applications. As a result, a lot of energy for producing CF can be saved. So, in order to realize chemical recycling of CFRP efficiently, a key scientific issue is to explore suitable reaction conditions under which both recovered CFs and the resultant decomposition products are highly useful and economical.

The main recycling method can be classified into three types: mechanical recycling [173–175], thermal processing [176–178], and chemical recycling [179–186]. Among them, chemical recycling is an attractive route for decomposing polymer matrices into useful chemicals and avoiding the damage to CFs.

6.3.4.1 Mechanical Recycling

Mechanical recycling techniques have been investigated for both glass-fiber- and CF-reinforced composites, but the most extensive research has been done on the former. The technique used is usually to reduce the scrap composite components in the primary crushing process. This would typically involve the use of a slow-speed cutting or crushing mill to reduce the material to pieces in the order of

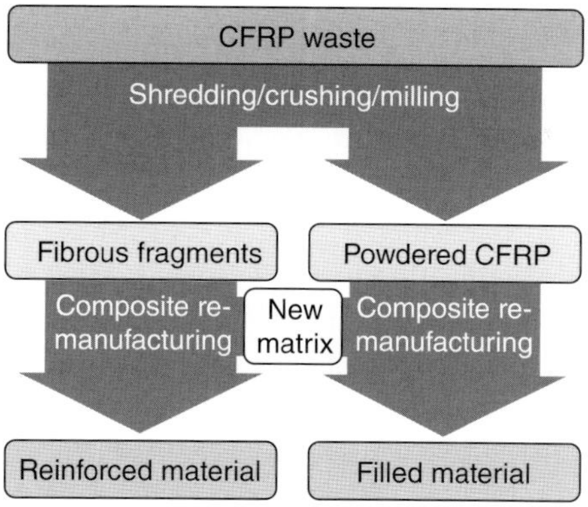

Figure 6.13 Mechanical recycling. (Pimenta and Pinho [188] reproduced with permission of Elsevier.)

50–100 mm in size. This facilitates the removal of metal inserts, and, if done in an initial stage where the waste arises, the volume reduction assists transport. The main size reduction stage would then be in a hammer mill or any other high-speed mill where the material is ground into a fine product ranging in size from typically 10 mm down to particles less than 50 mm. Then a classifying operation, typically comprising cyclones and sieves, would be employed to grade the resulting recyclates into fractions of different sizes [174, 175, 187]. In the mechanical recycling process, all constituents of the original composite appear in the resulting recyclates, which are mixtures of the polymer, fiber, and filler. Typically, the fine graded fractions are powders and contain a higher proportion of filler and polymer than in the original composite. The coarser fractions tend to be of a fibrous nature where the particles have a high aspect ratio and have higher fiber content.

Mechanical recycling processes are suitable for scrap composite material that is relatively clean and uncontaminated and from a known origin. There are several technologies developed for powder recyclates and fibrous recyclates, which have the potential for reuse. However, the powder recyclates have limited potential for reuse back into the thermoset compounds from which they originated. Although they are of lower density, there are other drawbacks for their use in terms of ease of processing and lower mechanical properties in the products in which they are used. The fibrous recyclates have some potential as reinforcement materials, but they are not as good as the virgin reinforcement, and there are problems associated with the bonding of the recyclate with polymers and the tendency for the larger pieces of recyclate to be stress raisers and to act as failure initiation sites. The process is shown in Figure 6.13.

6.3.4.2 **Thermal Recycling**

6.3.4.2.1 **Pyrolysis** Pyrolysis, as a common thermal process, gives chemical products, fibers, and fillers. It has also been investigated for the recycling of CF composites. The work on CF composites with epoxy and phenolic resin were investigated in Japan. The work mainly focused on the properties of the CF following heating in air and pyrolysis. Pyrolysis was unusually described as taking place in a stream of air in the experimental method. Temperatures of 400 °C, 500 °C and 600 °C were used to heating for an extended period. Tensile strength measurements were made and compared with virgin fiber. Results shows that for a CF composite heated under pyrolysis conditions at 500 °C, there was little deterioration of the tensile strength of the CF. On the other hand CF alone was heated in air, the tensile strength decreased by about 25%, which was interpreted as due to more severe oxidation as the fibers did not have a protective layer of resin. At 600 °C more severe oxidation takes place in CF and under pyrolysis conditions the tensile strength of the fiber reduced by over 30%. Gasification process developed for CF composites [189] includes the scrap is heated in a controlled flow of oxygen at a temperature of 600 °C. Short chain hydrocarbons, gases (H2 and CO), and the CFs can be recovered from polymer conversion and reuse. Char residue formed from the polymer remains on the fiber and this is generally less than 10%. Improved tensile strength (of over 25%) relative to glass fiber is obtained for the recycled fibers and is used as a substitute for glass fibers in a bulk molding compound. The advantage of thermal recycling processes is that it is able to tolerate more contaminated scrap materials. Fluidized-bed process produces a very clean fiber product, but it is not in the same form as the existing virgin fiber products. The reprocessing of materials into a cost effective new products needs to identify and more development requires in this field. It is same as in the case of fiber products developed from the pyrolysis processes. The limitation of this process includes varying degrees of char on the recycled fibers and this requires further processing for reuse. Pyrolysis processes are more complex than the fluidized process but produce potentially useful organic products from the polymer. It needs further processing to separate them from the mixture of products produced and it seems that it would only be cost effective on a large scale (Figure 6.14).

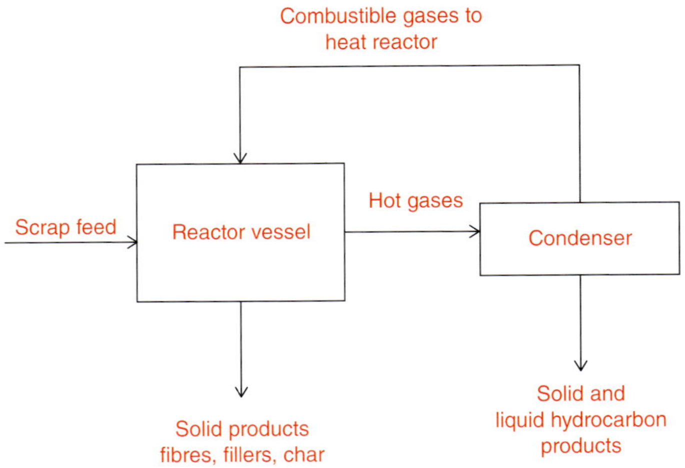

Figure 6.14 Pyrolysis process [36].

6.3.4.2.2 **Combustion or Incineration with Energy Recovery** Like all organic materials thermosetting polymers can be burned as a source of energy. During combustion some mineral fillers decomposes and absorb energy and fire retardants are used for this purpose. Fire retardants are used to reduce the initial ignition and flame spread, the amount of energy absorbed is small compared to the calorific value of the resin. For example, the energy absorbed by alumina trihydrate is 1000 kJ kg^{-1}. If there is as much alumina trihydrate is present in the composite the calorific value will decrease to 3.3%. Burning scrap composites in cement kilns is an effective route to recover some value from the incombustible material. Glass reinforcement and mineral fillers commonly used in composites contain minerals that can be incorporated in cement. The effect of these minerals on the cement production process has been investigated [190].

6.3.4.2.3 **Fluidized-Bed Thermal Process** Fluidized bed process is a thermal process used to produce clean fibers and fillers with energy recovery. The most recoverable value in the composite is fiber reinforcement. Over the past ten years the theme of research at university of Nottingham is that to develop a fluidized bed process to recover high grade glass and CF reinforcement from scrap glass and CF reinforced composites [177, 191]. Initially the scrap composites are reduced in size to about 25 mm and fed into a fluidized bed. The bed consists of silica sand with a particle size of about 0.85 mm. Typical fluidizing velocities used to fluidized the sand with a stream of hot air is 0.4–1.0 m/s at temperatures in the range of 450–550 °C. As a result the polymer volatilizes from the composite and releases the fibers and fillers which are carried out of the bed as individual particles in the gas stream. Which is then passed into a high temperature secondary combustion chamber where the polymer is fully oxidizes. Consequently energy may be recovered from these hot combustion products. A diagram showing this process is shown in Figure 6.15. This process has been used for the recovery of both glass fibers and CFs. The fiber product is in a feathery form contains individual fiber filaments typically of mean length (by weight) from 6 mm to more than 10 mm. The recovered fibers are clean and show very little surface contamination. Glass reinforced PES composites at 450 °C leads to volatilization of polymer and releases the fibers into the gas stream. Rapid volatilization polymer in epoxy resins needs 550 °C.

Pickering *et al.* [177, 192, 193] successfully recycled glass fibers by the fluidized-bed process from composites at a bed temperature of 450 °C and a fluidizing velocity of 1.3 m s^{-1}. Later, Zheng *et al.* [194] recycled glass fibers from printed circuit boards at a temperature ranging from 400 to 600 °C. The fluidized-bed process really gives clean fibers and other products, but the reclaimed fibers would be highly oxidized, causing reduction in their properties.

6.3.4.3 **Chemical Recycling**

Chemical recycling includes liquid-phase cracking, chemolysis, supercritical fluid (SCF) process, and so on. Compared to the mechanical and thermal processes mentioned above, chemical recycling is not effective to recycle composites because not all solutions are feasible to all composites. Developing a method for

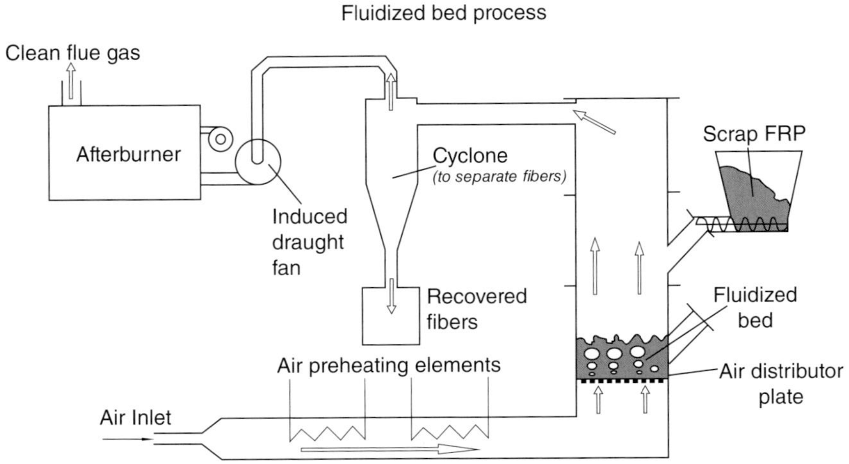

Figure 6.15 The fluidized-bed process.

the chemical recycling of thermoset composites to useful CFs and organic compounds is also an important issue. SCFs, which have the combined characteristics of liquid-like density, dissolving power, and gas-like viscosity, and diffusivity, have been used to recycle composites. Several types of fluids were studied, for example, water, methanol, and propanol.

6.3.4.3.1 **Supercritical Fluids** SCFs are fluids at temperatures and pressures (typically just) above a critical point; at this stage, the fluid presents itself in one single supercritical phase while having combined characteristics: liquid-like density and dissolving power and gas-like viscosity and diffusivity [195]. SCFs can therefore penetrate porous solids and dissolve organic materials while still being relatively innocuous under atmospheric conditions.

SCFs also posses unique properties intermediate between those of gases and liquids; thus they have low viscosities, a high mass transport coefficient, high diffusivity, and pressure-dependent solvent power [196]. Once in the supercritical state, a homogeneous phase is obtained, which is able to penetrate into porous materials and completely dissolve all gases. These properties, when coupled with the high solubilities of organic materials, make supercritical alcohols ideal candidates for extraction and pyrolysis reactions. The substrate materials can be accessed, expanded or loosened, pyrolyzed, and extracted in batch [197–198]. Supercritical water, carbon dioxide, and hydrocarbons have been previously used for a wide range of academic and commercial projects to perform reaction chemistry [199] and processing of materials [200]. A supercritical propanol flow system is shown in Figure 6.16.

Several types of SCFs (usually coupled with alkali catalysts) have been used for CF recycling, such as water, methanol, ethanol, and propanol [182]. Chemical recycling with SCFs is a more recent approach; it is nevertheless already recognized for producing recycled carbon fibers (rCFs) with virtually no mechanical

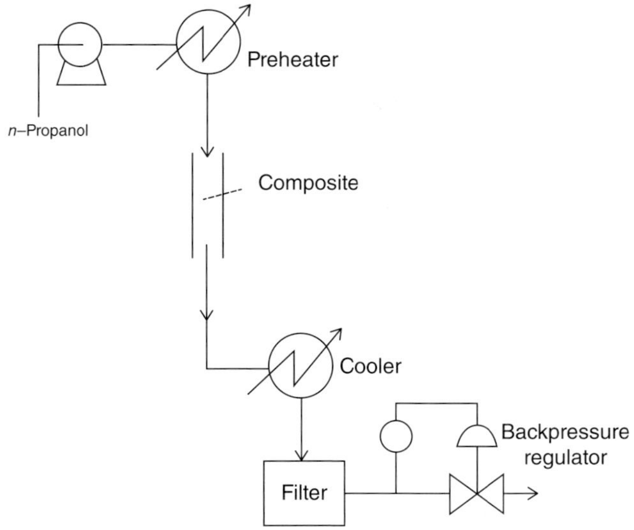

Figure 6.16 A schematic of the supercritical propanol flow system. (Hyde *et al.* [201] reproduced with permission of Elsevier.)

degradation – especially when using propanol – and for allowing the recovery of useful chemicals from the matrix [202, 203].

In their recent work, Piñero-Hernanz *et al.* [204, 205] tested SCFs at temperatures 250–400 °C and pressures 4–27 MPa for water, and at temperatures 300–450 °C and pressures 5–17 MPa for alcohols (methanol, ethanol, 1-propanol, and acetone) to recycle CF-reinforced composites. It was found that the use of an alkaline catalyst (e.g., KOH) gave a resin elimination efficiency of over 90% in supercritical water, and the recovered CF had only 2–10% degradation in mechanical strength. In the case of supercritical alcohol, a resin elimination efficiency of 98% was reported at 350 °C and the recovered CFs retained 85–99% of the strength of the virgin fibers.

Hyde *et al.* used supercritical propanol to extract and remove the epoxy resin from the surface of a CF composite material. The process appeared to be effective when operating above 450 °C and above 50 bar. The recovered fibers were found to be virtually as strong, in terms of tensile strength, as the virgin fibers, indicating that little damage was done to their structural integrity. One advantage of this method is that the polymer as well as the fiber material can be recovered. The recovered fibers are largely free from residual epoxy resin, and therefore are comparable in distribution to other extraction methodologies. The method is quick and simple, and can be operated in a semicontinuous mode, which is highly desirable for this type of process [201]. Figure 6.17a,b shows representative high-magnification SEM images of the virgin and recovered CFs, respectively. As can be seen, the fibers appear to be completely free from the epoxy resin and have surfaces that are clean and smooth. There is no apparent reduction in diameter or surface scratches on the fibers, indicating that the extraction process does not

Figure 6.17 SEM micrograph of (a) virgin carbon fibers and (b) carbon fibers obtained after scPrOH treatment. (Hyde *et al.* [201] reproduced with permission of Elsevier.)

damage the fibers. There are some very small deposits of residual epoxy resin remaining on the surface of the fibers. The fibers are 7 – 8 μm in diameter.

Palmer reported the development of a new SMC using low-cost CF recyclate as the reinforcing component in place of conventional glass fibers [206]. A newly designed manufacturing system allows large, coarse recyclate fibrous materials to be utilized within existing SMC production lines, and with careful control of virgin and recyclate fiber distribution, a viable SMC grade utilizing recycled CF can be obtained. The advantage of the sheet and dough molding processes is that they can utilize short fiber fractions, and therefore provide a potential application for CF recyclate derived from cheaper extraction processes. Short fibers in a dispersed form such as in bulk molding compounds or nonwoven veil or tissue products finds promising applications of recovered fibers.

6.4 Conclusion

Natural-fiber-reinforced composites face higher risk of degradation when subjected to outdoor applications as compared to composites with synthetic fibers. This is attributed to the characteristics of natural fibers, which are susceptible to biodegradation.

References

1. Carvalho, F.R. (2005) Composites of sisal fibers for use in reinforcement of wooden structures. PhD Thesis in Material Science and Engineering. São Carlos: Polytechnical School, Institutes of Physics and Chemistry. University of São Paulo (USP)
2. Saville, B.P. (2007) *Physical Testing of Textiles*, The Textile Institute, Woodhead Publishing Limited, Cambridge.
3. Ricken, L.C., Pozza, P.P., and Teixeira, J.A. (2008) Ecodesign: choice of materials aiming sustainability and manufacturing processes. Proceedings of the 8th Brazilian Congress of Research and Development in Design (P&D).

4. Scarlato, F.C. and Pontin, J.A. (1992) *From the Cradle to Grave: Society, Environmental and Education*, 18th edn, Atual Editora, São Paulo.
5. Alves, C., Ferrão, P.M.C., Silva, A.J., Reis, L.G., Freitas, M., Rodrigues, L.B. *et al.* (2010) Eco-design of automotive components making use of jute fiber composites. *J. Cleaner Prod.*, **18**, 313–327.
6. John, M.J. and Thomas, S. (2008) Biofibres and biocomposites. *Carbohydr. Polym.*, **71**, 343–364.
7. Iwatake, A., Nogi, M., and Yano, H. (2008) Cellulose nanofiber-reinforced polylactic acid. *Compos. Sci. Technol.*, **68**, 2103–2106.
8. Akil, H.M., Omar, M.F., Mazuki, A.A.M., Safiee, S., Ishak, Z.A.M., and Abu Bakar, A. (2011) Kenaf fiber reinforced composites: a review. *Mater. Des.*, **32**, 4107–4121.
9. Pappas, C., Tarantilis, P.A., Daliani, I., Mavromoustakos, T., and Polissiou, M. (2002) Comparison of classical and ultrasound-assisted isolation procedures of cellulose from kenaf (*Hibiscus cannabinus* L.) and eucalyptus (*Eucalyptus rodustrus Sm.*) *Ultrason. Sonochem.*, **9** (1), 19–23.
10. Wang, D., Shang, S., Song, Z., and Lee, M. (2010) Evaluation of microcrystalline cellulose prepared from kenaf fibers. *J. Ind. Eng. Chem.*, **16**, 152–156.
11. Kargarzadeh, H., Ahmad, I., Abdullah, I., Dufresne, A., Zainudin, S.Y., and Sheltami, R.M. (2012) Effects of hydrolysis conditions on the morphology, crystallinity, and thermal stability of cellulose nanocrystals extracted from kenaf bast fibers. *Cellulose*, **19**, 855–866.
12. Ralph, J. (1996) An unusual lignin from kenaf. *J. Nat. Prod.*, **59**, 341–342.
13. Xu, J., Sugawara, R., Widyorini, R., Han, G., and Kawai, S. (2004) Manufacture and properties of low-density binderless particle board from kenaf core. *J. Wood Sci.*, **50**, 62–67.
14. Kalaycıoglu, Ḧ and Nemli, Ğ (2006) Producing composite particleboard from kenaf (*Hibiscus cannabinus L.*) stalks. *Ind. Crops Prod.*, **24**, 177–180.
15. Shibata, S., Cao, Y., and Fukumoto, I. (2006) Lightweight laminate composites made from kenaf and polypropylene fibres. *Polym. Test.*, **25**, 142–148.
16. MacNeish, R.S. (1964) *Ancient Mesoamerican Civilization*, Bobbs-Merrill, Indianapolis, IN.
17. Huckell, L.W. (1993) Plant remains from the Pinaleno cotton cache, Arizona. *Kiva*, **59** (2), 147–203.
18. Fuller, D.Q. (2008) The spread of textile production and textile crops in India beyond the Harappan zone: an aspect of the emergence of craft specialization and systematic trade, in *Linguistics, Archaeology and the Human Past*, Industrial Project (eds T. Osada and A. Uesugi), Research Institute of Humanity and Nature, Kyoto, pp. 1–26.
19. Basra, A.S. and Malik, C.P. (1984) Development of cotton fiber. *Int. Rev. Cytol.*, **89** (1), 65–113.
20. Mwaikambo, L.Y. and Bisanda, E.T. (1999) The performance of cotton-kapok fabric-polyester composites. *Polym. Test.*, **18** (3), 181–198.
21. Mwaikambo, L.Y., Martuscelli, E., and Avella, M. (2000) Kapok/cotton fabric-polypropylene composites. *Polym. Test.*, **19** (8), 905–918.
22. Holt, G.A., Barker, G.L., Baker R.V., and Brashears A. (2000) Various parameters of cotton gin by-products produced from the gin processing machinery. Proceedings of the Belt Wide Cotton Conference, pp. 1595–1602.
23. Poore, M.H. and Rogers, G. (1995) Feeding whole cotton seed and other cotton by-products to beef cattle. *Vet. Med.*, **90** (11), 1077–1087.
24. Truhett, C. (1994) Developing markets for composted gin waste. Proceedings of the Beltwide Cotton Conference, Vol. 1, p. 609.
25. Ayers, V. (1997) Farmers composting of cotton gin waste. Proceedings of the Beltwide Cotton Conference, Vol. 2, pp. 1615–1616.
26. Ververis, C., Georghiou, K., Christodoulakis, N., Santas, P., and Santas, R. (2004) Fiber dimension, lignin and cellulose content of various plant materials and their suitability for

paper production. *Ind. Crops Prod.*, **19**, 245–254.

27. Coates, W. (2000) Using cotton plant residue to produce briquettes. *Biomass Bioenergy*, **18**, 201–208.
28. Holt, G.A., Simonton, J.L., Beruvides, M.G., and Canto, A. (2004) Utilization of cotton gin by-products for the manufacturing of fuel pellets: an economic perspective. *Appl. Eng. Agric.*, **20** (4), 423–430.
29. Kamath, M.G., Bhat, G.S., Parikh, D.V., and Mueller, D. (2005) Cotton fiber nonwovens for automotive composites. *Int. Nonwovens J.*, **14** (1), 34–40.
30. DEFRA (2006) *Recycling of Low Grade Clothing Waste*, Department for Environment, Food and Rural Affairs.
31. DEFRA (2009) *Maximising Reuse and Recycling of UK Clothing and Textiles*, Department for Environment, Food and Rural Affairs.
32. Ali, M. and Courtenay, P. (2014) Evaluating the progress of the UK's material recycling facilities: a mini review. *Waste Manage. Res.*, **32** (12), 1149–1157.
33. Hong, F., Guo, X., Zhang, S., Han, S.-F., Yang, G., and Jönsson, L.J. (2012) Bacterial cellulose production from cotton-based waste textiles: enzymatic saccharification enhanced by ionic liquid pretreatment. *Bioresour. Technol.*, **104**, 503–508.
34. Shen, F., Xiao, W., Lin, L., Yang, G., Zhang, Y., and Deng, S. (2013) Enzymatic saccharification coupling with polyester recovery from cotton-based waste textiles by phosphoric acid pretreatment. *Bioresour. Technol.*, **130**, 248–255.
35. Jeihanipour, A., Karimi, K., Niklasson, C., and Taherzadeh, M.J. (2010) A novel process for ethanol or biogas production from cellulose in blended-fibers waste textiles. *Waste Manage.*, **30** (12), 2504–2509.
36. Jeihanipour, A. and Taherzadeh, M.J. (2009) Ethanol production from cotton-based waste textiles. *Bioresour. Technol.*, **100** (2), 1007–1010.
37. Zou, Y., Reddy, N., and Yang, Y. (2011) Reusing polyester/cotton blend fabrics for composites. *Composites Part B*, **42** (4), 763–770.
38. DeVallance, D.B., Gray, J., and Lentz, H. (2012) Properties of wood/recycled textile composite panels. *Wood Fiber Sci.*, **44** (3), 310–318.
39. Ahmed, E.M., Sahari, B., and Pedersen, P. (1999) Non-linear behaviour of unidirectional filament wound COTFRP, CFRP, and GFRP composites. Proceedings of World Engineering Congress, WEC 99, Mechanical and Manufacturing Engineering, Kuala Lumpur, pp. 537–543.
40. Khalid, A.A., Sahari, B., and Khalid, Y.A. (1998) Environmental effects on the progressive crushing of cotton and glass fibre/epoxy composite cones. Proceedings of the Fourth International Conference on Advances in Materials and Processing Technologies, 98, Kuala Lumpur, pp. 680–689.
41. Chuayjuljit, S., Su-uthai, S., and Charuchinda, S. (2010) Poly(vinyl chloride) film filled with microcrystalline cellulose prepared from cotton fabric waste: properties and biodegradability study. *Waste Manage. Res.*, **28**, 109–117.
42. Bairagi, N. (2014) Recycling of textiles in India. *J. Text. Sci. Eng.*, **S2**, 003.
43. Kalliala, E.M. and Nousiainen, P. (1999) Life cycle assessment environmental profile of cotton and polyester-cotton fabrics. *Autex Res. J.*, **1** (1), 8–20.
44. Firgo, H., Eichinger, D., and Eibl, M. (1997) Process for the production of a cellulose moulded body. US Patent 5,601,767.
45. Haule, L.V., Rigout, M., Carr, C.M., and Jones, C.C. (2012) Surface and bulk chemical analysis of the durability of an easy care finish on cotton. *Cellulose*, **19** (3), 1023–1030.
46. Woodings, C.R. (1995) The development of advanced cellulosic fibres. *Int. J. Biol. Macromol.*, **17** (6), 305–309.
47. Chanzy, H. (1982) Cellulose amine oxide systems. *Carbohydr. Polym.*, **2**, 229–231.
48. Chanzy, H., Paillet, M., and Hagege, R. (1990) Spinning of cellulose from N-methyl morpholine N-oxide in the presence of additives. *Polymer*, **31** (3), 400–405.

49. Kim, D.B., Pak, J.J., Jo, S.M., and Lee, W.S. (2005) Dry jet-wet spinning of cellulose/N-methylmorpholine N-oxide hydrate solutions and physical properties of lyocell fibers. *Text. Res. J.*, **75** (4), 331–341.
50. Huong, M.B., Anelise, E., and Thomas, B. (2008) Pilling in man-made cellulosic fabrics, part 1: assessment of pilling formation methods. *J. Appl. Polym. Sci.*, **110** (1), 531–538.
51. Zhang, W., Okubayashi, S., and Bechtold, T. (2005) Fibrillation tendency of cellulosic fiberse—part 4. Effects of alkali pretreatment of various cellulosic fibers. *Carbohydr. Polym.*, **61** (4), 427–433.
52. Bates, I., Maudru, E., Phillips, D.A.S., Renfrew, A.H.M., Su, Y., and Xu, J. (2004) Cross-linking agents for the protection of lyocell against fibrillation: synthesis, application and technical assessment of 2,4-diacrylamidobenzenesulphonic acids. *Color. Technol.*, **120** (6), 293–300.
53. Bates, I., Ibbett, R., Reisel, R., and Renfrew, A.H.M. (2008) Protection of lyocell fibres against fibrillation; influence of dyeing with bis-monochloro-s-triazinyl reactive dyes. *Color. Technol.*, **124** (4), 254–258.
54. Goswami, P., Blackburn, R.S., Taylor, J., Westland, S., and White, P. (2007) Dyeing behaviour of lyocell fabric: effect of fibrillation. *Color. Technol.*, **123** (6), 387–393.
55. Farahbakhsh, N., Venditti, R.A., and Jur, J.S. (2014) Mechanical and thermal investigation of thermoplastic nanocomposite films fabricated using micro- and nano-sized fillers from recycled cotton T-shirts. *Cellulose*, **21**, 2743–2755.
56. Barbier, J. *et al.* (2011) Use of recycled cotton for producing ethanol and production method. US Patent 0236945 A1.
57. Alwood, J., Laursen, S.E., Rodríguez, C.M., and Bocken, N.M.P. (2006) *Well Dressed? The Present and Future Sustainability of Clothing and Textiles in the United Kingdom*, University of Cambridge, Institute for Manufacturing, London.
58. Sen, T. and Reddy, H.N.J. (2011) Application of sisal, bamboo, coir and jute natural composites in structural upgradation. *Int. J. Innovation Manage. Technol.*, **2**, 186–191.
59. Debnath, M., Pandey, M., Sharma, R., Thakur, G.S., and Lal, P. (2010) Biotechnological intervention of *Agave sisalana*: a unique fiber yielding plant with medicinal property. *J. Med. Plants Res.*, **43**, 177–187.
60. Wambua, P., Ivens, J., and Verpoest, I. (2003) Natural fibres: can they replace glass in fibre reinforced plastics? *Compos. Sci. Technol.*, **63**, 1259–1264.
61. Fung, K.L., Xing, X.S., Li, R.K.Y., Tjong, S.C., and Mai, Y.-W. (2003) An investigation on the processing of sisal fibre reinforced polypropylene composites. *Compos. Sci. Technol.*, **63**, 1255–1258.
62. Silva, O.R.R. and Beltrão, N.E.M. (1999) *O Agronegócio do sisal no Brasil*, Embrapa-CNPA, Brasília.
63. Botura, M.B., dos Santos, J.D.G., da Silva, G.D., de Lima, H.G., de Oliveira, J.V.A., de Almeida, M.A.O., Batatinha, M.J.M., and Branco, A. (2013) *In vitro* ovicidal and larvicidal activity of *Agave sisalana* Perr. (Sisal) on gastrointestinal nematodes of goats. *Vet. Parasitol.*, **192**, 211–217.
64. Bourmaud, A. and Baley, C. (2009) Rigidity analysis of polypropylene/vegetal fibre composites after recycling. *Polym. Degrad. Stab.*, **94**, 297–305.
65. Bourmaud, A. and Baley, C. (2007) Investigations on the recycling of hemp and sisal fibre reinforced polypropylene composites. *Polym. Degrad. Stab.*, **92**, 1034–1045.
66. Rao, S., Bhattacharyya, D., Jayaraman, K., and Fernyhough, A. (2009) Manufacturing and recycling of sisal-polypropylene composites. *Polym. Polym. Compos.*, **17**, 467–479.
67. Fávaro, S.L., Ganzerli, T.A., de Carvalho Neto, A.G.V., da Silva, O.R.R.F., and Radovanovic, E. (2010) Chemical, morphological and mechanical analysis of sisal fiber-reinforced recycled high-density polyethylene composites. *eXPRESS Polym. Lett.*, **4**, 465–473.

68. Maran, J.P. and Priya, B. (2015) Ultrasound-assisted extraction of pectin from sisal waste. *Carbohydr. Polym.*, **115**, 732–738.
69. Advertising Hazardous Waste Disposal – Asbestos, Recycle for Dorset, 2015. https://www2.barnsley.gov.uk/services/environment-and-planning/pollution/asbestos
70. Plescia, P., Gizzi, D., Benedetti, S., Camilucci, L., Fanizza, C., De Simone, P., and Paglietti, F. (2003) Mechanochemical treatment to recycling asbestos-containing waste. *Waste Manage. (Oxford)*, **23**, 209–218.
71. Foltz, A.D. and Sanders, M.L. (1999) Destruction of asbestos containing material and PCB's into non-regulated recyclables. WM'99 Conference.
72. Fuller, J. (2010), Fiberglass and Asbestos, Is insulation dangerous?.
73. Taylor, S. (1994) Asbestos removal system proves successful on 27-mile gas line. *Pipe Line Ind.*, **77**, 7.
74. The Asbestos Conversion Process, Regency International Group, Tecnosupplies. http://www.regency-uk.com/html/asbestos.html
75. The Ultimate Solution for Asbestos Disposal, HAZ Asbestos, Tetronics International. http://tetronics.com/our-solutions/hazardous-material-treatment/asbestos/
76. Smith, R.A. and Cracon, B.E. (1995) Polyamide 66 and 6 chemical recycling. Proceedings of the Recycle'95, Davos, Switzerland.
77. McKinney, R.J. and E. I. DuPont de Nemours (1995) Lewis acid catalyzed ammonolysis of nylon. US Patent 5,395,974.
78. Sifniades, S., Levy, A.B., and AlliedSignal Inc. (1995) Monomer recovery from multi-component materials. US Patent 5,457,197.
79. Peterson W.R., Production of polymer. US Patent 2,364,387 (to Du pont) Dec. 5, 1944; *Chem. Abstr.*, **39** (1945) 8403; 42658.
80. Dmitrieva, L.A. *et al.* (1975) Catalytic depolyamidation of cyclic oligomers of epsilon-Caprolactam. *Khim. Volokna*, **17**, 34.
81. Datye, K.V. (1991) Recycling processes and products in nylon 6 fibre industry. *Indian J. Fiber Text. Res.*, **16**, 46–51.
82. Kohan, M.I. (1995) Commercial nylon plastics, in *Nylon Plastics Handbook* (ed. M.I. Kohan), Hanser Gardner Publications, New York, p. 6.
83. Subramanian, P.M. (1995) Recovery of polyamide using a solution process. US Patent 5,430,068.
84. Papaspyrides, C.D. and Kartalis, C.N. (2000) A model study for the recovery of polymides using the dissolution/reprecipitation technique. *Polym. Eng. Sci.*, **40** (4), 979.
85. Eriksson, P.A., Albertsson, A.C., Boydell, P., Eriksson, K., and Manson, J.A.E. (1996) Reprocessing of fiberglass reinforced polyamide 66: influence on short term properties. *Polym. Compos.*, **17** (6), 823.
86. Claverie, A.L., Valdez, J.O., Hernandez, E.G., Zizumbo, A., Castillo, A.A., and Castano, V.M. (2002) Reprocessing effects on the properties of a hybrid Nylon 6,6-composite reinforced with short glass and carbon fibers. *J. Reinf. Plast. Compos.*, **21**, 847.
87. Moran, E.F. (1993) Nylon component reclamation. US Patent 5,266,694.
88. Sifniades, S., Levy, A.B., and Hendrk, J.A.J. (1997) Process for depolymerizing nylon-containing waste to form caprolactam. US Patent 5,681,952.
89. Datta, R.K., Polk, M.B., and Kumar, S. (1995) Reactive compatibilization of polypropylene and nylon. *Polym. Plast. Technol. Eng.*, **34** (4), 551.
90. Leichtnam, J.N., Schwartz, D., and Gadiou, R. (2000) The behaviour of fuel-nitrogen during fast pyrolysis of polyamide at high temperature. *J. Anal. Appl. Pyrolysis*, **55**, 255.
91. Mihut, C., Captain, D.K., Maria, F.G., and Amifudis, M.D. (2001) Review: recycling of nylon from carpet waste. *Polym. Eng. Sci.*, **41** (9), 1457.
92. Maier, S., Loontjens, T., Scholtens, B., and Mulhaupt, R. (2003) Carbonylbiscaprolactam: a versatile reagent for organic synthesis and isocyanate-free urethane chemistry. *Angew. Chem. Int. Ed.*, **42**, 5094.

93. Maier, S., Loontjens, T., Scholtens, B., and Mulhaupt, R. (2003) Isocyanate-free route to caprolactam-blocked oligomeric isocyanates via carbonylbiscaprolactam- (CBC-) mediated end group conversion. *Macromolecules*, **36**, 4727.
94. Zahr, G.E. (2003) Polyamide chain extension process and related polyamide product. US Patent 6,504,004.
95. Nery, L., Lefebvre, H., and Fradet, A. (2003) Kinetic and mechanistic studies of carboxylic acid–bisoxazoline chain-coupling reactions. *Macromol. Chem. Phys.*, **204**, 1755.
96. Scaffaro, R., Mantia, F.P.L., Canfora, L., Polacco, G., Filippi, S., and Magagnini, P. (2003) Reactive compatibilization of PA6/LDPE blends with an ethylene–acrylic acid copolymer and a low molar mass bis-oxazoline. *Polymer*, **44**, 6951.
97. Jeziorska, R. (2005) Studies on reactive compatibilisation of polyamide 6/poly(butylene terephthalate) blends by low molecular weight bis-oxazoline. *Polym. Degrad. Stab.*, **90**, 224.
98. Aharoni, S.M., Hammond, W.B., Xzobota, J.S., and Masilamani, D. (1984) Reactions in the presence of organic phosphites, I: High temperature amidation in the absence of solvents. *J. Polym. Sci., Part A: Polym. Chem.*, **22** (10), 2567.
99. Perry, E. and Savory, J. (1967) Modification of nylon 66 with diisocyanates and diacid chlorides. I. Chemical reaction. *J. Appl. Polym. Sci.*, **11**, 2473.
100. Ding, X., Rong, J., Xu, R., Yu, D., and Yang, Z. (2004) Effect of a chain extender on the crystallization behavior of nylon 66. *J. Appl. Polym. Sci.*, **91**, 68.
101. Villalobos, M., Awojulu, A., Greeley, T., Turco, G., and Deeter, G. (2006) Oligomeric chain extenders for economic reprocessing and recycling of condensation plastics. *Energy*, **31** (15), 3227.
102. Hong, J.H., Choi, C.W., Ramasundaram, S., Prabu, A.A., Lee, J.S., Kim, K.J., Yang, J.H., and Lee, D.J. (2008) Studies on the recycling of glycolyzed nylon 66 using novel chain extenders. *Polym. Degrad. Stab.*, **93**, 392–400.
103. Reisch, M.S. (1997) Moving toward proof positive: cultivating third-party verification of industry's environmental performance, *Chem. Eng. News*, **75** (15), 21–23.
104. Kotlair, A.M., Fountain, D.P., and Georgia Tech. (1997) Synthetic wood from waste fibrous materials. US Patent 5,626,939.
105. Hagewood, J. (2004) Production of polymeric nanofibers. *Int. Fiber J.*, **19** (1): 48–50.
106. Young, D.C., Chlystek, S.J., Malloy, R., Rios, I., and Lear Corporation (1999) Recycling of carpet scrap. US Patent 5,859,071.
107. David, D.J., Dickerson, J.L., and Sincock, T.F. (1994) Thermoplastic composition and method for producing thermoplastic composition by melt blending carpet. US Patent 5,294,384.
108. Hagberg, C.G. and Dickerson, J.L. (1997) Recycling nylon carpet via reactive extrusion. *Plast. Eng.*, **53**, 41–43.
109. Subramanian, P.M. and E. I. DuPont de Nemours and Company (1995) Recovery of polyamide using a solution process. US Patent 5,430,068.
110. Booij, M., Hendrix, J.A.J., Frentzen, Y.H., and Beckers, N.M.H., and DSM N.V. (1998) Process for recovering and recycling polyamide from carpet waste. US Patent 5,840,773.
111. Stefandl, R.E. (1998) Carpet recycling process for nylon containing carpeting. WO Patent 98/35998.
112. Sikorski, M.E. and Georgia Tech Research Corporation (1993) Recycling of polymeric materials from carpets and other multi-component structures by means of supercritical fluid extraction. US Patent 5,233,021.
113. Griffith, A.T., Park, Y., and Roberts, C.B. (1999) Separation and recovery of nylon from carpet waste using a supercritical fluid antisolvent technique. *Polym. Plast. Technol. Eng.*, **38** (3), 411.
114. Dagli, S.S., Xanthos, M., and Biesenberger, J.A. (1992) Effects of reactive extrusion variables on blend

characteristics, in *Emerging Technologies in Plastics Recycling*, American Chemical Society, (eds. G.D. Andrews and P.M. Subramania), pp. 240–257.
115. David, D.J., Dickerson, J.L., and Sincock, T.F. (1997) Thermoplastic composition and method for producing thermoplastic composition by melt blending carpet. US Patent 5,591,802.
116. Lemieux, P., Stewart, E., Realff, M., and Mulholland, J.A. (2004) Emissions study of co-firing waste carpet in a rotary kiln. *J. Environ. Manage.*, **70**, 27–33.
117. Zhang, Y., Muzzy, J.D., and Kumar, S. (1999) Recycling carpet waste by injection and compression molding. *Polym. Plast. Technol. Eng.*, **38** (3), 485–498.
118. Bockhorn, H., Donner, S., Gernsbeck, M., Hornung, A., and Hornung, U. (2001) Pyrolysis of polyamide 6 under catalytic conditions and its application to reutilization of carpets. *J. Anal. Appl. Pyrolysis*, **58–59**, 79–94.
119. Wang, Y. (2006) in *Recycling in Textiles* (ed. Y. Wang), Woodhead Publishing, Cambridge, pp. 213–224.
120. Ucar, M. and Wang, Y. (2011) Utilization of recycled post consumer carpet waste fibers as reinforcement in lightweight cementitious composites. *Int. J. Clothing Sci. Technol.*, **23**, 242–248.
121. Wang, Y., Zureick, A.H., Cho, B.S., and Scott, D. (1994) Properties of fibre reinforced concrete using recycled fibres from carpet industrial waste. *J. Mater. Sci.*, **29**, 4191–4199.
122. Miraftab, M. and Lickfold, A. (2008) Utilization of carpet waste in reinforcement of substandard soils. *J. Ind. Text.*, **38**, 167–174.
123. Sotayo, A., Green, S., and Turvey, G. (2015) Carpet recycling: a review of recycled carpets for structural composites. *Environ. Technol. Innovation*, **3**, 97–107.
124. Incornato, L., Scarfato, P., Di Maio, L., and Acierno, D. (2000) Structure and rheology of recycled PET modified by reactive extrusion. *Polymer*, **41**, 6825–6831.
125. Pacia (2002) Pacia National Plastics Recycling Survey 2002, http://www.pacia.org.au (accessed 13 May 2016).
126. Miller, C. (2002) Polyethylene terephthalate. *Waste Age*, **33** (5), 102–106.
127. Al-Sabagh, A.M., Yehia, F.Z., Eshaq, G., Rabie, A.M., and ElMetwally, A.E. (2016) Greener routes for recycling of polyethylene terephthalate. *Egypt. J. Pet.*, **25** (1), 53–64.
128. Babinchak, S. (1991) Recycling must work. *Chem. Tech.*, **21** (12), 728.
129. Chen, C.H., Chen, C.Y., Lo, Y.W., Mao, C.F., and Liao, W.T. (2001) Studies of glycolysis of poly (ethylene terephthalate) recycled from postconsumer soft-drink bottles. I. Influences of glycolysis conditions. *J. Appl. Polym. Sci.*, **80** (7), 943–948.
130. Kosmidis, V.A., Achilias, D.S., and Karayannidis, G.P. (2001) Poly (ethylene terephthalate) recycling and recovery of pure terephthalic acid. Kinetics of a phase transfer catalyzed alkaline hydrolysis. *Macromol. Mater. Eng.*, **286** (10), 640–647.
131. Yamaye, M., Hashime, T., and Yamamoto, K. (2002) Chemical recycling of poly (ethylene terephthalate). 2. Preparation of terephthalohydroxamic acid and terephthalohydrazide. *Ind. Eng. Chem. Res.*, **41** (16), 3993–3998.
132. Goje, A.S. and Mishra, S. (2003) Chemical kinetics, simulation, and thermodynamics of glycolytic depolymerization of poly (ethylene terephthalate) waste with catalyst optimization for recycling of value added monomeric products. *Macromol. Mater. Eng.*, **288**, 326–336.
133. Carta, D., Cao, G., and D'Angeli, C. (2003) Chemical recycling of poly (ethylene terephthalate) (PET) by hydrolysis and glycolysis. *Environ. Sci. Pollut. Res.*, **10** (6), 390–394.
134. Pardal, F. and Tersac, G. (2007) Kinetics of poly (ethylene terephthalate) glycolysis by diethylene glycol. Part II: effect of temperature, catalyst and polymer morphology. *Polym. Degrad. Stab.*, **92** (4), 611–616.
135. Genta, M., Iwaya, T., Sasaki, M., and Goto, M. (2007) Supercritical methanol for polyethylene terephthalate depolymerization: observation using simulator. *Waste Manage.*, **27** (9), 1167–1177.
136. Kurokawa, H., Ohshima, M.A., Sugiyama, K., and Miura, H. (2003)

Methanolysis of polyethylene terephthalate (PET) in the presence of aluminium triisopropoxide catalyst to form dimethyl terephthalate and ethylene glycol. *Polym. Degrad. Stab.*, **79** (3), 529–533.

137. Shukla, S.R. and Harad, A.M. (2006) Aminolysis of polyethylene terephthalate waste. *Polym. Degrad. Stab.*, **91** (8), 1850–1854.

138. Park, S.H. and Kim, S.H. (2014) Poly (ethylene terephthalate) recycling for high value added textiles. *Fashion Text.*, **1**, 1.

139. Badia, J.D., Strömberg, E., Karlsson, S., and Ribes-Greus, A. (2012) The role of crystalline, mobile amorphous and rigid amorphous fractions in the performance of recycled poly (ethylene terephthalate) (PET). *Polym. Degrad. Stab.*, **97**, 98–107.

140. Kang, D.H., Auras, R., Vorst, K., and Singh, J. (2011) An exploratory model for predicting postconsumer recycled PET content in PET sheets. *Polym. Test.*, **30**, 60–68.

141. Torres, N., Robin, J., and Boutevin, B. (2000) Study of thermal and mechanical properties of virgin and recycled poly (ethylene terephthalate) before and after injection molding. *Eur. Polym. J.*, **36**, 2075–2080.

142. Raffa, P., Coltelli, M.-B., Savi, S., Bianchi, S., and Castelvetro, V. (2012) Chain extension and branching of poly(ethylene terephthalate) (PET) with di- and multifunctional epoxy or isocyanate additives: an experimental and modelling study. *React. Funct. Polym.*, **72**, 50–60.

143. Zhang, Y., Zhang, C., Li, H., Du, Z., and Li, C. (2010) Chain extension of poly (ethylene terephthalate) with bisphenol-A dicyanate. *J. Appl. Polym. Sci.*, **117**, 2003–2008.

144. Nascimento, C.R., Azuma, C., Bretas, R., Farah, M., and Dias, M.L. (2010) Chain extension reaction in solid-state polymerization of recycled PET: the influence of 2, 2′-bis-2-oxazoline and pyromellitic anhydride. *J. Appl. Polym. Sci.*, **115**, 3177–3188.

145. Duarte, I.S., Tavares, A.A., Lima, P.S., Andrade, D.L.A.C.S., Carvalho, L.H., Canedo, E.L., and Silva, S.M.L. (2016) Chain extension of virgin and recycled poly (ethylene terephthalate): effect of processing conditions and reprocessing. *Polym. Degrad. Stab.*, **124**, 26–34.

146. Shen, L., Worrell, E., and Patel, M.K. (2010) Open-loop recycling: a LCA case study of PET bottle-to-fibre recycling. *Resour. Conserv. Recycl.*, **55**, 34–52.

147. Arena, U., Mastellone, M., and Perugini, F. (2003) Life cycle assessment of a plastic packaging recycling system. *Int. J. Life Cycle Assess.*, **8** (2), 92–98.

148. Detzel, A., Giegrich, J., Krüger, M., Möhler, S., and Ostermayer, A. (2004) *Life Cycle Assessment of PETOW Systems taking into Account Secondary Products*, IFEU GmbH, Heidelberg.

149. Dimitrov, N., Kratofil Krehula, L., Pticek Sirocic, A., and Hrnjak-Murgic, Z. (2013) Analysis of recycled PET bottles products by pyrolysis-gas chromatography. *Polym. Degrad. Stab.*, **98**, 972–979.

150. Lu, M. and Kim, S.W. (2001) Unsaturated polyester resins based on recycled PET: preparation and curing behavior. *J. Appl. Polym. Sci.*, **80**, 1052–1057.

151. Wallenberger, F.T. (1994) Melt viscosity and modulus of bulk glasses and fibers: challenges for the next decade, in present state and future prospects of glass science and technology. Proceedings of the Nobert Kreidl Symposium (Triesenberg, Liechtenstein), pp. 63–78.

152. Lopez, F.A., Martin, M.A., Alguacil, F.J. *et al.* (2012) Thermolysis of fibreglass polyester composite and reutilisation of the glass fibre residue to obtain a glass ceramic material. *J. Anal. Appl. Pyrolysis*, **93**, 104–112.

153. Erden, S., Sever, K., Seki, Y. *et al.* (2010) Enhancement of the mechanical properties of glass/polyester composites via matrix modification glass/polyester composite siloxane matrix modification. *Fibers Polym.*, **11**, 732–737.

154. Kim, J.K., Song, J.H., Chung, S.T., and Kwon, T.H. (1997) Morphology and mechanical properties of injection molded articles with weld-lines. *Polym. Eng. Sci.*, **37** (1), 228–241.

155. Lindhagen, J. and Berglund, L. (1998) Microscopical damage mechanisms in glass fiber reinforced polypropylene. *J. Appl. Polym. Sci.*, **69** (7), 1319–1327.
156. Soh, S.K., Lee, D.K., Cho, Q., and Rag, Q. (1994) Low temperature pyrolysis of SMC scrap. Proceedings of 10th Annual ASM/ESD Advanced Composites Conference, Dearborn, MI, November 7–10, 1994, pp. 47–52.
157. Roy, C. and Mallette, P. (1991) *Vacuum Pyrolysis of Automotive Shredder Waste*, Institute of Gas Technology, Chicago, IL.
158. de Marco, I., Legretta, J.A., Laresgoiti, M.F., Torres, A., Cambra, J.F., Chomon, M.J., Caballero, B., and Gondra, K. (1997) Recycling of the products obtained in the pyrolysis of fibre-glass polyester SMC. *J. Chem. Technol. Biotechnol.*, **69**, 187–192.
159. Torres, A., de Marco, I., Caballero, B.M., Laresgoiti, M.F., Legarreta, J.A., Cabrero, M.A., Gonzalez, A., Chomon, M.J., and Gondra, K. (2000) Recycling by pyrolysis of thermoset composites: characteristics of the liquid and gaseous *fuels* obtained. *Fuel*, **79**, 897–902.
160. Watt, D.F. and Xu, J.Z. (1991) On the use of pyrolysed SMC as a concrete additive. Proceedings of the 7th Annual ASM/ESD Conference, pp. 615–619.
161. Hudec, P.P. (1991) Reduction of alkali reactivity and de-icing salt damage in concrete by additions of pyrolysed sheet moulding compound. Proceedings of the Advanced Composite Materials, ASM International, pp. 611–614.
162. Kamimura, A., Yamada, K., Kuratoni, T., Taguchi, Y., and Tomonaga, F. (2006) Effective depolymerization waste FRPs by treatment with DMAP and supercritical alcohol. *Chem. Lett.*, **35**, 586.
163. Suyama, K., Kubota, M., Shirai, M., and Yoshida, H. (2006) Effect of alcohols on the degradation of crosslinked unsaturated polyester in sub-critical water. *Polym. Degrad. Stab.*, **91**, 983–986.
164. Suyama, K., Kubota, M., Shirai, M., and Yoshida, H. (2007) Degradation of crosslinked unsaturated polyesters in sub-critical water. *Polym. Degrad. Stab.*, **92**, 317–322.
165. Oliveux, G., Bailleul, J.-L., and La Salle, E.L.G. (2012) Chemical recycling of glass fibre reinforced composites using subcritical water. *Composites Part A*, **43**, 1809–1818.
166. Asokan, P., Osmani, M., and Price, A.D.F. (2010) Improvement of the mechanical properties of glass fibre reinforced plastic waste powder filled concrete. *Constr. Build. Mater.*, **24**, 448–460.
167. Conroy, A. and Halliwell, S. (2006) Tim reynolds composite recycling in the construction industry. *Composites Part A*, **37**, 1216–1222.
168. Shah, S.P. (1992) *Fibre-Reinforced Cement Composites* (eds P.N. Balaguru and S.P. Shah), McGraw-Hill, p. 110.
169. Kwolek, S.L. and Yang, H.H. (1993) *Manmade Fibers: Their Origin and Development* (eds R.B. Seymour and R.S. Porter), Elsevier Applied Science Publishers Ltd., Barking, pp. 315–336.
170. Zheng, Q. and Chung, D.D.L. (1989) Carbon fiber reinforced cement composites improved by using chemical agents. *Cem. Concr. Res.*, **19**, 25–41.
171. Akihama, H., Suenaga, T., and Nakagawa, H. (1988) Application of three-dimensional fabric reinforced concrete to building panels. *Concrete*, **10** (l), 40–47.
172. Pickering, S.J. (2006) Recycling technologies for thermoset composite materials- current status. *Composites Part A*, **37**, 1206–1215.
173. Kouparitsas, C.E., Kartalis, C.N., Vareilidis, P.C., and Papaspyrides, C.D. (2002) Recycling of the fibrous fraction of reinforced thermoset composites. *Polym. Compos.*, **23**, 682–684.
174. Scheirs, J. (1998) *Polymer Recycling Science Technology and Applications*, John Wiley & Sons Ltd., London.
175. Curcuras, C.N., Flax, A.M., Graham, W.D., and Hartt, G.N. (1991) Recycling of Thermoset Automotive Components. SAE Technical Paper Series 910387, p. 16.
176. Nahil, M.A. and Williams, P.T. (2011) Recycling of carbon fibre reinforced polymeric waste for the production of activated carbon fibres. *J. Anal. Appl. Pyrolysis*, **91**, 67–75.

177. Yip, H.L.H., Pickering, S.J., and Rudd, C.D. (2002) Characterisation of carbon fibres recycled from scrap composites using fluidised bed process. *Plast. Rubber Compos.*, **31** (6), 278–282.
178. Cunliffe, A.M., Jones, N., and Williams, P.T. (2003) Recycling of fibre-reinforced polymeric waste by pyrolysis: thermogravimetric and bench-scale investigations. *J. Anal. Appl. Pyrolysis*, **70**, 315–338.
179. Braun, D., Gentzkow, W., and Rudolf, A.P. (2001) Hydrogenolytic degradation of thermosets. *Polym. Degrad. Stab.*, **74**, 25–32.
180. Liu, Y.Y., Meng, L.H., Huang, Y.D., and Du, J.J. (2004) Recycling of carbon/epoxy composites. *J. Appl. Polym. Sci.*, **94**, 1912–1916.
181. Dang, W.R., Kubouchi, M., Sembokuya, H., and Tsuda, K. (2005) Chemical recycling of glass fiber reinforced epoxy resin cured with amine using nitric acid. *Polymer*, **46**, 1905–1912.
182. Jiang, G., Pickering, S.J., Lester, E.H., Turner, T.A., Wong, K.H., and Warrior, N.A. (2009) Characterisation of carbon fibres recycled from carbon fibre/epoxy resin composites using supercritical *n*-propanol. *Compos. Sci. Technol.*, **69**, 192–198.
183. Liu, Y.Y., Shan, G.H., and Meng, L.H. (2009) Recycling of carbon fibre reinforced composites using water in subcritical conditions. *Mater. Sci. Eng., A*, **520**, 179–183.
184. Bai, Y.P., Wang, Z., and Feng, L.Q. (2010) Chemical recycling of carbon fibers reinforced epoxy resin composites in oxygen in supercritical water. *Mater. Des.*, **31**, 999–1002.
185. Fromenteil, C., Bardelle, P., and Cansell, F. (2000) Hydrolysis and oxidation of an epoxy resin in sub- and supercritical water. *Ind. Eng. Chem. Res.*, **39**, 922–925.
186. Goto, M. (2009) Chemical recycling of plastics using sub- and supercritical fluids. *J. Supercrit. Fluids*, **47**, 500–507.
187. Hartt, G.N. and Carey, D.P. (1992) *Economics of Recycling Thermosets*, SAE Technical Paper 920802, Society of Automobile Engineers, Warrendale, PA.
188. Pimenta, S. and Pinho, S.T. (2011) Recycling carbon fibre reinforced polymers for structural applications: technology review and market outlook. *Waste Manage. (Oxford)*, **31**, 378–392.
189. Unser, J.F., Staley, T., and Larsen, D. (1996) Advanced composites recycling. *SAMPE J.*, **32** (5), 52–57.
190. Pickering, S.J. and Benson, M. (1993) Recovery of materials and energy from thermosetting plastics, in *Sixth European Composite Materials Conferenced, Recycling Concepts and Procedures. Bordeaux, France, September*, Published by European Association for Composite Materials, Bordeaux, pp. 41–46.
191. Pickering, S.J., Kelly, R.M., Kennerley, J.R., and Rudd, C.D. (2000) A fluidised bed process for the recovery of glass fibres from scrap thermoset composites. *Compos. Sci. Technol.*, **60**, 509–523.
192. Kennerley, J.R., Fenwick, N.J., Pickering, S.J., and Rudd, C.D. (1997) The properties of glass fibres recycled from the thermal processing of scrap thermoset composites. *J. Vinyl Add. Tech.*, **3**, 58–63.
193. Jiang, G., Pickering, S.J., Walker, G.S., Wong, K.H., and Rudd, C.D. (2008) Surface characterisation of carbon fibre recycled using fluidised bed. *Appl. Surf. Sci.*, **254**, 2588–2593.
194. Zheng, Y.H., Shen, Z.G., Ma, S.L., Cai, C.J., Zhao, X.H., and Xing, Y.S. (2009) A novel approach to recycling of glass fibres from nonmetal materials of waste printed circuit boards. *J. Hazard. Mater.*, **170**, 978–982.
195. Eckert, C.A., Knutson, B.L., and Debenedetti, P.G. (1996) Supercritical fluids as solvents for chemical and materials processing. *Nature*, **383**, 313–318.
196. Wu, D., Fu, R., Zhang, S., Dresselhaus, M.S., and Dresselhaus, G. (2004) Preparation of low-density carbon aerogels by ambient pressure drying. *Carbon*, **42**, 2033–2039.
197. Goto, T. (2004) Scission of cross-linking element of cross-linked polymer using supercritical alcohol. *Soc. Rubber Ind. Jpn.*, **77** (10), 359–364.

198. Goto, T. (2003) Recycling of silane cross-linked polyethylene for insulation of cables by supercritical alcohol. Proceeding of the International Conference on Properties and Applications of Dielectric Materials, pp.1218–1221.
199. Licence, P., Ke, J., Sokolova, M., Ross, S.K., and Poliakoff, M. (2003) Chemical reactions in supercritical carbon dioxide: from laboratory to commercial plant. *Green Chem.*, **5**, 99.
200. Tiltscher, H. and Hofmann, H. (1987) Trends in high pressure chemical reaction engineering. *Chem. Eng. Sci.*, **42** (5), 959–977.
201. Hyde, J.R., Lester, E., Kingman, S., Pickering, S., and Wong, K.H. (2006) Supercritical propanol, a possible route to composite carbon fibre recovery: a viability study. *Composites Part A*, **37**, 2171–2175.
202. Marsh, G. (2009) Carbon recycling: a soluble problem. *Reinf. Plast.*, **53**, 22–23 and 25–27.
203. Warrior, N.A., Turner, T.A., and Pickering, S.J. (2009) AFRECAR and HIRECAR Project results. Carbon Fibre Recycling and Reuse 2009 Conference, IntertechPira, Hamburg, Germany.
204. Piñero-Hernanz, R., Dodds, C., Hyde, J., García-Serna, J., Poliakoff, M., Lester, E., Cocero, M.J., Kingman, S., Pickering, S., and Wong, K.H. (2008) Chemical recycling of carbon fibre reinforced composites in near critical and supercritical water. *Composites Part A*, **39**, 454–461.
205. Piñero-Hernanz, R., García-Serna, J., Dodds, C., Hydec, J., Poliakoff, M., Cocero, M.J., Kingman, S., Pickering, S., and Lester, E. (2008) Chemical recycling of carbon fibre composites using alcohols under subcritical and supercritical conditions. *J. Supercrit. Fluids*, **46**, 83–92.
206. Palmer, J., Savage, L., Ghita, O.R., and Evans, K.E. (2010) Sheet moulding compound (SMC) from carbon fibre recyclates. *Composites Part A*, **41**, 1232–1237.

7
Recycling of Polymer Blends and Composites (Epoxy Blends)

Jyothi V. Sunny

7.1
Introduction

Polymers are highly versatile materials with wide range of applications. Because of their wide range of properties, polymers are essential and have a unique role in everyday life. Mainly there are two classes of polymers: natural and synthetic (or manmade). Naturally occurring polymers are proteins, cellulose, latex, and so on, whereas synthetic polymers include polybutadiene, epoxy resin, polyester, and so on. Polymers are again classified as thermoplastics and thermosets, based on the their behaviors when exposed to heat. Thermoplastic materials can be heated, softened and reshaped, or recycled in to new articles. But thermosets, upon heating, become soft and cannot be reshaped into new articles. A variety of polymerization techniques can be employed to create polymeric materials with distinctive properties. In addition, there are several ways to combine the unique properties by blending two or more polymers.

7.2
Polymer Blends and Composites

Composites are materials comprised of two or more components with distinct performance properties compared to those of their individual components. One can easily identify the components of a composite because the components will not melt or dissolve in the composite. Composites can be designed in to different shapes and sizes depending on the application. The resulting materials can be lightweight and, at the same time, very tough and strong. Yang Xiang *et al.* [1] have provided a detailed description of the various classes of composites materials, as shown in Figure 7.1. As can be seen, there can be various groups of composite materials, which can be tailored for specific applications. The focus of this chapter will be mainly thermoset-based composites, their manufacturing, methods of recycling, characterization, and application of the recyclates.

Recycling of Polymers: Methods, Characterization and Applications, First Edition. Edited by Raju Francis.

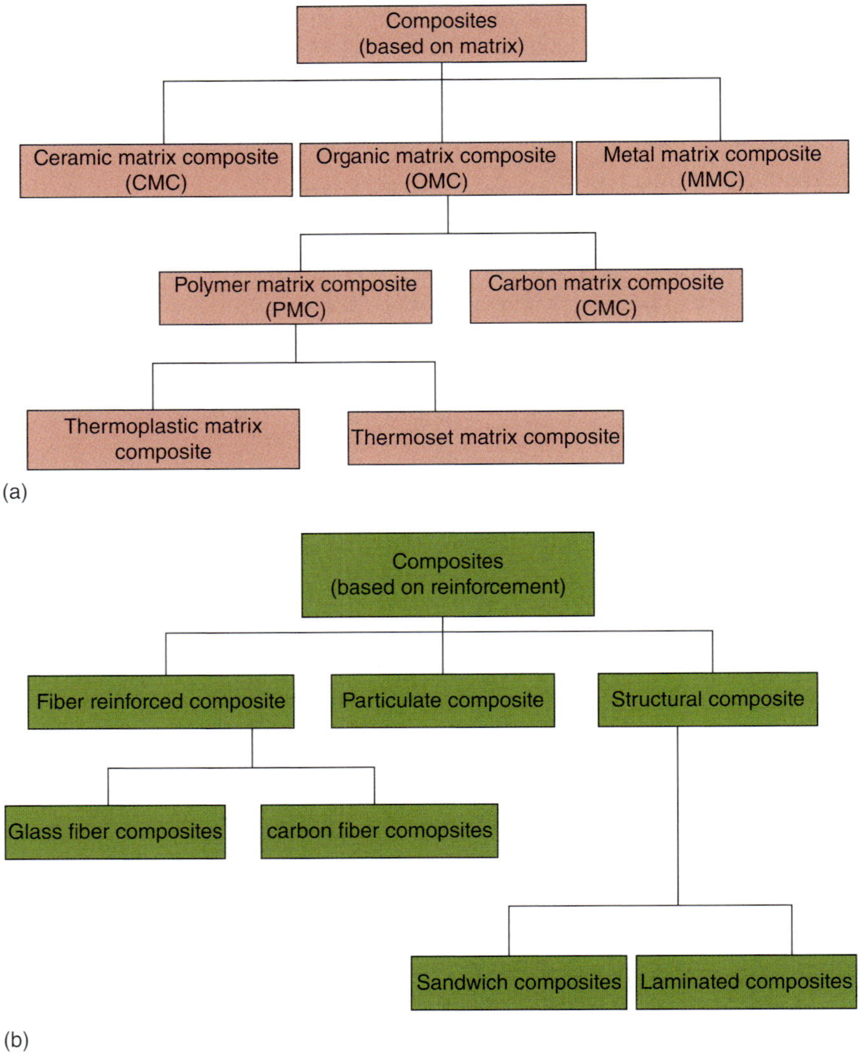

Figure 7.1 Classification of composite materials (a) based on matrix materials and (b) based on reinforcement materials.

Composite and blends are inevitable in the construction, OEM (original equipment manufacturer), and transportation industries. Composites are widely used for manufacturing wind turbine blades, medical devices, safety equipment, marine, aerospace, and automotive parts, and so on. Figures 7.2–7.5 show some of these applications of thermoset composites.

Polymer composites are primarily composed of a thermosetting resin, which is the matrix that holds the fibrous material and fillers. The main thermoset resins used in making blends or composites are epoxy, polyester, and vinyl ester. Some

Figure 7.2 (a) Structural composites made of fiber-reinforced thermosets as excellent replacements for metal [2]. (b) Safety helmets [3]. (c) Wind turbine blades [4]. (d) Carbon fiber reinforced polymer composites are used in aircraft wings [5].

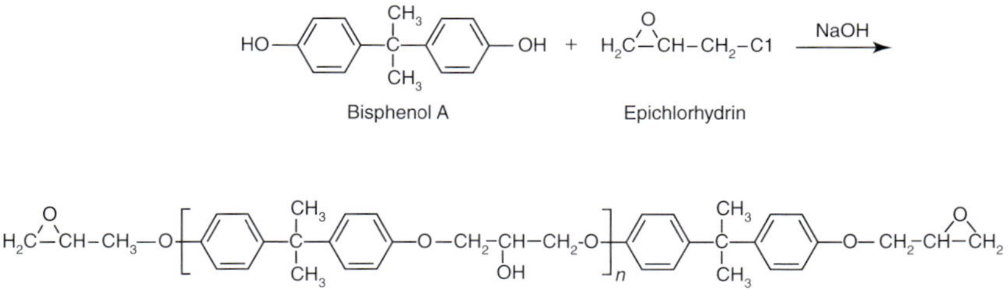

Figure 7.3 Epoxy resin formation.

of the excellent characteristics of thermosetting resins are resistance to harsh chemicals or solvents, adhesion to substrates, and high heat stability. For these reasons, products made out of thermosets can be used under extreme conditions without affecting the performance. However, when compared, epoxies have slightly better performance than polyester resin-based products.

A detailed description of the various thermosetting composites, the additives, manufacturing process, and so on, has been reported earlier [6]. Epoxy resins are used in a wide variety of applications including coatings and as adhesives because of their easy processability and curing characteristics [7, 8]. Because of their unique characteristics combined with their reasonable cost, epoxies are often well accepted for all areas of application from commodity to high-performance requirements. The key demand for epoxy resins are due to their increased usage in wind turbine blades and coatings. Once the epoxy is crosslinked, the resulting structure gains all high qualities such as heat and thermal stability, high modulus,

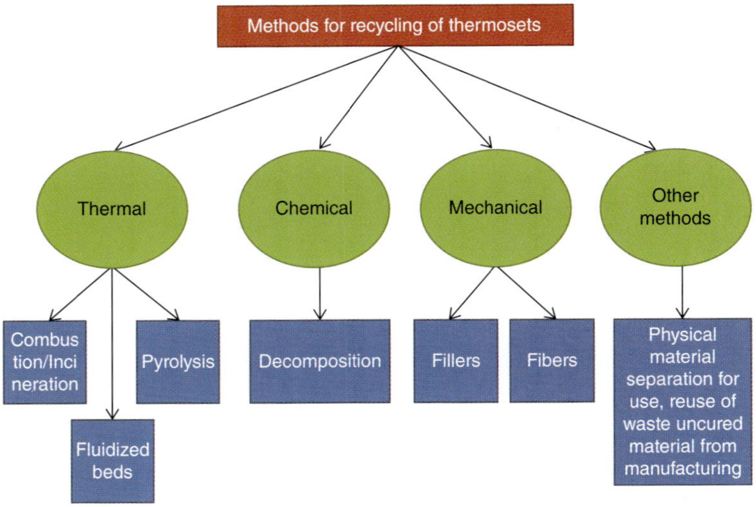

Figure 7.4 Methods of recycling taken from Pickering.

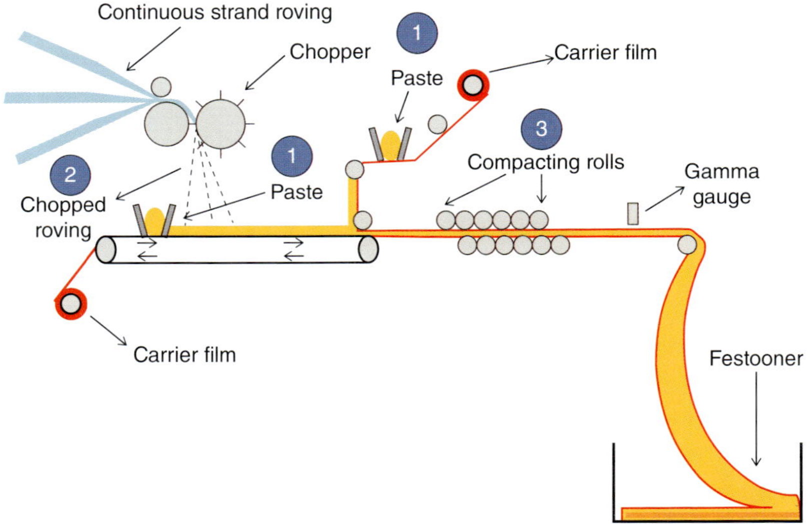

Figure 7.5 Schematic model of SMC manufacturing process. (Landis [17] http://www.designworldonline.com/reinforcement-strategies-for-high-performance-composites/#.)

low creep, and so on. Because of their hard-to-break network structure, epoxies and their blends often create environmental problems when they end in landfills. So, generally speaking, thermoset composites and blends are not recyclable. Most of the epoxy resins used around the world are derived from epichlorohydrin and bisphenol A. The formation of the diglycidyl ether of bisphenol A (DGEBA) resin is shown in Figure 7.3.

Extensive studies have been carried out over the years on the effect of various fillers [9–14] and fibers [15] on epoxy resins. So these additives can change the performance of the resulting composites significantly. Extensive use of these thermoset composites has led to the question of what would happen to these materials once they are not usable or need replacement. Environmental concerns are the main reasons for recycling, as there are growing concerns associated with landfills. Therefore, there is growing pressure to recycle thermoset-based materials. Polymer thermosets are very difficult to recycle; besides, the cost associated [16] with recycling is huge compared with the low-cost landfill option. According to Pickering, waste management in the European Union has laid out a well-established hierarchy starting from the prevention of waste generation during manufacturing, and clearly landfill is the last option.

7.2.1 *Methods of Recycling*

Several researchers have investigated the methods of recycling of thermoset resins and their blends. There are three main ways to recycle a thermoset, mainly in the case of epoxy resin-based material. Mechanical and thermal methods are the most investigated paths for recycling, whereas chemical recycling has been getting more attention recently. Methods of recycling of polymer composites are given in the following subsections.

7.2.1.1 Mechanical Recycling

There are several methods for manufacturing composites materials. Some of these techniques are hand lay-up, resin transfer molding (RTM), compression molding, sheet molding (SM), dough molding (DM), and bulk molding (BM). SM compounds (SMCs) and DM compounds (DMCs) are the major composite products widely used in large quantities for automotive applications. The main components of these two types of products are the resin, filler, and fiber. Figure 7.5 shows a typical SMC manufacturing process [17].

Studies [18] have been conducted to reuse the uncured materials from SMCs in new products. Automotive manufactures are under pressure to recover and reuse at least 95% of the weight cars by 2015 [19]. The methodology for the mechanical recycling is reducing the particle size of the material by mechanical force or simply by grinding. Mechanical breakdown is a more commercially viable technique compared to other recycling technologies. One has to keep in mind the tremendous amount of energy required for the mechanical breakdown of these composites, which often makes it a less attractive from the economic standpoint. In mechanical grinding, the initial step is breaking the material in to smaller pieces so that further breakdown into smaller particles can be carried out in specialized equipment. Inoh *et al.* [20] have given a detailed description of the various grinding mechanisms, such as compression, impact, attrition, and shearing. They also have provided the description of the different grinding machines (Table 7.1). The

Table 7.1 Various grinding equipment.

Type of grinding machine		Principle of grinding	Type
Roll mill	Roller type	Compression (between rollers)	A
High-speed revolution hammer mill	Screen classifier type	Impact (hammer and particles)	B
	Air classifier type		—
Ball mill	Rolling, vibration, epicyclic mill	Compression + impact (between balls)	C
Jet mill	Impact board type	Impact (particle board)	—
	Air impact type	Attrition (among particles)	E
Others	Mortar, stone mill	Attrition	—

reuse of the recyclates produced from DMCs and woven glass-reinforced phenolics (GWPs) using a Hammer mill was described by Bream and Hornsby [21]. The reduced-size particles were characterized for their physical properties and their behavior at elevated temperatures. In another interesting work [22], the authors claim two strategies. In the first strategy, pure mechanical grinding converts the SMCs to fine powders. In the second strategy, a mixed process is used in which first shredding and then partial dissolution of the filler in an acid solution takes place. Both processes were compared in terms of the performance of the new composites where the filler and fiber were incorporated. The most difficult part associated with polymer composite reclamation is separating the fiber from the polymer matrix. Figure 7.6 shows how tightly the filler and fibers associate with the matrix in SMCs [21].

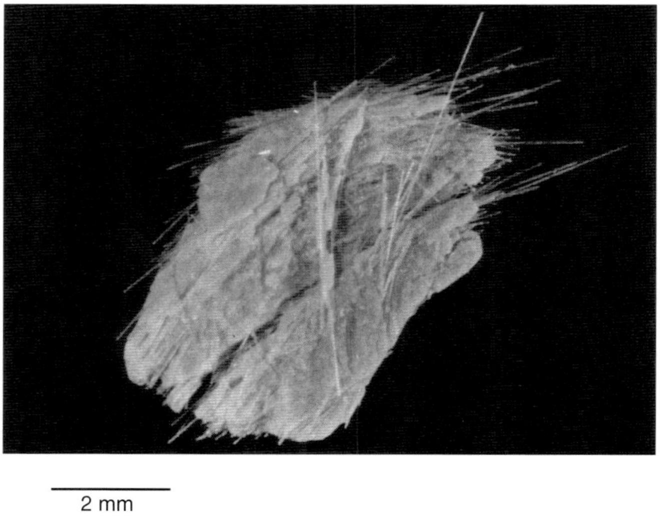

Figure 7.6 DMC recyclate fiber–particulate bundle. (Bream and Hornsby [21] reproduced with permission of Springer.)

There are no official classifications for these mixed materials. Sieve analysis can give an idea of particles and their separation. In general, these recycled materials can be classified in to coarse and fine particles, of which the coarse particles are processed again to smaller particles, whereas the fines can be used as such without any more processing.

7.2.1.2 Chemical Recycling

Recycling of thermosets is difficult due to their complex crosslink structure. Chemical recycling is a better way of breaking the structure and reversing the thermosetting resin to its original components. Chemical recycling can be used to degrade resin-based neat products and fiber-reinforced resin composites. Unfortunately, most of the thermoset materials used as composites utilize fibers, which make them more complex and difficult to recycle. Chemical recycling can therefore be categorized in to several groups based on the method and chemicals used to breakdown the structure. Figure 7.7 shows a schematic model of the different methods for chemical recycling.

The degradation of carbon-fiber-reinforced epoxy composites by ultraviolet radiation and condensation was reported by Kumar *et al.* [23]. Their observation showed the breakdown of the epoxy matrix under the severe physical and chemical environment without affecting the oriented fiber properties. When considering this factor, UV radiation can be used for recovering the fibers from a composite by matrix erosion.

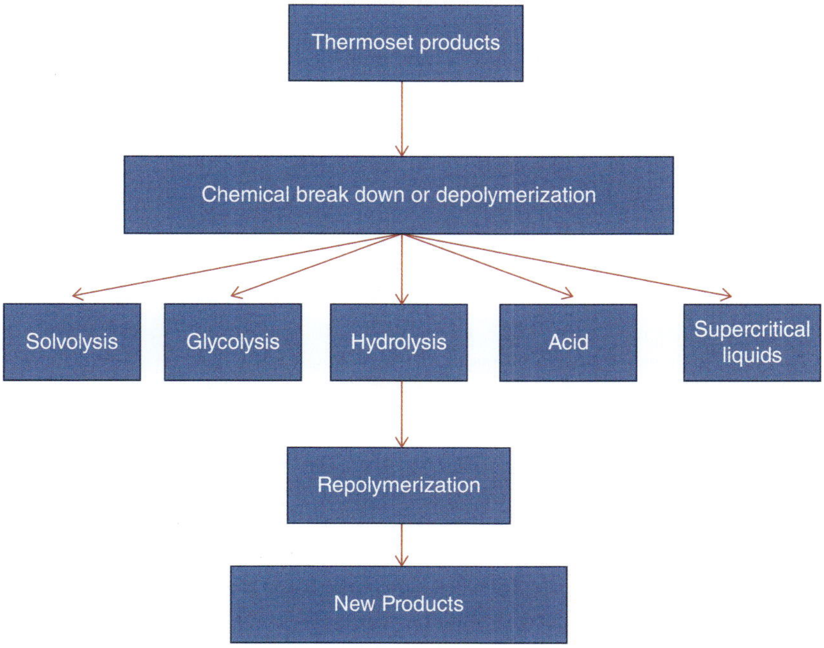

Figure 7.7 Schematic model for methods of chemical recycling.

Dang *et al.* [24, 25] have carried out extensive research on the chemical recycling of epoxy resins and their composites. They successfully decomposed the epoxy resin using nitric acid and then repolymerized the decomposed products with a curing agent. The cured epoxy resin had low resistance to concentrated nitric acid. This disadvantage of low resistance to nitric acid can be very well turned in to advantage for recycling these resins.

Solvolysis involves the breakdown of the network structure using solvents [26, 27] and heat. The solvents used for this purpose are different alcohols [28], acetone at super critical conditions, and so on. Mormann and Frank [29] demonstrated the use of supercritical ammonia for the chemical breakdown of thermosets.

Several other studies has been carried out to show the effect of acids on the cured epoxy resin and how it can imitate the breakdown or degradation of the cured structure [7, 30–32]. Decomposition of the cured epoxy system using nitric acid and microwave energy was studied by Bolasodun *et al.* [33]. Among the several methods mentioned in schematic model, glycolysis was successfully applied for recycling of polymers by several researchers [34–37]. Use of glycol to degrade unsaturated polyester was reported, as shown in Figure 7.7 [38]. The interesting fact from this study was that the curing reaction of the recycled resin was faster than the neat resin.

Other methods of chemical depolymerization have also been described [39, 40]. One approach to reuse a thermoset consists in creating molecules with cleavable [7] epoxide groups and synthesis of biodegradable thermosets [41, 42]. The use of enzymes is another option to break down thermosets, as enzymes can help in the cleavage of certain ester groups. This will be an advantage for polyester-based thermosets. It has been recently reported that researchers [43] have developed new classes of thermosets that can be hydrolyzed in to monomers and can be repolymerized to create polymers of the desired properties.

7.2.1.3 Thermal Recycling

Thermal recycling involves the application of heat. Thermal methods can be classified in to three groups, which will be discussed below.

7.2.1.3.1 Combustion or Incineration for Energy Recovery In this recycling route, no material is recovered but the composites are combusted to recover heat energy. The energy recovered is mainly from the resin part of the composite, which is usually the smaller part of the total composition compared to the filler and fiber parts. The fillers and fibers are combustible and contribute to the highest percentage of the composites. Studies have been conducted to measure the calorific values of the several thermosetting resins 70. All thermosetting resins have a certain calorific value, which is in the order of 30 000 kJ kg^{-1} with some exception. The inorganic residues resulting from the combustion can be used in the cement industry as an additive for cement [44].

7.2.1.3.2 Combustion for Fiber Recovery Using Fluidized Beds Use of fluidized beds is another thermal degradation method to recover the fibers or fillers to be reused

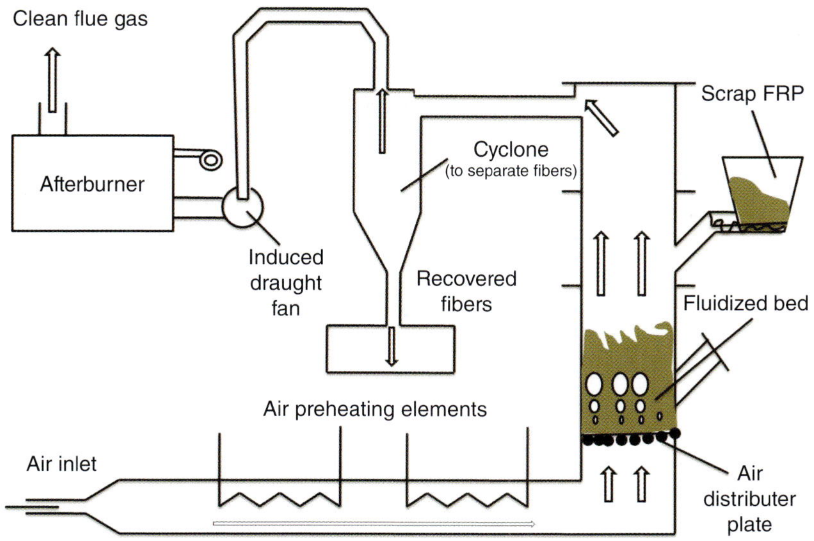

Figure 7.8 Fluidized-bed recycling process.

in other composite materials. This technology was investigated in the recovery of glass and carbon fibers as well as energy [45, 46]. The fluidized-bed recycling process is illustrated in Figure 7.8. These investigations on glass and carbon reinforced composites have led to some interesting facts, such as the effect of temperature on the recovered material, which is determined by the resin type. For polyester resins, the process temperature was 4500 °C, and for epoxy the temperature was 550 °C; there was a significant reduction in strength of the recovered fibers as the temperature was increased.

7.2.1.3.3 **Pyrolysis** In pyrolysis, the composite is heated at very high temperatures in the absence of oxygen to break it down into smaller molecules. The thermally broken down organic part of the composite such as oils and gases can be used as fuel, whereas the inorganic part of the fibers can be reused for fabricating composite materials. So pyrolysis can be effectively used to recover the matrix part as well as the fibers. This marks the main advantage of pyrolysis over other thermal processes: it breaks down the polymer network with energy recovery, which can be reused as feedstock in chemical processing. Several studies have been conducted on the application of pyrolysis for the recycling of thermoset composites as well as their characterization and possible reuse in other products [47–50].

Pyrolysis is now accepted as a standard method for recycling polymer-based products. Pyrolysis can be carried out in different types of reactors [51–53]. The resulting products from these different reactors will be different, which makes it difficult from an industrial standpoint and reusing these materials. There are several facilities for pyrolysis established around the world for various materials including tires and other thermoplastic materials. Microwave pyrolysis allows even distribution of heat and can achieve much higher temperatures within a short time. Tech-En Ltd have developed the microwave-induced pyrolysis technology [54]. So, in general, several manufactures of composite materials have already started looking into the recycling of composites materials. The first commercial recycling activity was carried out by ERCOM in Rastatt, Germany [55]. Their success in establishing a mechanical recycling method inspired several companies around the world, who have developed recycling methods for composite materials [56–59].

7.3 Characterization and Application of Recyclates

The materials – both fiber and the polymer – that are recycled by any of the above-mentioned methods need to be characterized before they can be successfully used in new formulations. As mentioned earlier, sieve analysis will be an acceptable method in particular for identifying the different sizes of the materials. The material can be sieved and weighed, and the fractions can be used to find their distribution. Both optical microscopy and scanning electron microscopy can be used to further characterize the particulates. Longer fibers [60–62] can be incorporated in to virgin thermoset composites. The majority of the recycling and reuse work has been done on carbon and armid fiber-based composites due to the cost advantage of reusing these fibers along with neat fibers [63–65]. Studies conducted on different fiber-reinforced composites have shown the recovery of fibers and their characterization using mechanical grinding. Table 7.2 shows [67] the percentage of aramid fibers recovered form aramid epoxy composites.

Studies have been conducted with different levels of recovered fiber incorporation in neat composites [17, 20]. It has been found that the properties of the composites with recycled fibers deteriorate after certain levels of the recycled fibers. One of the factors for this reduction in performance is the cleanness of the fibers that are recovered from the matrix. The presence of cured resin in the recovered fiber can lead to poor adhesion [66].

It has been found that incorporation of the recycled materials not only acts as low filler but also imparts some sort of reinforcement. The results of the studies conducted by Isabel de Marco *et al.* [49] showed the reinforcement provided by the recyclates in virgin composite materials. Table 7.3 shows a comparison of mechanical properties of the virgin composite with composites containing recyclates.

Table 7.2 Aramid fiber recovered from aramid–epoxy composites [66].

Screen size (mm)	Reground fraction obtained	%w/w (%)
>3.5	Aramid short fibers	53.2
>2	Coarse epoxy pieces and aramid fibrils	14.8
>2	Fine epoxy powder	32

Table 7.3 Mechanical properties of bulk molding compounds [50].

Property	Virgin BMC-1	BMC with recycled SMC (BMC-R)	Virgin BMC-II
Tensile strength ($MN\,m^{-2}$)	19.1 ± 4.0	24.0 ± 5.0	22.9 ± 7.0
Flexural strength ($MN\,m^{-2}$)	64.9 ± 11.0	68.9 ± 17.0	85 ± 12.0
Flexural modulus ($GN\,m^{-2}$)	10.9 ± 0.9	11.1 ± 1.0	11.5 ± 1.0
Barcol hardness	68 ± 5.0	66 ± 3.2	65 ± 5.3
Unnotched charpy impact strength ($kJ\,m^{-2}$)	16.4 ± 3.3	16.5 ± 4.8	18.5 ± 5.2

A detailed study on the mechanical recycling of automotive composites for use as reinforcement in thermoset composites was conducted by Palmer [68]. His extensive report includes the effect of different levels of recyclates in to the composites and their impact on properties. But the performance decreases when the weight percentage of the recyclates increases in the virgin composites. The addition of chemically extracted resin to the neat composition shows that at lower levels the flexural strength of the composite is higher than that of the neat composition but as the content of recyclates increases, the strength decreases dramatically [25]. This again depends of the end application of the composites. So it is necessary to carry out detailed investigation on the recycled material before incorporating it into a virgin composite with regard to the levels and the performance of the resulting material. a Another industry where the recycled powder can be used is the cement and concrete industry.

7.4 Conclusions

Recycling of thermoset composites are receiving more and more attention due to the enormous amounts of waste generated and also used composites. Because of the chemical nature of thermosets, they cannot be reprocessed to create new products. This chapter gave a detailed description of the various recycling methods for recycling thermoset composites. It also covered the properties of the recycled composites.

References

1. Yang, Y., Boom, R., Irion, B., van Heerden, D.J., Kuiper, P., and de Wit, H. (2012) Recycling of composite materials. *Chem. Eng. Process.*, **51**, 53–68.
2. SRI International Fiber Reinforced Composites to Replace Metals, http://www.sri.com/engage/products-solutions/ceramic-composites (accessed 11 May 2016).
3. SMC Products http://www.idicomposites.com/overview-products.php (accessed 11 May 2016).

4. Eco Chunk Wind Turbine Blades, http://www.ecochunk.com/tag/wind-turbine-blades (accessed 11 May 2016).
5. Kennametal Fiber Reinforced Composites, http://www.kennametal.com/hi/industry-solutions/aerospace/machining-cfrp-composites.html (accessed 11 May 2016).
6. Waldemar, S. (2000) Thermosets. *Encycl. Ind. Chem.*, **36**, 493–507.
7. Buchwalter, S.L. and Kosbar, L.L. (1996) Cleavable epoxy resins: design for disassembly of a thermoset. *J. Polym. Sci., Part A: Polym. Chem.*, **34** (2), 249–260.
8. McAdams, L.V. and Gannon, J.A. (0000) *Encyclopedia of Polymer Science and Engineering*, 2nd edn, John Wiley & Sons, Inc., New York, pp. 323–382.
9. Park, J.M., Kim, D.S., Lee, J.R., and Kim, T.W. (2003) Nondestructive damage sensitivity and reinforcing effect of carbon nanotube/epoxy composites using electro-micromechanical technique. *Mater. Sci. Eng., C*, **23**, 971.
10. Sandler, J.K.W., Kirk, J.E., Kinloch, I.A., Shaffer, M.S.P., and Windle, A.H. (2003) Ultra-low electrical percolation threshold in carbon-nanotube-epoxy composites. *Polymer*, **44**, 5893.
11. Miyagawa, H. and Drzal, L.T. (2004) Thermo-physical and impact properties of epoxy nanocomposites reinforced by single-wall carbon nanotubes. *Polymer*, **45**, 5163.
12. Jan, I.N., Lee, T.M., Chiou, K.C., and Lin, J.J. (2005) Comparisons of physical properties of intercalated and exfoliated clay/epoxy nanocomposites. *Ind. Eng. Chem. Res.*, **44**, 2086.
13. Nigam, V., Setua, D.K., Mathur, G.N., and Kar, K.K. (2004) Epoxy-montmorillonite clay nanocomposites: synthesis and characterization. *J. Appl. Polym. Sci.*, **93**, 2201.
14. Ratna, D., Manoj, N.R., Varley, R., Singh, R.R.K., and Simon, G.P. (2003) Clay-reinforced epoxy nanocomposites. *Polym. Int.*, **52**, 1403.
15. Di Landro, L. and Janszen, G. (2014) Composites with hemp reinforcement and bio-based epoxy matrix. *Composites Part B*, **67**, 220–226.
16. Pickering, S. (2006) Recycling technologies for thermoset composite materials—current status. *Composites Part A*, **37**, 1206–1215.
17. Landis, L. (2013) Reinforcement Strategies for High Performance Composites, Design World, March 4, 2013.
18. Michaeli, W. and Oelgarth, A. (1996) *International SAMPE Symposium and Exhibition*, Vol. 41, Issue: Materials and Process Challenges: Aging Systems, Affordability, Alternative Applications, Book 2, pp. 1551–1562, ISSN: 0891-0138.
19. Palmer, J., Ghita, O.R., Savage, L., and Evans, K.E. (2009) Successful closed-loop recycling of thermoset composites. *Composites Part A*, **40**, 490–498.
20. Inoh, T., Yokoi, T., and Sekiyama, K.-I. (1994) SMC recycling technology. *J. Thermoplast. Compos. Mater.*, **7** (1), 42–55.
21. Bream, C.E. and Hornsby, P.R. (2001) Comminuted thermoset recyclate as a reinforcing filler for thermoplastics, part I characterisation of recyclate feedstocks. *J. Mater. Sci.*, **36** (12), 2965–2975.
22. Perrin, D., Guillermain, C., Bergeret, A., Lopez-Cuesta, J.-M., and Tersac, G. (2006) SMC composites waste management as reinforcing fillers in polypropylene by combination of mechanical and chemical recycling processes. *J. Mater. Sci.*, **41** (12), 3593–3602.
23. Kumar, B.G., Singh, R.P., and Nakamura, T. (2002) Degradation of carbon fiber-reinforced epoxy composites by ultraviolet radiation and condensation. *J. Compos. Mater.*, **36** (24), 2713–2733.
24. Dang, W., Kubouchi, M., Sembokuya, H., and Tsuda, K. (2005) Chemical recycling of glass fiber reinforced epoxy resin cured with amine using nitric acid. *Polymer*, **46**, 1905–1912.
25. Dang, W., Kubouchi, M., Sembokuya, H., and Tsuda, K. (2002) An approach to chemical recycling of epoxy resin cured with amine using nitric acid. *Polymer*, **43**, 2953–2958.
26. Winter, H., Mostert, H.A.M., Smeets, P.J.H.M., and Paas, G. (1995) Recycling of sheetmolding compounds by chemical routes. *J. Appl. Polym. Sci.*, **57** (11), 1409–1417.

27. Patel, S.H. and Gonsalves, K.E. (1993) Alternative procedures for the recycling of sheet molding compounds. *Adv. Polym. Tech.*, **12** (1), 35–45.
28. Pinero-Hernanz, R., Garcıa-Serna, J., Dodds, C., Hyde, J., Poliakoff, M., Cocero, M.J., Kingman, S., Pickering, S., and Lester, E. (2008) Chemical recycling of carbon fibre composites using alcohols under subcritical and supercritical conditions. *J. Supercrit. Fluids*, **46**, 83–92.
29. Mormann, W. and Frank, P. (2006) (Supercritical) ammonia for recycling of thermoset polymers. *Macromol. Symp.*, **242**, 165–173.
30. Chen, M.C., Hourston, D.J., and Schafer, F.-.U. (1995) Miscibility and fracture behaviour of epoxy resin-nitrated polyetherimide blends. *Polymer*, **36** (17), 3287–3293.
31. Kubouchi, M., Tsuda, K., Nishiyama, T., and Hojo, H. (1995) A study on an application of corrosion behavior to glass fiber composite disposal and recycling. *Adv. Compos. Lett.*, **4** (1), 13–15.
32. Chen, J.-S., Ober, C.K., Poliks, M.D., Zhang, Y., Wiesner, U., and Cohen, C. (2004) Controlled degradation of epoxy networks: analysis of crosslink density and glass transition temperature changes in thermally reworkable thermosets. *Polymer*, **45**, 1939–1950.
33. Bolasodun, B., Nesbitt, A., Wilkinson, A., and Day, R. (2013) Decomposition and chemical analysis of cured epoxy system using a microwave reaction system. *Int. J. Sci. Technol. Res.*, **2** (2), 19–26.
34. Sugeno, T. and Tagaya, H. (2014) The effects of solvents on the chemical decomposition of foamed phenol resin in high-temperature conditions. *J. Mater. Cycles Waste Manage.*, **17**, 453–458.
35. Campanelli, J.R., Kamal, M.R., and Cooper, D.G. (1994) Kinetics of glycolysis of poly(ethylene terephthalate) melts. *J. Appl. Polym. Sci.*, **54**, 1731.
36. Chen, J.Y., Ou, C.F., Hu, Y.C., and Lin, C.C. (1991) Depolymerization of poly(ethylene terephthalate) resin under pressure. *J. Appl. Polym. Sci.*, **42**, 1501.
37. Oku, A., Hu, L.-C., and Yamada, E. (1997) Alkali decomposition of poly(ethylene terephthalate) with sodium hydroxide in nonaqueous ethylene glycol: a study on recycling of terephthalic acid and ethylene glycol. *J. Appl. Polym. Sci.*, **63**, 595.
38. Yoon, K.H., DiBenedettot, A.T., and Huang, S.J. (1997) Recycling of unsaturated polyester resin using propylene glycol. *Polymer*, **38** (9), 2281–2285.
39. Campanelli, J.R., Kamal, M.R., and Cooper, D.G. (1993) A kinetic study of the hydrolytic degradation of polyethylene terephthalate at high temperatures. *J. Appl. Polym. Sci.*, **48**, 443.
40. Yoshioka, T., Sato, T., and Okuwaki, A. (1994) Hydrolysis of waste PET by sulfuric acid at 150 °C for a chemical recycling. *J. Appl. Polym. Sci.*, **52**, 1353.
41. Mohantya, A.K., Misraa, M., and Hinrichsen, G. (2000) Biofibres, biodegradable polymers and biocomposites: an overview. *Macromol. Mater. Eng.*, **276/277**, 1–24.
42. Datta, J. and Włoch, M. (2014) Selected biotrends in development of epoxy resins and their composites. *Polym. Bull.*, **71**, 3035–3049.
43. García, J.M., Jones, G.O., Virwani, K., McCloskey, B.D., Boday, D.J., ter Huurne, G.M., Horn, H.W., Coady, D.J., Bintaleb, A.M., Alabdulrahman, A.M.S., Alsewailem, F., Almegren, H.A.A., and Hedrick, J.L. (2014) Recyclable, strong thermosets and organogels via paraformaldehyde condensation with diamines. *Science*, **344**, 732–735.
44. Asokan, P., Osmani, M., and Price, A.D.F. (2009) Assessing the recycling potential of glass fibre reinforced plastic waste in concrete and cement composites. *J. Cleaner Prod.*, **17**, 821–829.
45. Kennerley, J.R., Kelly, R.M., Fenwick, N.J., Pickering, S.J., and Rudd, C.D. (1998) The characterization and reuse of glass fibers recycled from scrap composites by the action of a fluidized bed process. *Composites Part A*, **29** (7), 839–845.
46. Pickering, S.J., Kelly, R.M., Kennerley, J.R., Rudd, C.D., and Fenwick, N.J. (2000) A fluidized-bed process for the recovery of glass fibers from scrap thermoset composites. *Compos. Sci. Technol.*, **60** (4), 509–523.

47. Onwudili, J.A., Insura, N., and Williams, P.T. (2013) Autoclave pyrolysis of carbon reinforced composite plastic waste for carbon fibre and chemicals recovery. *J. Energy Inst.*, **86** (4), 227–232, Journal; Online Computer File.
48. Torres, A., de Marco, I., Caballero, B.M., Laresgoiti, M.F., Legarretaa, J.A., Cabreroa, M.A., González, A., Chomón, M.J., and Gondra, K. (2000) Recycling by pyrolysis of thermoset composites: characteristics of the liquid and gaseous fuels obtained. *Fuel*, **79**, 897–902.
49. de Marco, I., Legarreta, J.A., Laresgoiti, M.F., Torres, A., Cambra, J.F., Chomón, M.J., and Caballero, B. (1997) Recycling polymeric wastes by means of pyrolysis. *J. Chem. Technol. Biotechnol.*, **69**, 187–192.
50. Cunliffe, A.M. and Williams, P.T. (2003) Characterisation of products from the recycling of glass fibre reinforced polyester waste by pyrolysis. *Fuel*, **82**, 2223–2230.
51. Blazsó, M. (2010) Pyrolysis for recycling waste composites, in *Management, Recycling and Reuse of Waste Composites* (ed. V. Goodship), WP and CRC Press, Cambridge, pp. 102–121.
52. Henshaw, J.M., Han, W., and Owens, A.D. (1996) An overview of recycling issues for composite materials. *J. Thermoplast. Compos. Mater.*, **9**, 4–20.
53. Yun, Y.M., Seo, M.W., Koo, G.H., Ra, H.W., Yoon, S.J., Kim, Y.K., Lee, J.G., and Kim, J.H. (2014) Pyrolysis characteristics of GFRP (Glass Fiber Reinforced Plastic) under non-isothermal conditions. *Fuel*, **137**, 321–327.
54. Holland, K.M. (1995) Apparatus for waste pyrolysis. US Patent 5,387,321 and (1994) Process of destructive distillation of organic material. US Patent 5,330,623.
55. Nomaguchi, K. (2005) Recycling Trends in the Composites Industry. JEC Magazine #17 (May 2005).
56. Composite recycling—raw material for the future. ERCOM composite recycling GmbH.
57. Allred, R.E. (1996) Recycling process for scrap composites and prepregs. *SAMPE J.*, **32** (5), 46–51.
58. Harth, T. (1985) Proceedings of Auto Recycle '95, p. 195.
59. Conroy, A., Halliwell, S., Reynolds, T., and Waterman, A. (2004) *Recycling Fibre Reinforced Polymers in Construction: a Guide to Best Practicable Environmental Option*, Building Research Establishment, BREPress. ISBN: 1 86081 689
60. (2002) Recycling takes off at MCR. *Compos. Int.*, **48**, 27.
61. 1999) Thermo setting materials: recycling and applications. *Composites (Paris)*, **35**, 49.
62. Monce, C. (2002) Recyclage Recuperation Magazine, 2002,16,19.
63. Mlecnik, E. (1997) Proceedings of the 3rd International Congress, on Recovery, Recycling and Re-Intergratwn, Geneva, February 4–7, 1997, vol. 111, p. 165.
64. Papaspyrides, C.D., Poulakis, J.G., and Arvanitopoulos, C.D. (1995) Recycling of glass fiber reinforced thermo-plastic composites. I. Ionomer and low density poly ethylene based composites. *Resour. Conserv. Recycl.*, **14**, 91.
65. Poulakis, J.G., Varelidis, P.C., and Papaspyrides, C.D. (1997) Recycling of polypropylene- based composites. *Adv. Polym. Technol.*, **16**, 313.
66. Åkesson, D., Foltynowicz, Z., Christéen, J., and Skrifvars, M. (2012) Recycling of thermoset composites by microwave pyrolysis. 8th International Conference On Composite Materials, September, 2012, vol. 31, no. 17, pp. 1136–1142.
67. Kouparttsas, C.E., Kawtalis, C.N., Varelidis, P.C., Tsenoglou, C.J., and Paf'aspyrides, C.D. (2002) Recycling of the fibrous fraction of reinforced thermoset composites. *Polym. Compos.*, **23** (4), 682–689.
68. Palmer, J.A.T. (2009) Thesis submitted to the University of Exeter as a thesis for the degree of Doctor of Philosophy in Engineering, May 2009.

8
Recycling of Other Layered Mixed Plastics or Resins: Polyurethanes

Jyothi V. Sunny

8.1
Introduction

Polyurethanes (PURs) are versatile and durable, show a variety of properties, and are derived from reacting di or polyisocyantes with a polyalcohol. The characteristic of polyurethanes is the presence of the urethane (–{–NH–CO–O–}–) group. Scheme 8.1 shows the formation polyurethanes.

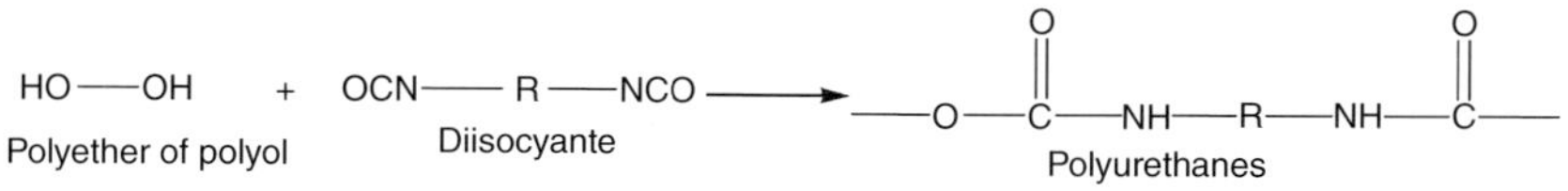

Scheme 8.1 Schematic diagram of urethane formation.

The properties of PURs are derived from those of their precursors, that is, the polyols and isocyanates. There are several kinds of polyols and polyisocyantes that are used to create polyurethanes. The majority of the polyurethanes are derived from aromatic polyisocyantes such as toluene diisocyanate (TDI) and methylene diphenyl diisocyanate (MDI), both polymeric and monomeric (Figures 8.1 and 8.2).

In addition to above aromatic isocyanates, there are different aliphatic and cyclo aliphatic isocyanates also used in various UV-stable product applications. Examples are hexamethylene diisocyanate (HDI) and isophorone diisocyanate (IPDI). Additionally, several dimers, derivatives of MDIs, and blocked isocyanates are used for developing polyurethanes.

The molecular weights, functionality, and structure of the polyols are very important for the development of PURs. Polyether polyols are the most common and largest group of polyols used for PURs. Polyester polyols are also used for certain specific applications. Polyether polyols are made by the reaction of either ethylene oxide (EO) or propylene oxide (PO) with an initiator and in the presence of a catalyst. Polyols have two or more hydroxyl groups as end groups that can

Recycling of Polymers: Methods, Characterization and Applications, First Edition. Edited by Raju Francis.

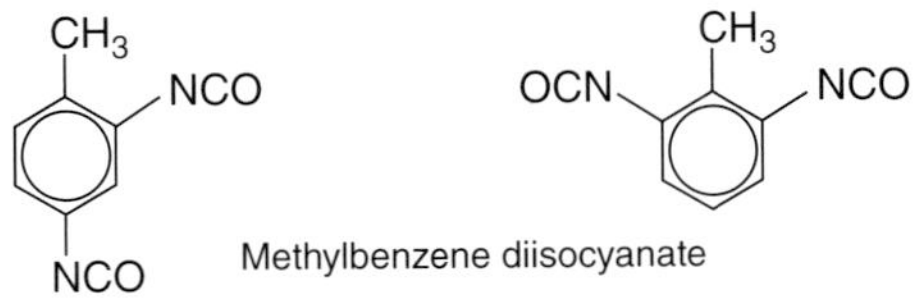

Figure 8.1 Toluene diisocyante (TDI) structure.

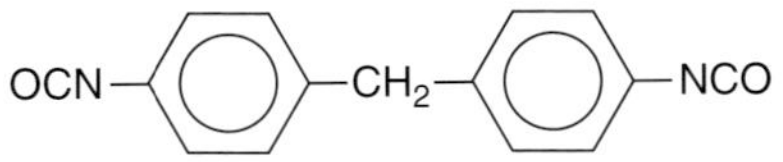

Figure 8.2 Methylene diphenyl diisocyanate (MDI).

react with other reactants. Various series of polyols can be synthesized depending on the starter molecule. The properties and reactivity depend on the structure of the polyols. Polyols containing terminal EO groups react faster because of the primary hydroxyl groups compared to polyols with PO end groups due to the secondary hydroxyl groups. Polyols can be distinguished on the basis of their structure [1]. Long-chain polyols are used for flexible foams and short-chain polyols for rigid foams. The properties of long-chain and short-chain polyols are given in Tables 8.1 and 8.2 [2]. The use of polyols is shown in the pie chart [1]. PURs are formed by the exothermic reaction of polyisocyanates and polyol molecules. Many different types of PUR materials are produced from a few types of isocyanates and a range of polyols with different functionality and molecular weights. Some of the diversity of functionality depends on whether the polyols are based on polyethers or polyesters. Most polyether polyols are produced for PUR applications; however, other end uses range from synthetic lubricants and functional fluids to surface-active agents.

PURs are used in a variety of applications starting from commodity to high-performance products to provide comfort, convenience, and safety to everyday activities of modern life. Because of the wide range of isocyanates and polyols availabille, a variety of materials can be created for target performance or applications. The main difference between PUR and polyurethane foam (PUF) is the incorporation of gas in the foam. This is achieved by reacting isocyanates with

Table 8.1 Properties of long-chain polyether polyols.

Starter	Alkylene oxide	Functionality	Mn (g mol^{-1})	OH no (mg KOH g^{-1})	Viscosity (25 °C) (mPa s)
PG	PO/EO	2	4000	28	870
PG	PO	2	2000	56	310
TMP/PG	PO	2.8	3750	42	630
GLY	PO/EO	3	4800	35	860

PG: 1,2-propanediol, TMP: Trimethylolpropane, GLY: Glycerol, PO: Propylene Oxide, EO: Ethylene Oxide.

Table 8.2 Properties of short-chain polyether polyols.

Starter	Alkylene oxide	Functionality	Mn	OH no (mg KOH g^{-1})	Viscosity (25 °C) (~mPa s)
TMP	PO	3	435	385	600
TMP	PO	3	305	550	1 800
EDA	PO	4	350	620	19 200
TDA	PO/EO	4	475	470	8 000
Sucrose/PG	PO	3.1	385	450	15 000

Figure 8.3 Examples of flexible polyurethane foam applications.

Figure 8.4 Rigid polyurethane foam applications.

water, forming CO_2 and amine. There are other chemicals that can be used as blowing agents to produce foams.

Most polyurethane applications comprise rigid and flexible foams. Flexible foams are mainly used in automobile cushions, furniture, carpet, clothing, packaging, and so on, whereas the rigid polyurethanes foams are used in mainly insulation applications. Examples [3–5] of flexible PUFs are shown in Figure 8.3, and those of rigid [6, 7] PUF are shown in Figure 8.4. Generally speaking, open-cell forms are flexible and closed-cell foams are rigid.

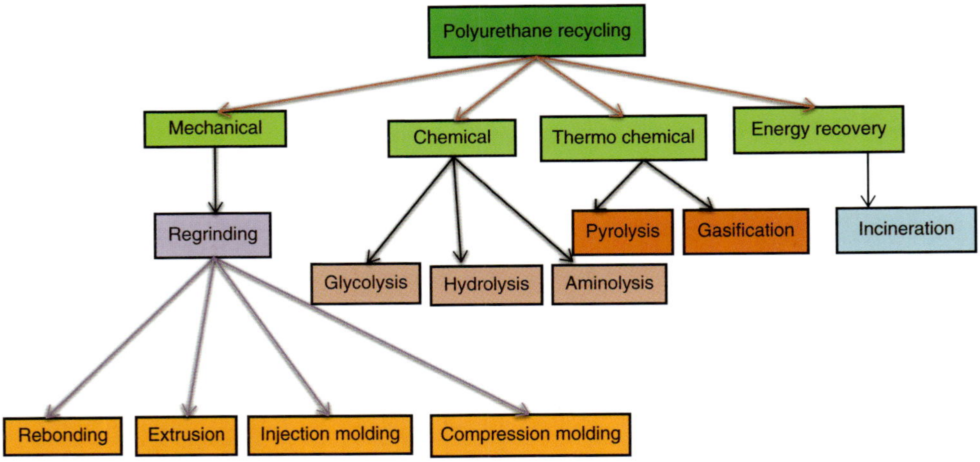

Figure 8.5 Summary sheet of polyurethane recycling.

The coatings, sealant, adhesives, and elastomer (CASE) field is the next big category of PUR applications. Because of their strong adhesion, desirable mechanical properties, and weathering characteristics, there is a high demand for PURs in CASE applications.

Recycling of PUR is receiving more and more attention worldwide now days due to excessive product development involving PUR for various applications. In addition, the use of isocyanates in product development is also being subjected to stringent environmental regulations. Therefore, the need for more environmentally friendly product development or reuse of these materials is necessary. Rigid PUFs mainly end up in landfills and for incineration. Landfilling has serious environmental consequences, so it is the least favorable. A detailed study of on the disposal PUF by incineration and heat recovery was conducted by Hilyard *et al.* [8] Even though the primary methods for recycling flexible PUR are either mechanical or chemical, generally they can be divided into four. The methods for recycling polyurethane are shown in Figure 8.5.

8.2
Mechanical Recycling

In mechanical recycling, the material is ground and rebonded, and this is oldest method for recycling PURs. Some examples of mechanical recycling of polyurethane include the carpet industry and its reuse in mattresses and automobile cushions. In other words, the material is reused in different shapes or forms: that is, only physical change to the material occurs. The majority of flexible foams generated is recycled in the carpet industry to be used as underlay by mixing the ground foam with a polymer binder. Also, the ground foam or

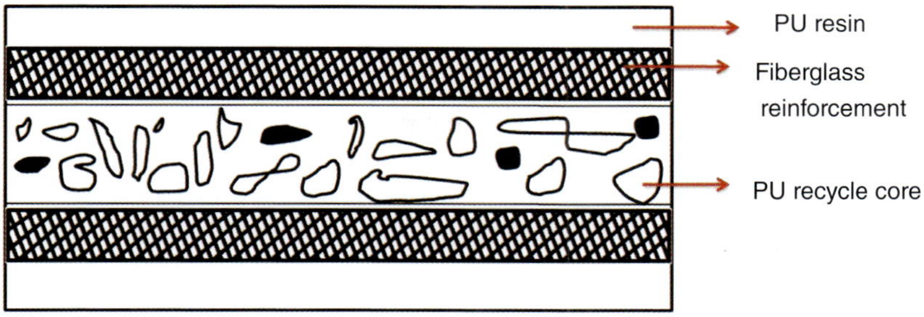

Figure 8.6 Structure reaction injection molding (SRIM) recycling of PU scrap [9].

Table 8.3 Polyurethane particle size distribution after granulation [12].

Size/range	wt%
<150	6.26
150–300 μm	3.94
300–600 μm	8.71
600 μm–1.18 mm	8.41
1.18–2.36 mm	24.5
2.36–5 mm	48.16

composite particles can be mixed with neat materials and can be processed in to new products by extrusion, injection molding, and compression molding. Figure 8.6 shows a schematic of structural reaction injection molding of recycled PUF [9].

A new method of recycling PUF was proposed by Quadrini *et al.* [10], where scrap pulverization and subsequent compression molding of the resulting particles were done. Bledzki *et al.* [11] reported a detailed study on the uses of rigid PUF and soft PUFs. Another option for recycling PUFs is to grind and reuse them as fillers [12] in new materials. The particles size distribution is shown in Table 8.3.

Some of the associated technologies are two-roll milling, extrusion, cryogenic grinding, and so on. A detailed review on the various methods of recycling polyurethanes was provided by Khalid *et al.* [13].

8.3 Chemical Recycling

In chemical recycling, the breaking down of PURs in to different raw materials is done by means of chemical treatments, that is, recovering polyols from polyurethanes [14]. Some of the chemical recycling methods used for recycling polyurethane are glycolysis, hydrolysis, and aminolysis.

8.3.1
Glycolysis

Glycolysis is the most widely used chemical recycling method for the recovery of polyols from rigid and flexible PUFs. In single-phase glycolysis, PUF can be degraded to low molecular weight components in the presence of a catalyst and lower molecular weight alcohol at high temperatures.

Chemical recycling by glycolysis was extensively studied by Molero *et al.* [15–17]. They pointed out that the glycolysis progresses in two phases: an upper phase where high-quality polyols are generated and a lower phase where side reaction products are generated. The influence of the glycolysis conditions such as the type of glycol (polyol), type of catalyst, catalyst concentration, reaction temperature, and the particle size of PUF on the dissolution rate of rigid PUF in the reaction was studied by Murai *et al.* [18].

The steps involved in glycolysis are detailed in Schemes 8.2 and 8.3 by, which are from the studies conducted by Vahid [19]. In their study, scrap PUF was subjected

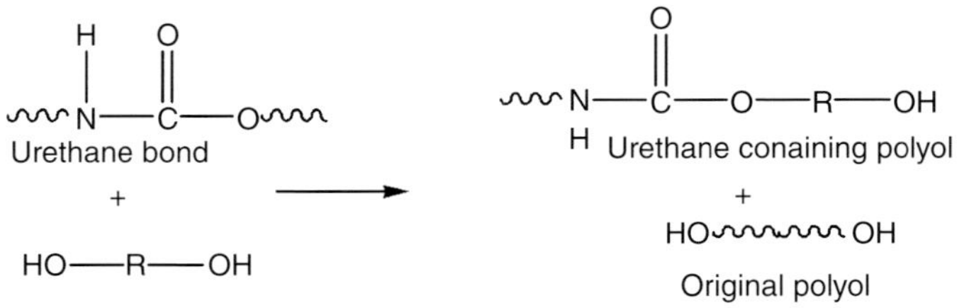

Scheme 8.2 First step in glycolysis of polyurethane foam with glycerol to form a mixture hydroxyl oligomers [19].

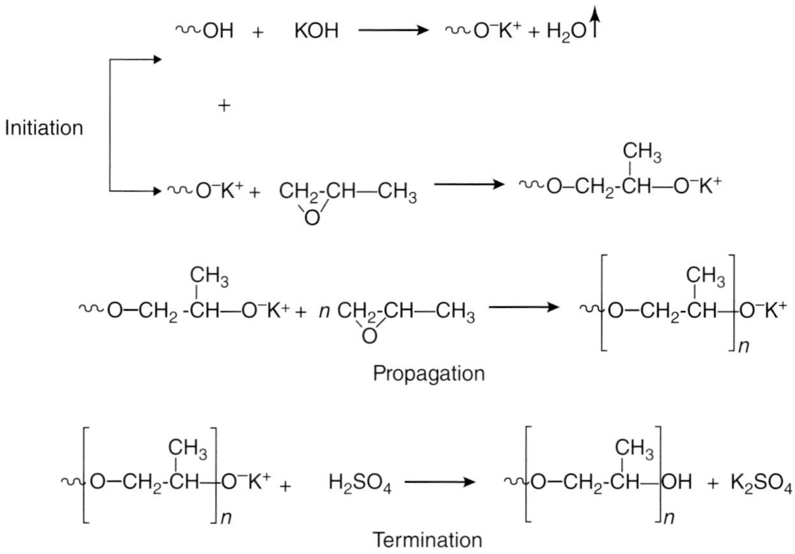

Scheme 8.3 Second step in the glycolysis process: the glycolyzed products are used as initiators in a reaction with alkylene oxide to produce oxyalkylated polyols [19].

Table 8.4 Recycling of polyurethane [26].

Austria	Elastomeric foams/ instrument panels (IP)	Elastomeric foams/ IP components
France	Rigid foam	Rigid foam
Germany	RIM	RIM/integral skin form
Germany	Shoe soles	Shoe soles
Italy	Shoe soles	Rigid foam
Italy	Shoe soles	Shoe soles
England	Flexible slab stock	Flexible and rigid foam

to glycolysis and the glycolyzed products were formed first, which were then used as initiator in a reaction with alkylene oxide to produce oxyalkylated polyols.

Various studies [20–25] have been conducted of the use of glycolysis for recycling PUF to obtain polyol. Glycolysis is more appropriate for recycling production waste than post-consumer waste, and several companies are involved, as indicated [26] in Table 8.4.

8.3.2 Hydrolysis

Hydrolysis involves the use of an alkali metal hydroxide as catalyst in the presence of water and temperatures above 200 °C. The hydrolysis process is represented in [27] Scheme 8.4.

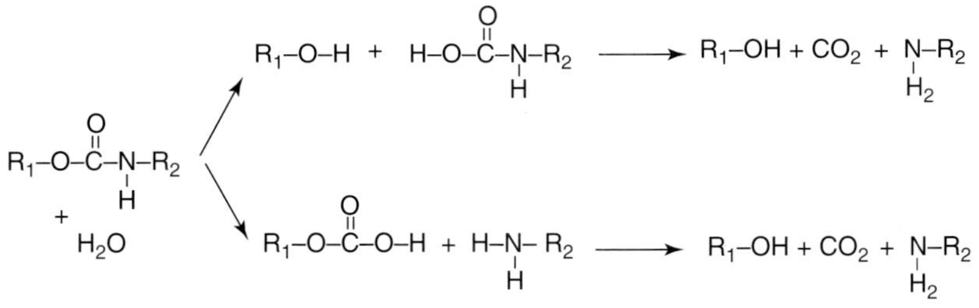

Scheme 8.4 Hydrolysis of polyurethane [27].

As a result of hydrolysis, the resulting products are a mixture of polyamines and polyols. Several studies [28–31] have reported on hydrolysis for the recycling of polyurethanes.

8.3.3 Aminolysis

The reaction between urethane and amines is well known in the PUR chemistry, and it can be used in recycling PURs. In the aminolysis of PUR, a substituted urea is formed [32] (Scheme 8.5).

$$\text{Ar-NHCOORx+RyNH}_2\text{–} \longrightarrow \text{Ar-NHCONHRy +RxOH}$$

Scheme 8.5 Aminolysis of polyurethane [32].

Studies have been conducted on the aminolysis of PUR, which is still in the research phase [28, 32, 33]. In aminolysis, PUF is reacted with amines under pressure and at elevated temperatures. The amines used in the process include dibutylamine and ethanolamine, among others. The reaction of polyurethane with alkanolamine is slower compared to the reaction with amines. For improving the purity of the resulting polyols, a blend of mono and polyamines can be used [34].

8.4 Thermochemical methods

Thermochemical methods include pyrolysis, gasification, and hydrogenation, which are used as efficient recycling methods for polyurethanes. Several studies have been conducted on the methods for their practical applicability.

8.4.1 Pyrolysis

Pyrolysis is the breaking down of polyurethane in the absence of oxygen and in presence of nitrogen to generate gaseous and liquid products. Several studies have been conducted on pyrolysis of polyurethane foams and adhesives [35–40]. One of the disadvantages of pyrolysis is the toxicity of the gases produced during pyrolysis. The liquid components are extremely viscous and solidify over time.

8.4.2 Gasification

In gasification, polyurethane is heated at very high temperatures under limited amount of oxygen to form carbon monoxide, carbon dioxide, and hydrogen. Gasification has been in use worldwide for more than 60 years for recycling carbonaceous materials [41]. Gasification uses only small fraction of the amount of oxygen used in combustion. It is less expensive, utilizes less energy, and generates no hazardous heavy metal. Figure 8.7 shows a flowchart of the gasification process [42].

There has been a significant amount of research conducted to modify or improve the process of gasification [43–45].

8.4.3 Hydrogenation

Polymers, especially the C–C bonds of polyurethanes, can be degraded by hydrogenation, which makes hydrogenation a recycling method for polyurethanes.

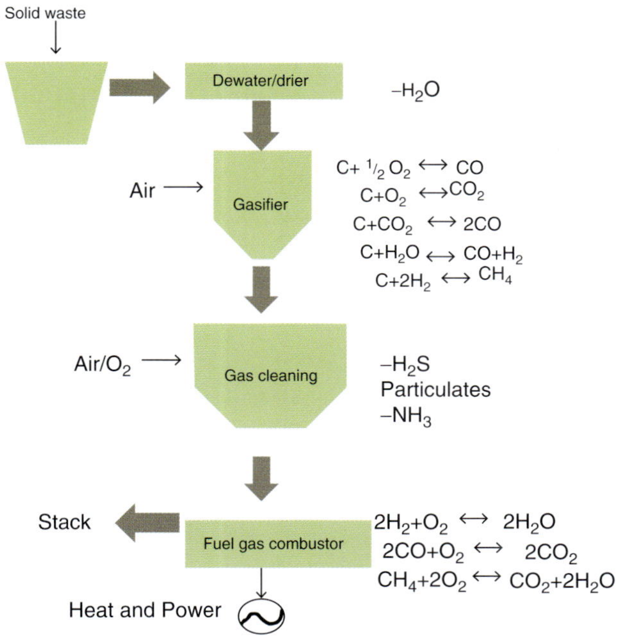

Figure 8.7 Gasification process flowchart [42].

Hydrogenation is carried out at high temperatures and pressures in the presence of a metal as catalyst [46]. Polyurethane is heated in the absence of oxygen and at high levels of hydrogen. The recycled products include solids, gases, and liquids. In other words, hydrogenation is a combination pyrolysis and gasification.

8.5
Energy Recovery by Incineration

Energy recovery by incineration is applied when the recovered product can no longer be used in different products. There is no recovery of products from incineration or combustion; only energy is recovered. Toxic gases are generated and the heat generated is very high during the incineration of polyurethanes [47, 48]. The presence of flame retardants in polyurethane form makes it harder to recycle and reuse. Incineration will reduce the volume of foam in landfills significantly. Thermal degradation of polyurethane starts above 110 °C, and the typical degradation temperatures for urethane-related linkages are shown in Table 8.5 [49].

Detailed reports [50] on energy recovered compared to actual recycling show that recycling saves more energy than the energy recovered by incineration. Other reports on the energy recovery from incineration or combustion of polyurethanes are available in the literature [51–53].

Table 8.5 Thermal dissociation temperature for various linkages in polyurethanes [49].

Linkage		On set of dissociation (°C)
Allophante	Aliphatic/aromatic	85–105/100–120
Biuret	Aliphatic/aromatic	100–110/115–125
Urea	Aliphatic/aromatic	140–180/160–200
Urethane	Aliphatic/aromatic	160–180/180–200
Disubstituted urea		235–250

References

1. (2015) *Polyether Polyols for Urethanes, Chemical Economics Handbook.*
2. Adam, N., Avar, G., Blankenheim, H., Friederichs, W., Giersig, M., Weigand, E., Halfmann, M., Wittbecker, F.-W., Larimer, D.-R., Maier, U., Meyer-Ahrens, S., Noble, K.-L., and Wussow, H.-G. (2005) Polyurethanes, in *Ullmann's Encyclopedia of Industrial Chemistry.*
3. Upholstery http://www.decsignco.com/main.asp?page=%7B5EA8D29E-5755-4ABA-B065-4C12019DF861%7D (accessed 12 May 2016).
4. DesignNews http://www.designnews.com/document.asp?doc_id=229183 (accessed 12 May 2016).
5. Foam Factory http://www.foamdistributing.com/polyurethane-foam-sheets.html (accessed 12 May 2016).
6. Treehugger http://www.treehugger.com/green-architecture/spray-polyurethane-foam-manufacturer-may-face-class-action-lawsuit.html (accessed 12 May 2016).
7. Building System Design https://www.bsdsoftlink.com/manufacturers/mfrpages/demilecusa.htm (accessed 12 May 2016).
8. Hilyard, N.C.; Kinder, A.I.; Axelby, G.L., A techno-economic study of the disposal of polyurethane foam waste by incineration and heat recovery, *Cell. Polym.*, **4** (5), 367-384.
9. Scheirs, J. (1998) *Polymer Recycling*, Chapter 10, John Wiley & Sons, Ltd., Chichester.
10. Quadrini, F., Bellisario, D., and Santo, L. (2013) Recycling of thermoset polyurethane foams. *Polym. Eng. Sci.*, **53** (7), 1357–1363.
11. Bledzki, A.K., Zicans, J., Merijs Meri, R., and Kardasz, D. (2008) Some physical and mechanical properties of recycled polyurethane foam blends. *Mech. Compos. Mater.*, **44** (5), 515–522.
12. Hulme, A.J. and Goodhead, T.C. (2003) Cost effective reprocessing of polyurethane by hot compression moulding. *J. Mater. Process. Technol.*, **139** (1–3), 322–326.
13. Khalid, M.Z., Haq, N.B., and Bhatti, I.A. (2007) Methods for polyurethane and polyurethane composites, recycling and recovery: a review. *React. Funct. Polym.*, **67**, 675–692.
14. American Chemistry Council http://polyurethane.americanchemistry.com/Sustainability/Recycling (accessed 12 May 2016).
15. Molero, C., de Lucas, A., and Rodriguez, J.F. (2009) Activities of octoate salts as novel catalysts for the trans-esterification of flexible polyurethane foams with diethylene glycol. *Polym. Degrad. Stab.*, **94**, 533–539.
16. Molero, C., de Lucas, A., and Rodriguez, J.F. (2008) Recovery of polyols from flexible polyurethane foam by "split-phase" glycolysis: study on the influence of reaction parameters. *Polym. Degrad. Stab.*, **93**, 353–361.
17. Molero, C., de Lucas, A., and Rodrıguez, J.F. (2006) Recovery of polyols from flexible polyurethane foam by "split-phase" glycolysis: glycol influence. *Polym. Degrad. Stab.*, **91**, 221–228.

18. Murai, M., Sanou, M., Fujimoto, T., and Baba, F. (2003) Glycolysis of rigid polyurethane foam under various reaction conditions. *J. Cell. Plast.*, **39**, 15–27.
19. Vahid, S. (2007) Chemical recycling of mixed polyurethane foam stream recovered from shredder residue into polyurethane polyols. *J. Cell. Plast.*, **43**, 31–46.
20. Borda, J., Pásztor, G., and Zsuga, M. (2000) Glycolysis of polyurethane foams and elastomers. *Polym. Degrad. Stab.*, **68**, 419–422.
21. Hekmatjoo, N., Ahmadi, Z., Afshar Taromi, F., Rezaee, B., Hemmati, F., and Saeb, M.R. (2015) Modeling of glycolysis of flexible polyurethane foam wastes by artificial neural network methodology. *Polym. Int.*, **64** (9), 1111–1120.
22. Wu, C.-H., Chang, C.-Y., Cheng, C.-M., and Huang, H.-C. (2003) Glycolysis of waste flexible polyurethane foam. *Polym. Degrad. Stab.*, **80**, 103–111.
23. Modesti, M. and Simioni, F. (1996) Chemical recycling of reinforced polyurethane from the automotive industry. *Polym. Eng. Sci.*, **36** (17), 2173–2178.
24. Huang, H.-C., Wu, C.-H., and Chang, C.-Y. (2000) Recycling of wasted flexible polyurethane foam by chemolysis. Proceedings of the Air & Waste Management Association's Annual Conference & Exhibition, 93rd, Salt Lake City, UT, pp: 6071–6079.
25. Wang, X., Chen, H., Chen, C., and Li, H. (2011) Chemical degradation of thermoplastic polyurethane for recycling polyether polyol. *Fibers Polym.*, **12** (7), 857–863.
26. ISOPA (2001) Fact Sheet – Recycling and Recovering Polyurethane, June 2001.
27. Wang, X., Chen, H., Chen, C., and Li, H. (2013) Description and modeling of polyurethane hydrolysis used as thermal insulation in oil offshore conditions. *Polym. Test.*, **32** (8), 1588–1593.
28. Behrendt, G. and Naber, B.W. (2009) The chemical recycling of polyurethanes. *J. Univ. Chem. Technol. Metall.*, **44** (1), 3–23.
29. Campbell, G.A. and Meluch, W.C. (1976) Polyurethane foam recycling superheated steam hydrolysis. *Environ. Sci. Technol.*, **10** (2), 182–185.
30. Ge, J.-J. and Sakai, K. (1998) Decomposition of polyurethane foams derived from condensed tannin II: hydrolysis and aminolysis of polyurethane foams. *J. Wood Sci.*, **44**, 103–105.
31. Mahoney, L.R., Weiner, S.A., and Ferris, F.C. (1974) Hydrolysis of polyurethane foam waste. *Environ. Sci. Technol.*, **8** (2), 135–139.
32. Kanaya, K. and Takahashi, S. (1994) Decomposition of polyurethane foams by alkanolamines. *J. Appl. Polym. Sci.*, **51**, 675–682.
33. Scheirs, J. (1998) *Polymer Recycling*, John Wiley & Sons, Chichester, chapter 10.
34. Chuayjuljit, S., Norakankorn, C., and Pimpan, V. (1994) Chemical recycling of rigid polyurethane foam scrap via base catalyzed aminolysis. *J. Appl. Polym. Sci.*, **51**, 675–682.
35. Chambers, J., Jiricny, J., and Reese, C.B. (1981) The thermal decomposition products of rigid polyurethanes and polyisocyanurates. *Fire Mater.*, **5**, 133–141.
36. Napier, D.H. and Wong, T.W. (1972) Toxic products from the combustion and pyrolysis of polyurethane foams. *Br. Polym. J.*, **4**, 45–52.
37. Takamoto, D.Y. and Petrich, M.A. (1994) Effect of heterogeneous secondary pyrolysis reactions on the thermal decomposition of polyurethane scrap. *Ind. Eng. Chem. Res.*, **33**, 3004–3009.
38. Terakado, O., Yanase, H., and Hirasawa, M. (2014) Pyrolysis treatment of waste polyurethane foam in the presence of metallic compounds. *J. Anal. Appl. Pyrolysis*, **108**, 130–135.
39. Jomaa, G., Goblet, P., Coquelet, C., and Morlot, V. (2015) Kinetic modeling of polyurethane pyrolysis using non-isothermal thermogravimetric analysis. *Thermochim. Acta*, **612**, 10–18.
40. Rafael, F., Andres, F., Jose, C., Candela, J., and Garcia, A. (2001) Pyrolysis study of polyurethane. *J. Anal. Appl. Pyrolysis*, **58**, 63–77.
41. Guo, X., Wang, L., Li, S., Tang, X., and Hao, J. (2015) Gasification of waste rigid

polyurethane foam: optimizing operational conditions. *J. Mater. Cycles Waste Manage.*, **17** (3), 560–565.
42. W2E http://w2es.com/gasification.php (accessed 12 May 2016).
43. WasteSyn http://www.wastesyn.com/tech_wg.html (accessed 12 May 2016).
44. Schingnitz, M., Gaudig, M., McVey, I.E., and Wood, K. (2000) Gasifier to convert nitrogen waste organics at Seal Sands, UK. 2000 Gasification Technologies Conference, San Francisco, CA, October 8–11, 2000.
45. Aznar, M.P., Caballero, M.A., Sancho, J.A., and Francé, E. (2006) Plastic waste elimination by co-gasification with coal and biomass in fluidized bed with air in pilot plant. *Fuel Process. Technol.*, **87** (5), 409–420.
46. K.C. Frisch, D. Klempner, *Advances in Urethane: Science & Technology*, Vol. 13, p 244.
47. Landry, T.D., Daems, D., Pauluhn, J., and Reimann, K.A. (2002) *Polyurethane Products in Fires: Acute Toxicity of Smoke and Fire Gasses*, American Chemistry Council.
48. Paabo, M. and Levin, B.C. (1987) A review of to be literature on to be gaseous products and toxicity generated from the pyrolysis and combustion of rigid polyurethane foams. *Fire Mater.*, **11**, 1–29.
49. Isopa Fact sheets, Energy Recovery, 2001.
50. Szycher, M. (1999) *Szycher's Handbook of Polyurethanes*, CRC Press, Boca Raton, FL.
51. Morris, J. (1996) Recycling versus incineration: an energy conservation analysis. *J. Hazard. Mater.*, **47**, 277–293.
52. Perry, M.J., Lambach, J.L., and Critchfield, F.E. (2000) Energy management polyurethane rigid foams with high recovery. US Patent US6127443 A. PCT Int. Appl.
53. Peter, S. and David, P. (2013) Method of recovering/enhancing the energy from municipal solid waste. UK Patent GB 2494532 A 20130313.

9
Ecoprofiles of Recycled Polymers at a Glance

9.1
Advantages of Recycled Polymers on the Environment

Raju Francis and Anjaly Sivadas

9.1.1
Introduction

Plastic materials comprise a steadily increasing proportion of the municipal and industrial waste going into landfill. Owing to the huge amount of plastic wastes and environmental pressures, recycling of plastics has become a predominant subject in today's plastics industry. Once a material enters the waste stream, recycling is the process of using recovered material to manufacture a new product. For organic materials such as plastics, the concept of recovery can also be expanded to include energy recovery, where the calorific value of the material is utilized by controlled combustion as a fuel, although this results in a lesser overall environmental performance than material recovery, as it does not reduce the demand for new (virgin) material. Development of technologies for reducing plastic waste, which are acceptable from the environmental standpoint and are cost effective, has proven to be a difficult challenge, because of the complexities inherent in the reuse of polymers. Establishing optimal processes for the reuse/recycling of plastic materials thus remains a worldwide challenge in the new century. Plastic materials find applications in agriculture as well as in plastic packaging which is a high-volume market owing to the many advantages of plastics over other traditional materials. However, such materials are also the most visible in the waste stream and have received a great deal of public criticism as solid materials have comparatively short life-cycles and usually are nondegradable. The growth of plastics waste has a great impact on landfilling and incineration [1] (Figure 9.1.1).

Plastics do not decompose naturally, and so alternative methods need to be implemented in order to recycle such materials. Plastics recycling is concerned with the production of second-grade pellets of a single-type polymer. Virgin materials are difficult to find and are uneconomical when used as a modifier [2]. Many

Recycling of Polymers: Methods, Characterization and Applications, First Edition. Edited by Raju Francis.

Figure 9.1.1 Plastic pollution.

of the waste polymers could be hazardous and pose an environmental burden if these are not effectively recycled or reused. There are several kinds of polymers that can be recycled in bitumen [3, 4] such as polypropylene (PP) used in straw, furniture, and wrapping industries; high-density polyethylene (HDPE) used in plastic bottles and packaging; low-density polyethylene (LDPE) used widely in soft drink and mineral water bottles [5]; poly(vinyl chloride) (PVC) used in fittings and plumbing pipes; and poly(ethylene terephthalate) (PET) widely used in soft drink and water bottles.

9.1.2 Poly(ethylene terephthalate) (PET)

PET is widely used to make containers for carbonated and noncarbonated soft drinks. It is an eco-friendly, recyclable polymer, which on burning generates only heat, CO_2, and water, without any toxic gases. It has several advantages over glass, such as consumer safety (as when dropped) and weight savings, which benefit customers and reduce the cost and environmental impacts of transport. Recycling PET has the added advantage that the operation contributes to national municipal waste recycling targets and to achieve the Packaging Directive's objectives. However, compared with thermal recovery, there may be problems with product quality; for example, beverage manufacturers consider the presence of black specks in premium bottled water and cola products to be unacceptable. However, this should not be a problem if the PET is used in colored bottles for non-food uses such as motor oils (Figure 9.1.2).

It has been estimated that PET bottle recycling gives a net benefit in greenhouse gas emissions of 1.5 t of CO_2-e per ton of recycled PET [6] as well as reduction in landfill and net energy consumption. An average net reduction of 1.45 t of CO_2-e per ton of recycled plastic has been estimated as a useful guideline to policy [7]; one basis for this value appears to have been a German life-cycle analysis (LCA) study [8], which also found that most of the net energy and emission benefits arise from the substitution of virgin polymer production. A recent LCA specifically for PET bottle manufacture calculated that the use of 100% recycled PET

Figure 9.1.2 Recycled polyethylene terephthalate used for making bottles.

instead of 100% virgin PET would reduce the full life-cycle emissions from 446 to 327 g CO_2 per bottle, resulting in a 27% relative reduction in emissions [9]. Mixed plastics, the least favored source of recycled polymer, could still provide a net benefit in the vicinity of 0.5 t of CO_2-e per ton of recycled product [10]. The higher eco-efficiency for bottle recycling is due to both the more efficient process for recycling bottles as opposed to mixed plastics and the particularly high emissions profile of virgin PET production. However, the mixed plastics recycling scenario still has a positive net benefit, which is considered superior to the other options studied, namely landfills and energy recovery as solid refuse fuel, as long as it can be a substitute for the virgin polymer [11]. Recycling of PET serves not only as a partial solution to the solid-waste problem but also as a source of raw material to some industries and contributes to the conservation of high cost raw petrochemical products and energy, which are of great importance in today's world [12].

9.1.3 High-Density Polyethylene (HDPE)

HDPE or polyethylene high density (PEHD) is a polyethylene thermoplastic made from petroleum. It takes 1.75 kg of petroleum (in terms of energy and raw materials) to make 1 kg of HDPE. It is the third largest commodity plastic material in the world in terms of volume. Large containers for liquids, mainly made of HDPE, are very good candidates for recycling. Indeed, they are very easy to collect separate, and HDPE undergoes, in general, little degradation both during its lifetime and during the recycling operations. However, the extent of thermomechanical

Figure 9.1.3 Various applications of recycled HDPE.

degradation and the properties of the recycled material are strongly dependent on the structure of the HDPE and the processing conditions [13]. Chemical recycling of HDPE involves the successful degradation of waste HDPE to petrol/diesel-like liquid fuels by suitable processes using different catalysts and advanced reactors [14] (Figure 9.1.3).

The amount of plastic used in plastic bags has reduced by around 70% in the last 20 years by the introduction of reusable canvas bags and other biodegradable materials, but the majority of bags are still produced from HDPE. Furthermore, there is a growing market for HDPE containers in China and India due to increased standards of living, as well as a higher demand for HDPE pipes and cables due to rapidly growing industries. HDPE is nonbiodegradable and can take centuries to decompose, so it is imperative that these bags and containers are recycled and used again. Recycling HDPE has many benefits. For example, it is more cost efficient to produce an article from recycled HDPE than it is to manufacture from "virgin" plastic. HDPE, like many plastic polymers, is produced using considerable amounts of fossil fuels, and it takes a total of 1.75 kg of oil to manufacture just 1 kg of HDPE. Many new products can be manufactured using recycled HDPE, as shown in Figure 9.1.3.

Figure 9.1.4 Recycled PVC used for making plastic floor tiles.

9.1.4 Poly(vinyl chloride) (PVC)

PVC packaging materials are used in the packaging of food, cleaning materials, textiles, beverages, and medical devices, and also in long-life products such as cable insulation, window frames, pipes, floors coverings, and roofing sheets [15–21]. PVC can be successfully recycled into a variety of products such as bottles, pipes, pipe fittings, and other articles with good appearance and properties [8, 22–26]. It is also claimed that the properties of the manufactured articles are unaffected when virgin PVC is replaced by PVC recyclates. Manufacturing window profiles with 70% recycled PVC rather than virgin PVC results in savings of up to 50% in energy, over 60% in air emissions, and more than 60% in water emissions. In addition, because it is a thermoplastic, PVC can be recycled several times without significant loss of performance. Using recycled PVC helps to meet the resource efficiency objectives and allows the preservation of raw materials. The use of recycled PVC reduces emission and landfill requirements (Figure 9.1.4).

Figure 9.1.5 Recyclable polypropylene.

9.1.5 Polypropylene (PP)

PP is one of the high-volume commodity plastics and is used extensively in appliances such as refrigerators, washing machines, and air-conditioners. The closed-loop recycling of PP is therefore crucial in waste appliances. PP products degrade very slowly in landfills, taking about 20–30 years to fully decompose. This poses a serious problem because of the additives present, which include stabilizers, colorants, and plasticizers. These additives can contain a range of toxins including cadmium and lead. According to various studies, plastics can account for about 28% of the total cadmium in solid municipal wastes in landfills, which has the potential to leach into other systems with potentially harmful consequences for a range of biosystems. Moreover, burning of thermoplastics can release toxins such as vinyl chloride and dioxins, and this makes atmospheric pollution a real issue to consider.

PP has diverse applications in a range of single-use packaging products such as bottles, yogurt containers, medicine bottles, caps, straws, plastic cups, and food packaging, leading to steadily increasing PP waste (Figure 9.1.5). Recycling of PP in a cost-effective and eco-friendly way is very much needed for managing the plastic waste. Also, since PP is a petroleum-based product, recycling of products made of PP will help us preserve our oil/petroleum resources. Recycled PP is currently used in the manufacture of the following products:

1. Battery cables
2. Brushes
3. Ice scrapers
4. Pallets and trays
5. Landscape borders
6. Signal lights
7. Brooms
8. Auto battery cases
9. Bicycle racks
10. Rakes and bins.

Figure 9.1.6 Recyclable EPS forms.

9.1.6 Polystyrene (PS)

The amount of waste polystyrene (PS) foams is increasing year by year due to the increase in their use in isolation, protection, and storage of many different food products. Continuous accumulation of waste plastics has led to serious problems all over the world [27]. PS is generally not a high-profile target in recycling terms (unlike glass, paper, aluminum cans, batteries, tires, etc.), but many companies, local authorities, and individuals may not have considered the implications of the quantity of PS they are dumping. Recycling of this waste has recently received significant attention all over the world due to changes in both regulatory and environmental issues.

Increasing landfill costs and decreasing landfill space are forcing the consideration of alternative options for the disposal of PS materials. Expanded polystyrene (EPS) is another similar foam material that has found even more uses. Recycled EPS may be used to create packaging or other materials, but new EPS is always needed for coffee cups and plates (Figure 9.1.6) [28]. Using recyclables rather than fresh raw material to make new products has the following merits:

1. Uses less energy
2. Results in less pollution
3. Saves natural resources

4. Saves scarce landfill space
5. Creates extra income and jobs
6. Reduces littering.

9.1.7 Other Polymers

By incorporating waste polymers in paving applications, pavement performance improves, environmental pollution reduces, and the use of virgin polymers reduces, thus leading cost savings and pollution abatement [29]. Recycling reduces the quantity of residues in landfills and those indiscriminately discarded in the environment. Thus, it also leads to a reduction in the spread of diseases and the contamination of soil, air, and water bodies [30]. Waste that is not properly packed and discarded can attract some vectors such as rats and mosquitoes. This latter can transmit dengue, yellow fever, and many others. Socially, recycling has advantages, because it generates direct and indirect jobs and transforms waste in to artifacts. The generation of jobs is an important factor for developing countries. Recycled polymers have been found to show similar results in improving road performance as when using virgin polymers. Therefore using waste plastic bottles as a modifier in road surfacing can potentially help reduce material wastage and improve the performance of road surfaces at the same time [31–35]. The use of waste polymers in asphalt paving has also been shown to improve certain properties of the pavement to a level similar to that achieved when using virgin polymers. From an economical point of view, using waste material in road construction and paving is beneficial in different ways. It can improve the performance of pavements, reduce landfills, and solve the waste disposal problem [36]. An investigation on the rheology of modified bitumen with recycled polymers has been done by Garcia-Morales *et al.* They studied the flow behavior of bitumen which was modified with 5% and 9% waste poly(ethylene-*co*-vinyl acetate) (EVA)/LDPE at high temperature and linear viscoelasticity at low and intermediate temperature. In their study, waste polymers were mixed with 60/70 penetration grade bitumen with a four-blade propeller. The test results showed that the performance of modified [37] bitumen was improved. Casey *et al.* found that a binder with 4% waste HDPE gave better results than other modified binders. The optimum mixing process was chosen according to the type of modifier, mixing time, and mixing temperature. Results of this study were used to compare the performance of modified binders with recycled polymers with that of the traditional binders already in use in road construction. Fatigue and wheel tracking test results showed that the polymer-modified binder was better than the traditional binders used in asphalt. From their study results, it was concluded that modified recycled EVA/LDPE bitumen had better mechanical properties and that the polymer improved the performance of the road surface. It also contributes to solving the problem of disposal of waste plastic [38]. In Western countries, the market for most recycled products (i.e., low-pressure pipes, traffic barriers, outdoor furniture, dustbins) is becoming very competitive [39], and new applications with higher "added value" are being found for recycled thermoplastics. Low-cost composites with improved structural properties – in particular stiffness, heat deflection temperature, and creep resistance – can be obtained from recycled plastics by the addition of rigid fillers or reinforcements. The most widely reported inorganic fillers for recycled polymers are $CaCO_3$, talc, and wollastonite [40–46]. Glass fibers (GFs) are one of several low-cost rigid filler options used to upgrade thermoplastic waste. When reinforcing recycled thermoplastics with virgin GF, an increase in impact strength is also generally observed, while elongation at break is only marginally reduced [47]. The combustion of tires in cement kilns is environmentally safe because of the much lower emission of dust, carbon dioxide, nitrogen oxides, and heavy metals (except zinc) compared to coal combustion [48]. Apart from their use in the cement industry, used tires are also used as a fuel for the production of steam, electrical energy, paper, lime, and steel. This is because the co-combustion of coal with ground rubber waste improves the thermal efficiency of steam boilers and furnaces and the amounts of exhaust gases and dusts do not exceed the permissible limits [49–51]. Product recycling is a separate form of material recycling that is based on the recycling of entire used tires in their original form, without any physical or chemical treatment. Because of their shape and size, high elasticity, and good damping properties against vibration, noise, and shock, tires are used as a cheap material in construction engineering. They can be used to form protective barriers along roads and highways and to protect sloping waterfront banks and roadsides. They can also be used as fenders for boats, artificial reefs offering protection to marine organisms, as a material for road substrates, as insulation for the foundations of buildings [52–55], and for footwear with soles made from suitably shaped pieces of tire treads. Product recycling of tires is of marginal importance, and so contributes little to solving the problem of their management. Grinding with a water jet was developed for the recycling of highly resistant, large-sized tires from trucks, construction vehicles, and farm tractors. Ground tires are also used in the building industry as filler for cement mortar, which enables concrete compositions to be obtained that are more resistant to bending and with better thermal insulation and acoustic properties as well as improved resistance to dynamic loading and cracking. The presence of a rubber material in concrete structures also reduces moisture absorption and permeability to chloride ions, thus offering steel structures better protection from corrosion [56–58]. Biocomposites with a biopolymer matrix are the future of "green composites", and their recycling makes it possible to retain the whole carbon content and save primary resources. Furthermore, they offer reductions in weight and cost and lead to less reliance on oil resources [59]. Biocomposites are currently used mainly in the automotive, construction, furniture, and packaging industries, where renewable products are attracting great attention because of the driving forces of increasing environmental awareness and depletion of fossil resources [60]. Retreading is economically very profitable: it requires only 30% of the energy and 25% of the raw materials needed to produce new tires [61].

Moreover, it is a practically waste-free process, the only byproduct of retreading being pulp rubber, which can be used for manufacturing polymer composites and in the construction industry. One of the basic ways of recovering used tires or other used rubber products is to use them as an energy raw material. Fuel consisting of shredded tires is called TDF (tire-derived fuel) in the international classification. Used tires have a calorific value of 32 MJ kg^{-1}, which makes them competitive with other types of fuel, especially with coal, which has a far lower calorific value [62]. The cement industry is one of the largest consumers of shredded tires, which uses them as an alternative fuel co-combusted with coal; their management is thus waste-free. Cement plants are now able to use as a fuel only whole tires. This is possible because of the high temperatures in cement kilns (>1200 °C), which ensure the complete combustion of all tire components. The ash and steel cord are permanently bound to the clinker, but this does not seriously impair its physicochemical properties apart from a slightly longer cement binding time and a greater water demand [63]. Moreover, the combustion of tires in cement kilns is environmentally safe because of the much lower emission, compared to coal combustion, of dusts, carbon dioxide, nitrogen oxides, and heavy metals (except zinc) [48]. Apart from the cement industry, used tires are also used as a fuel for the production of steam, electrical energy, paper, lime, and steel. This is because the co-combustion of coal with ground rubber wastes improves the thermal efficiency of steam boilers and furnaces, and the amounts of exhaust gases and dusts do not exceed any permissible limits[49–51]. Plastic solid waste (PSW) is derived from oil and has a recoverable energy, in some cases comparable to other energy sources. Direct incineration via one- or two-stage combustion technologies can certainly reduce the volume of PSW as well as the dependence on fossil fuels, which as a result can lead to a better conservation of natural resources and integrated waste management schemes. It is very important to consider recycling and energy recovery methods in plastic manufacturing and converting facilities [64]. About 33% of the material in electronic devices is plastic, however just 25% of that is clean, homogeneous, and free from contamination [65]. Plastics recovered from electronics are used for making outdoor furniture, plastic lumber and road materials in United States [50, 66]. Reused acrylonitrile/butadiene/styrene (ABS) can be used for compact disk trays, battery boxes, and camera casings. Recycled plastics can be used in new equipment such as HP printer. ABS and HIPS (high-impact polystyrene) like mixed plastic resins can substitute for stone and gravel, due to its enhanced bond strength. Thermoplastic resins are used for hot-mix asphalt concrete.

Tufts University and IBM have demonstrated that reused HIPS could be utilized as a part of the assembling of some high-end consumer electronic products at no expense to the manufacturer [67]. Constructing laminated floor is another area for marketing reused plastics [50]. Another is making Pallets, it made from recycled plastics have longer life, are cost effective, and are a convenient alternative to the traditional wooden pallets. Recently various research works are reported to develop new applications for EOL (end of life) electronic plastics. In Western Europe, electrical and electronic industries consumed about 2.78 million tons of plastics [68]. If one-third of the resins can be recycled, 0.83 million ton will be available annually. With the end goal plastics should be reused mindfully and utilized aggressively with virgin plastics as a part of new items, utilized plastics must be appropriately gathered, sorted, and cleaned. The APME (Association of Plastics Manufacturers in Europe) has proposed that energy recovery might be the most valuable alternative for managing EOL electronic plastics in Europe. In the United States, however, industry keeps on favoring an integrated way to deal with taking care of plastics from EOL electronics. The exact balance of management options, obviously, contrasts as per the assets, technology, and supplies accessible in various districts of the nation.

9.1.8 Conclusion

This chapter outlined the recycling benefits of some important recyclable polymers for the environment. In general, recycling reduces the amount of waste sent to landfilling and incinerators. It prevents pollution by reducing the need to collect new raw materials. Recycling helps save energy. It also reduces greenhouse gas emissions that contribute to global climate change. It helps in creating new well-paying jobs in the recycling and manufacturing industries.

References

1. Shent, H., Pugh, R.J., and Forssberg, E. (1999) A review of plastics waste recycling and the flotation of plastics. *Resour. Conserv. Recycl.*, **25**, 85–109.
2. Adhikari, B. and de Maiti, D.S. (2000) Reclamation and recycling of waste rubber. *Prog. Polym. Sci.*, **25** (7), 909–948.
3. Murphy, M., O'Mahony, M., Lycett, C., and Jamieson, I. (2001) Recycled polymers for use as bitumen modifiers. *J. Mater. Civ. Eng.*, **13** (4), 306–314.
4. Satapathy, S., Nag, A., and Nando, G.B. (2010) Thermoplastic elastomers from waste polyethylene and reclaim rubber blends and their composites with fly ash. *Process Saf. Environ. Prot.*, **88** (2), 131–134.
5. Zheng, Y., Shen, Z., Cai, C., Ma, S., and Xing, Y. (2009) The reuse of nonmetals recycled from waste printed circuit boards as reinforcing fillers in the polypropylene composites. *J. Hazard. Mater.*, **163** (2–3), 600–606.
6. Department of Environment and Conservation (NSW) (2005) Benefits of Recycling. Australia: Parramatta.
7. ACRR (2004) *Good Practices Guide on Waste Plastics Recycling*, Association of Cities and Regions for Recycling, Brussels.
8. Patel, M., von Thienen, N., Jochem, E., and Worrell, E. (2000) Recycling of plastics in Germany. *Resour. Conserv. Recycl.*, **29** (1–2), 65–90.
9. WRAP (2008) Study Reveals Carbon Impact of Bottling Australian Wine in the UK in PET and Glass Bottles.
10. WRAP (2008) *LCA of Management Options for Mixed Waste Plastics*, Waste & Resources Action Programme, London.
11. Hopewell, J., Dvorak, R., and Kosior, E. (2009) Plastics recycling: challenges and opportunities. *Philos. Trans. R. Soc. London, Ser. B*, **364**, 2115–2126.
12. Sabnis, A.S., Kathalewar, M.S., Mare, S., Bhave, V.G., and Raut, P.P. (2012) New polyester polyol derived from recycled

poly (ethylene terephthalate) for coating application. *Arch. Appl. Sci. Res.*, **4** (1), 85–93.
13. Kostadinova Loultcheva, M., Proietto, M., Jilov, N., and La Mantia, F.P. (1997) Recycling of high density polyethylene containers. *Polym. Degrad. Stab.*, **57**, 77–81.
14. Kumar, S. *et al* (2011) A review on tertiary recycling of high-density polyethylene to fuel. *Resour. Conserv. Recycl.*, **55**, 893–910.
15. Garcia, D., Balart, R., Crespo, J.E., and Lopez, J. (2006) Mechanical properties of recycled PVC blends with styrenic polymers. *J. Appl. Polym. Sci.*, **101**, 2464–2471.
16. La Mantia, F.P. (1996) *Recycling of PVC and Mixed Plastic Waste*, ChemTec Publishing.
17. Braun, D. (2001) PVC-origin, growth, and future. *J. Vinyl Add. Tech.*, **7**, 168–176.
18. Wenguang, M. and La Mantia, F.P. (1996) Processing and mechanical properties of recycled PVC and of homopolymer blends with virgin PVC. *J. Appl. Polym. Sci.*, **59** (5), 759–767.
19. Matuschek, G., Milanov, N., and Kettrup, A. (2000) Thermoanalytical investigations for the recycling of PVC. *Thermochim. Acta*, **361**, 77–84.
20. Garcia, D., Balart, R., Sanchez, L., and Lopez, J. (2007) Compatibility of recycled PVC/ABS blends effect of previous degradation. *Polym. Eng. Sci.*, **47**, 789–796.
21. Burat, F., Güney, A., and Olgaç Kangal, M. (2009) Selective separation of virgin and post consumer polymers (PET and PVC) by flotation method. *Waste Manage.*, **29**, 1807–1813.
22. Sombatsompop, N. and Thongsang, S. (2001) Rheology, morphology, and mechanical and thermal properties of recycled PVC pipes. *J. Appl. Polym. Sci.*, **82**, 2478–2486.
23. Hegberg, B.A., Hallenbeck, W.H., and Brenniman, G.R. (1993) Plastics recycling rates. *Resour. Conserv. Recycl.*, **9**, 89–107.
24. Carroll, W.F. Jr., (1994) Recycling post-consumer PVC bottles: technology and market development. *J. Vinyl Add. Tech.*, **16** (3), 169–176.
25. Thomas, N.L. and Quirk, J.P. (1995) Use of bottle recyclate in rigid PVC foam. *Plast. Rubber Compos.*, **24** (2), 89–96.
26. Kameda, T., Fukuda, Y., Grause, G., and Yoshioka, T. (2010) Chemical modification of rigid poly(vinyl chloride) by the substitution with nucleophiles. *J. Appl. Polym. Sci.*, **116**, 36–44.
27. Seung-Soo, K. and Seungdo, K. (2004) Pyrolysis characteristics of polystyrene and polypropylene in a stirred batch reactor. *Chem. Eng. J.*, **98**, 53–60.
28. García, M.T. *et al.* (2009) Study of the solubility and stability of polystyrene wastes in a dissolution recycling process. *Waste Manage. (Oxford)*, **29**, 1814–1818.
29. Kalantar, Z.N., Karim, M.R., and Mahrez, A. (2012) A review of using waste and virgin polymer in pavement. *Constr. Build. Mater.*, **33**, 55–62.
30. Lavee, D. (2007) Is municipal solid waste recycling economically efficient? *Environ. Manage.*, **40** (6), 926–943.
31. Huang, Y., Bird, R.N., and Heidrich, O. (2007) A review of the use of recycled solid waste materials in asphalt pavements. *Resour. Conserv. Recycl.*, **52** (1), 58–73.
32. Sengoz, B. and Topal, A. (2005) Use of asphalt roofing shingle waste in HMA. *Constr. Build. Mater.*, **19**, 337–346.
33. Xue, Y., Hou, H., Zhu, S., and Zha, J. (2009) Utilization of municipal solid waste incineration ash in stone mastic asphalt mixture: pavement performance and environmental impact. *Constr. Build. Mater.*, **23**, 989–996.
34. Arabani, M., Mirabdolazimi, S.M., and Sasani, A.R. (2010) The effect of waste tire thread mesh on the dynamic behaviour of asphalt mixtures. *Constr. Build. Mater.*, **24**, 1060–1068.
35. Huang, B., Dong, Q., and Burdette, E.G. (2009) Laboratory evaluation of incorporating waste ceramic materials into Portland cement and asphaltic concrete. *Constr. Build. Mater.*, **23**, 3451–3456.
36. Gonzalez, O., Pena, J.J., Munoz, M.E., Santamaria, A., Perez-Lepe, A., Martinez-Boza, F. *et al.* (2002) Rheological techniques as a tool to analyze

polymer–bitumen interactions: bitumen modified with polyethylene and polyethylene-based blends. *Energy Fuels*, **16**, 1256–1263.

37. Casey, D., McNally, C., Gibney, A., and Gilchrist, M.D. (2008) Development of a recycled polymer modified binder for use in stone mastic asphalt. *Resour. Conserv. Recycl.*, **52**, 1167–1174.
38. Garcia-Morales, M., Partal, P., Navarro, F.J., and Gallegos, C. (2005) Effect of waste polymer addition on the rheology of modified bitumen. *Fuel*, **85** (11), 936–943.
39. Pringle, R.T. and Barker, M.B. (2004) Starting a Waste Plastics Recycling Business. Report Prepared For Aberdeen Forward and Aberdeenshire Council, Edinburgh.
40. Sanchez-Soto, M., Rossa, A., Sanchez, A.J., and Gamez-Perez, J. (2008) Blends of HDPE wastes: study of the properties. *Waste Manage. (Oxford)*, **28**, 2565–2573.
41. Putra, H.D., Ngothai, Y., Ozbakkaloglu, T., and Seracino, R. (2009) Mineral filler reinforcement for commingled recycled-plastic materials. *J. Appl. Polym. Sci.*, **112** (6), 3470–3481.
42. Tall, S., Albertsson, A., and Karlsson, S. (2001) Enhanced rigidity of recycled polypropylene from packaging waste by compounding with talc and high-crystallinity polypropylene. *Polym. Adv. Technol.*, **12** (5), 279–284.
43. Sahnoune, F., Cuesta, J.M., and Crespy, A. (2004) Improvement of the mechanical properties of an HDPE/PS blend by compatibilization and incorporation of $CaCO_3$. *Polym. Eng. Sci.*, **43** (3), 647–660.
44. Brachet, P., Høydal, L., Hinrichsen, E., and Melum, F. (2008) Modification of mechanical properties of recycled polypropylene from post-consumer containers. *Waste Manage.*, **28** (12), 2456–2464.
45. Elloumi, A., Pimbert, S., Bourmaud, A., and Bradai, C. (2010) Thermomechanical properties of virgin and recycled polypropylene impact copolymer/CaCO3 nanocomposites. *Polym. Eng. Sci.*, **50** (10), 1904–1913.
46. Tzankova Dintcheva, N., Jilovand, N., and La Mantia, F.P. (1997) Recycling of plastics from packaging. *Polym. Degrad. Stab.*, **57**, 191–203.
47. Hugo, A.-M., Scelsi, L., Hodzic, A., Jones, F.R., and Dwyer-Joyce, R. (2011) Development of recycled polymer composites for structural applications. *Plast. Rubber Compos.*, **40** (6–7), 317–323.
48. Conesa, J.A., Galvez, A., Mateos, F., Ignacio Martin-Gullon, I., and Font, R. (2008) Organic and inorganic pollutants from cement kiln stack feeding alternative fuels. *J. Hazard. Mater.*, **158**, 585–592.
49. Levendis, Y.A., Atal, A., Carlson, J., Dunayevskiy, J., and Vouros, P. (1996) Comparative study on the combustion and emissions of waste tire crumb and pulverized coal. *Environ. Sci. Technol.*, **30** (9), 2742–2754.
50. Courtemanche, B. and Levendis, Y.A. (1998) A laboratory study on the NO, NO2, SO2, CO and CO2 emissions from the combustion of pulverized coal, municipal waste plastics and tires. *Fuel*, **77** (3), 183–196.
51. Singh, S., Nimmo, W., Gibbs, B.M., and Williams, P.T. (2009) Waste tyre rubber as a secondary fuel for power plants. *Fuel*, **88**, 2473–2480.
52. Murugan, S., Ramaswamy, M.C., and Nagarajan, G. (2008) The use of tyre pyrolysis oil in diesel engines. *Waste Manage. (Oxford)*, **28**, 2743–2749.
53. Bosscher, P.J., Tuncer, B.E., and Kuraoka, S. (1997) Design of highway embankments using tire chips. *J. Geotech. Geoenviron. Eng.*, **4**, 295–304.
54. Collins, K.J., Jensen, A.C., Mallinson, J.J., Roenelle, V., and Smith, I.P. (2002) Environmental impact assessment of a scrap tyre artificial reef. *J. Mar. Sci.*, **59**, 243–249.
55. Lee, H.J. and Roh, H.S. (2007) The use of recycled tire chips to minimize dynamic earth pressure during compaction of backfill. *Constr. Build. Mater.*, **21**, 1016–1026.
56. Piercea, C.E. and Blackwell, M.C. (2003) Potential of scrap tire rubber as lightweight aggregate in flowable fill. *Waste Manage. (Oxford)*, **23**, 197–208.
57. Benazzouk, A., Douzane, O., Langlet, T., Mezreb, K., Roucoult, J.M., and

Queneudec, M. (2007) Physico-mechanical properties and water absorption of cement composite containing shredded rubber wastes. *Cem. Concr. Compos.*, **29**, 732–740.
58. Oikonomou, N. and Mavridou, S. (2009) Improvement of chloride ion penetration resistance in cement mortars modified with rubber from worn automobile tires. *Cem. Concr. Compos.*, **31**, 403.
59. Mukherjee, T. and Kao, N. (2011) PLA based biopolymer reinforced with natural fibre: a review. *J. Polym. Environ.*, **19** (3), 714–725.
60. Fowler Paul, A., Hughes, J.M., and Elias, R.M. (2007) *Biocomposites from Crop Fibres and Resins*, BioComposites Centre, University of Wales, Bangor.
61. Ferrer, G. (1997) The economics of tire remanufacturing. *Resour. Conserv. Recycl.*, **19**, 221–255.
62. Gieré, A.,.R., Smith, K., and Blackford, M. (2006) Chemical composition of fuels and emissions from a coal + tire combustion experiment in a power station. *Fuel*, **85**, 2278–2285.
63. Pipilikaki, P., Katsioti, M., Papageorgiou, D., Fragoulis, D., and Chaniotakis, E. (2005) Use of tire derived fuel in clinker burning. *Cem. Concr. Compos.*, **27**, 843–847.
64. Al-Salem, S.M., Lettieri, P., and Baeyens, J. (2009) Recycling and recovery routes of plastic solid waste (PSW): a review. *Waste Manage. (Oxford)*, **29**, 2625–2643.
65. American Plastics Council (2000) Plastics from Residential Electronics Recycling. Report 2000, American Plastics Council, Inc.
66. Environmental and Plastics Industry Council [EPIC] (2003) Management of Plastics in EOL Electronics. Special News and Views Report.
67. American Plastics Council (2003) An industry Full of Potential: Ten Facts to Know about Plastics from Consumer Electronics.
68. Association of Plastics Manufacturers in Europe(APME) (2003) An Analysis of Plastics Consumption and Recovery in Europe.

9.2 Toxic or Environmental Effects of Recycled Polymers

Raju Francis, Nidhin Joy, and Anjaly Sivadas

9.2.1 Introduction

This chapter evaluates effect of recycled polymer on the basis of two parameters – environmental and human health. The recycling of plastics has been less successful due to difficulties in the identification and sorting as well as the presence of various other materials and additives such as fillers and plasticizers [1]. As a result of this and their longevity and widespread use, particularly in disposable products, managing plastic waste can be a major problem. Being organic materials, they have the potential to be used for energy production, but uncontrolled or poorly managed incineration can result in hazardous emissions [2], leading to considerable public opposition. Additionally, when plastic waste enters the natural environment as litter through poor waste management or incorrect disposal, this poses threats to wildlife [3]. The large volumes and bright colors of plastic waste make them highly visible in the waste stream and as litter. Such matters foster a negative emotional response and give plastics considerable bad press coverage. Plastic waste posses negative environmental externalities such as greenhouse gas emissions and ecological damage. It is usually nonbiodegradable and therefore can remain as waste in the environment for a very long time, it may pose risks to human health as well as the environment, and it can be difficult to reuse and/or recycle in practice. Plastic recycling needs to be carried out in a sustainable manner.

9.2.2
Will Recycling Reduce the Amount of Waste?

Recycling is clearly a waste management strategy, but it can also be seen as a current example of implementing the concept of industrial ecology, whereas in a natural ecosystem there are no wastes but only products [4, 5]. Recycling is associated with many difficulties, which are related, principally, to the lack of fiscal subsidies, precarious programs for selective garbage separation, and high electricity costs [6]. Recycling of plastics is one method for reducing environmental impact and resource depletion. Fundamentally, high levels of recycling, as with reduction in use, reuse, and repair or remanufacturing can allow for a given level of product service with lower material inputs than would otherwise be required. Recycling can therefore decrease energy and material usage per unit of output and so yield improved eco-efficiency [7]. It should be noted, however, that in spite of the ability to maintain whatever residual level of material and energy inputs, the effects of external impacts on ecosystems will decide the ultimate sustainability of the overall system. Recycled plastics can have negative impacts on human health. At a basic level, in facilities where manual sorting is still in place, workers may risk injury and disease while sorting materials. At times, consumers are unaware of what can and cannot be recycled, and items such as hypodermic needles and broken glass may get mixed in and potentially injure workers. There is a risk of plastic waste recycling affecting local populations in countries with less stringent regulations (than, say in the EU). Recycling techniques used to treat plastic waste can be relatively primitive in those countries, and in some cases there is a lack of appropriate facilities to safeguard environmental and human health. For example, chipping and melting of plastics in unventilated areas (a practice not seen in Europe) can have negative consequences on human health [8]. The common environmental impact categories are global warming (GW), acidification (AC), nutrient enrichment (NE), photochemical ozone formation (POF), human toxicity (HT), and persistent toxicity (PT), and their normalization references.

In the following section, the disadvantages of polymer recycling and how recycling affects humans and the environment are discussed.

9.2.2.1 Recycling of Waste Electrical and Electronic Equipment (WEEE)

Waste electrical and electronic equipment (WEEE) can contain over 1000 different substances, many of which are toxic and some which have a relatively high market value when extracted. Atmospheric pollution due to burning and dismantling activities seems to be the main cause for occupational and secondary exposure at WEEE recycling sites. Inadequate disposal and poor recycling practices to recover metals such as gold, copper, and silver contribute to potentially harmful impacts on the environment and pose health risks to the exposed individuals. Three main substances released during recycling can be recognized:

- *Original substances.* Which are constituents of electrical and electronic equipment
- *Auxiliary substances.* Utilized as a part of recycling techniques
- *Byproducts.* Formed by the transformation of primary constituents [9].

Dismantling activities release dust particles loaded with heavy metals and flame retardants into the environment. The dismantling and sorting activities contribute only between 40% (abiotic resources, climate change) and 10% (eutrophication) and to the total impact of the licensed area, whereas mechanical processing is responsible for more than 75%. These particles are either accumulated near or away from the emission source depending on their size (higher halogenated compounds often deposit near the sources of emission, while the lower halogenated compounds are typically transported over longer distances). Thermal or inadequate metallurgical treatment of WEEE can lead to the formation of polyhalogenated dioxins and furans like extremely hazardous anthropogenic pollutants. During WEEE burning process, chlorinated and brominated dioxins and furans (PCDD/Fs-Polychlorinated dibenzo-p-dioxins and dibenzofurans and PBDD/Fs-Polybrominated dibenzo-p-dioxins and dibenzofurans) and mixed halogenated compounds like the polybrominated-chlorinated dibenzo-p-dioxins (PBCDDs) and mixed halogenated compounds like the polybrominated-chlorinated dibenzo-p-dioxins (PBCDDs) and polybrominated-chlorinated dibenzo furans (PBCDFs) can be formed [10]. Once emitted into the climate, dioxins and furans are accumulating in organic rich media due to its semi volatile and hydrophobic properties [11, 12]. Higher chlorinated or brominated congeners degrade more slowly and tend to partition more into lipids [13]. Lower halogenated compounds are typically transported over longer distances, while the higher ones deposit near the sources of emission. Other emissions from WEEE recycling, such as leachates and toxic liquids, increase human risk of exposure through impacted natural materials such as soils, crops, drinking water, livestock, fish, and shell fish [14]. Wastewater containing dismantling and shredding residues and other toxic liquids from WEEE recycling activities (such as acid and cyanide leaching) represents a serious threat to ecosystems and human health [15].

9.2.2.2
Recycling of Tires

The technology of manufacturing rubber products is based primarily on the irreversible vulcanization reaction that takes place between natural and synthetic diene rubbers, sulfur, and a variety of auxiliary compounds. As a result, transverse bonds connect the elastomer chains to form the cross-linked structure of rubber. That is why rubber articles are elastic, insoluble, and infusible solids that cannot be reprocessed, as is the case with thermoplastic materials. Their recycling therefore requires a long time and high energy outlay and is based solely on the mechanical, thermal, or chemical destruction of the rubber product; recovery of the raw materials used to produce them is impossible [16]. Used tires are a category of waste whose recycling is exceedingly difficult. This is due to their highly complex structure, the diverse composition of the raw material, and the structure of the rubber from which the tire was made.

9.2.2.3
Recycling of Plastics

Some plastics, for example, PET (polyethylene terephthalate) or HDPE (high-density polyethylene), are increasingly being recycled back into food contact applications in the UK. However, recycled plastics are not commonly used in food packaging (although this segment is one of the biggest single markets for plastics) because of concerns about food safety and hygiene standards. Multilayered containers, which enclose the recycled plastic between layers of virgin plastic, are therefore being used in some drinks bottles, but recycling cannot eliminate the colors from plastics so they cannot be used in transparent or light-colored applications. A number of environmental concerns are associated with co-incinerating plastic solid waste (PSW), which are mainly the emission of certain air pollutants such as CO_2, NO_2, and SO_2. The combustion of PSW is also known to generate volatile organic compounds (VOCs), smoke (particulate matter), particulate-bound heavy metals, polycyclic aromatic hydrocarbons (PAHs), polychlorinated dibenzo furans (PCDFs), and dioxins.

9.2.2.4
Recycling of Polymers

Recycling of synthetic polymers such as PVC (poly(vinyl chloride)), PET, PS (polystyrene), and PE by incineration or combustion results in a large amount of carcinogenic substances like PAHs, nitro-PAHs, dioxins, and so on, which have been identified in the air. On the other hand, all factors – except climate change – are dominated by the secondary metals production step. This step is responsible for up to more than 90% of the total load of the secondary production. One of the main reasons for this is the fact that the fraction of metals accounts for more than 60% of the output of the dismantling step [17]. The combustion of PSW is also known to generate VOCs, smoke (particulate matter), particulate-bound heavy metals, PAHs, PCDFs, and dioxins.

9.2.2.5
Health Problems

Dumped materials containing heavy metals and brominated and chlorinated flame retardants can influence the nature of soil. The mobility of these substances toward other environmental compartments depends on diverse environmental parameters such as pH, organic matter content, uptake by biota, adsorption -desorption processes, temperature, degradation processes, complexation and the intrinsic chemical characteristics of the substance [18–22]. People are exposed to toxics compounds in several ways.

1. By breathing the air, which affects both workers in the plant and people who live nearby
2. By eating locally produced foods or water that has been contaminated by air pollutants from the incinerator
3. By eating fish or wildlife that have been contaminated by the air emissions.

Generally speaking, a growing body of epidemiological and clinical evidence has led to increased concerns about the potential damaging effects of ambient air pollution on health [23]. Combustion typically generates smaller particles [24] (PM 2.5 μm in diameter), and consequently fine particulate matter is strongly implicated in pulmonary and cardiovascular disease. Among the direct and indirect exposed groups to PM2.5, the more vulnerable are pregnant women and children. Eighty percent of children in Guiyu suffer from respiratory diseases and are particularly vulnerable to Pb poisoning [25–30]. Blood lead levels (BLLs) in children of Guiyu (15.3 $\mu g\, dl^{-1}$) exceed the Chinese mean (9.29 $\mu g\, dl^{-1}$), thus posing a potentially serious threat to children's health; air pollution probably is the cause for this [31, 32]. Human exposure to dioxins begins with atmospheric emissions [33], of which incineration releases the largest quantity [34]. Dioxin levels in hair reflect those in the atmosphere [35–38].

9.2.2.6
Recycling by Polymer Incineration

Incineration is done by a special type of reactors called *incinerators*, to burn wastes in the presence of air in a controlled manner to convert hydrocarbons of the plastic into carbon dioxide and water. A number of environmental concerns are associated with co-incinerating PSW, the main one being the emission of certain air pollutants such as HCl, CO_2, NO_2, and SO_2. Polymer incineration systems produce a wide variety of pollutants depending on the chemical composition of polymer, which are detrimental to human health. Such systems are expensive and does not eliminate or adequately control the toxic emissions from chemically complex MSW. Even new incinerators release toxic metals, dioxins, and acid gases. Far from eliminating the need for a landfill, waste incinerator systems produce toxic ash and other residues. Emissions from incinerators can include heavy metals, dioxins, and furans, which may be present in the waste gases, water, or ash. Plastic and metals are the major source of the calorific value of the waste. The combustion of plastics, such as PVC, gives rise to these highly toxic pollutants. Through complete combustion, PVC can be broken down into water, carbon dioxide, and hydrogen chloride (HCl). However, complete combustion is rarely feasible in reality, and dioxins can be unintentionally generated as byproducts according to incineration conditions.

9.2.3
Conclusion

Recycling is the only way to reduce plastic and polymeric waste from the earth. This chapter reviewed the negative effects of recycling in different manner. Recycling produces environmental, human health, and economic imbalance. So what can people do? Separate, avoid, reduce, and reuse the polymeric material, rather than recycling them!

References

1. Davis, G. and Song, J. (2006) Biodegradable packaging based on raw materials from crops and their impact on waste management. *Ind. Crops Prod.*, **23** (2), 147–161.
2. Hopewell, J., Dvorak, R., and Kosior, E. (2009) Plastics recycling: challenges and opportunities. *Philos. Trans. R. Soc. London, Ser. B*, **364** (1526), 2115–2126.
3. Rao, K. (2010) The Current State of Biopolymers and Their Potential Future. SpecialChem 2010, http://www.omnexus.com/resources/editorials.aspx?id=25041 (accessed 12 February 2012).
4. Frosch, R. and Gallopoulos, N. (1989) Strategies for manufacturing. *Sci. Am.*, **261**, 144–152.
5. McDonough, W. and Braungart, M. (2002) *Cradle to Cradle: Remaking the Way We Make Things*, North Point Press, New York, NY.
6. Pacheco, E.B.A.V., Ronchetti, L.M., and Masanet, E. (2012) An overview of plastic recycling in Rio de Janeiro. *Resour. Conserv. Recycl.*, **60**, 140–146.
7. WBCSD (2000) Eco-Efficiency:, http://www.wbcsd.org/web/publications/eco_efficiency_creating_more_value.pdf (accessed 1 November 2008).
8. Wong, M.H., Wu, S.C., Deng, W.J., Yu, X.Z., Luo, Q., Leung, A.O.W., Wong, C.S.C., Luksemburg, W.J., and Wong, A.S. (2007) Export of toxic chemicals – A review of the case of uncontrolled electronic-waste recycling. *Environ. Pollut.*, **149**, 131–140.
9. Sepúlveda, A. *et al.* (2010) A review of the environmental fate and effects of hazardous substances released from electrical and electronic equipments during recycling: examples from China and India. *Environ. Impact Assess. Rev.*, **30**, 28–41.
10. Söderström, G. (2003) *On the Combustion and Photolytic Degradation Products of Some Brominated Flame Retardants*, Department of Chemistry, Environmental Chemistry, University of Umeå, Umeå, Sweden, p. 54.
11. Adriaens, P., Fu, Q., and Grbic-Galic, D. (1995) Bioavailability and transformation of highly chlorinated dibenzo-*p*-dioxins and dibenzofurans in anaerobic soils and

sediments. *Environ. Sci. Technol.*, **29**, 2252–2261.

12. Smith, K.E.C. and Jones, K.C. (2000) Particles and vegetation: implications for the transfer of particle bound organic contaminants to vegetation. *Sci. Total Environ.*, **246**, 207–236.
13. Webster, E. and Mackay, D. (2007) Modelling the Environmental Fate of Dioxins and Furans Released to the Atmosphere during Incineration. CEMC Report No. 200701. Report to Environment Canada. Peterborough, Ontario. Canadian Environmental Modelling Centre, Trent University. p. 22.
14. Azuma, R. (2003) E-Waste in China. TED Case Studies Number 687, http://www.americanedu/TED/computer-recycling.htm (accessed 12 May 2016).
15. Wong, C.S.C., Duzgoren-Aydin, N.S., Aydin, A., and Wong, M.H. (2007) Evidence of excessive releases of metals from primitive e-waste processing in Guiyu, China. *Environ. Pollut.*, **148**, 62–72.
16. White, R. and De, S. (2001) *Rubber Technologist's Handbook*, vol. **1**, iSmithers Rapra Publishing.
17. Hischier, R., Wäger, P., and Gauglhofer, J. (2005) Does WEEE recycling make sense from an environmental perspective? *Environ. Impact Assess. Rev.*, **25**, 525–539.
18. Sauvé, S., Martínez, C.E., McBride, M., and Hendershot, W. (2000) Adsorption of free lead (Pb2+) by pedogenic oxides, ferrihydrite, and leaf compost. *Soil Sci. Soc. Am. J.*, **64**, 595–599.
19. Georgopoulos, A.R., Yonone-Lioy, M.J., Opiekun, R.E., and Lioy, P.J. (2001) *Copper: Environmental Dynamics and Human Exposure Issues*, 207 pp.
20. Hu, H. (2002) Human health and heavy metals exposure, in *Life Support: the Environment and Human Health* (ed. M. McCally), MIT Press.
21. Gouin, T. and Harner, T. (2003) Modelling the environmental fate of polybrominated diphenyl ethers. *Environ. Int.*, **29**, 717–724.
22. Qin, F., Shan, X.Q., and Wei, B. (2004) Effects of low-molecular-weight organic acids and residence time on desorption of Cu, Cd, and Pb from soils. *Chemosphere*, **57**, 253–263.
23. Brook, R.D., Franklin, B., Cascio, W., Hong, Y., Howard, G., Lipsett, M. *et al.* (2004) Air pollution and cardiovascular disease: a statement for healthcare professionals from the expert panel on population and prevention science of the American Heart Association. *Circulation*, **109**, 2655–2671.
24. Cormier, S.A., Lomnicki, S., Backes, W., and Dellinger, B. (2006) Origin and health impacts of emissions of toxic by-products and fine particles from combustion and thermal treatment of hazardous wastes and materials. *Environ. Health Perspect.*, **114** (6), 810–817.
25. Baghurst, P.A., McMichael, A.J., Wigg, N.R., Vimpani, G.V., Robertson, E.F., Roberts, R.J. *et al.* (1992) Environmental exposure to lead and children's intelligence at the age of seven years: the Port Pirie Cohort Study. *N. Engl. J. Med.*, **327**, 1279–1284.
26. Wasserman, G.A., Staghezza-Jaramillo, B., Shrout, P., Popovac, D., and Graziano, J. (1998) The effect of lead exposure on behaviour problems in preschool children. *Am. J. Public Health*, **88**, 481–486.
27. Guilarte, T.R., Toscano, C.D., McGlothan, J.L., and Weaver, S.A. (2003) Environmental enrichment reverses cognitive and molecular deficits induced by developmental lead exposure. *Ann. Neurol.*, **53**, 50–56.
28. Grigg, J. (2004) Environmental toxins: their impact on children's health. *Arch. Dis. Child.*, **89**, 244–250.
29. Needleman, H. (2004) Lead poisoning. *Annu. Rev. Med.*, **55**, 209–222.
30. Jain, N.B. and Hu, H. (2006) Childhood correlates of blood lead levels in Mumbai and Delhi. *Environ. Health Perspect.*, **114**, 466–470.
31. Wang, S. and Zhang, J. (2006) Blood lead levels in children, China. *Environ. Res.*, **101**, 412–418.
32. Huo, X., Peng, L., Xu, X., Zheng, L., Qiu, B., Qi, Z. *et al.* (2007) Elevated blood lead levels of children in Guiyu, an electronic waste recycling town in China. *Environ. Health Perspect.*, **115** (7), 1113–1117.

33. Beck, H., Dross, A., and Mathar, W. (1994) PCDD and PCDF exposure and levels in humans in Germany. *Environ. Health Perspect.*, **102**, 173–185.
34. WHO (2004) Fact Sheet: Dioxin and their Effects on Human Health, http://www.who.int/mediacentre/factsheets/fs225/en/ (accessed 12 May 2016).
35. Schramm, K.W., Keuttnear, T., Weber, S., and Lützke, K. (1992) Dioxin hair analysis as monitoring pool. *Chemosphere*, **24**, 351–358.
36. Tirler, W., Voto, G., and Donega, M. (2001) PCDD/F, PCB and hexachlorobenzene level in hair. *Organohalogen Compd.*, **52**, 290–292.
37. Nakao, T., Aozasa, O., Ohta, S., and Miyata, H. (2002) Assessment of human exposure to PCDDs, PCDFsand Co-PCBs. Using hair as a human pollution indicator sample I: development of analytical method for human hair and evaluation for exposure assessment. *Chemosphere*, **48**, 885–896.
38. Nakao, T., Aozasa, O., Ohta, S., and Miyata, H. (2005) Survey of human exposure to PCDDs, PCDFs, and coplanar PCBs using hairs an indicator. *Arch. Environ. Contam. Toxicol.*, **17**, 124–130.

Index

a
abrasion 112
acrylics 38
additives 6
– antibacterial additives 29
– antiblocking additives 39
– antifogging agents 41
– antioxidants
– – function of 21
– – primary antioxidants 22
– – secondary antioxidants 22
– antistatic agents 26
– blowing agent
– – expanded polystyrene 40
– – mechanism of 41
– categorization 13
– classes of 13
– in colouring agents 47
– compatibilizers 19
– coupling agents 29
– in fillers 48
– fillers and modifiers 25
– flame retardants 27, 46
– impact modifiers 23
– intermaterial competition 43
– lubricants 28
– multifunctional additives 49
– nucleating agents
– – coupled nucleating agents 33
– – defining 32
– – inorganic nucleating agents 32
– – phosphates 33
– – sorbitol compounds 33
– optical brighteners
– – chemical structure of 37
– – enhancing stability of 37
– – examples of 36
– – inorganic brighteners 36
– – organic brighteners 36
– – selection of 36
– plasticizers 28, 47
– purpose of 13
– reinforcing agents/fillers
– – carbon blacks and carbon fibers 35
– – carbon nanotubes 35
– – inorganic/organic 33
– – nanoclays 35
– role of 13
– stabilizing agents 46
– – combinations for applications of polymers 19
– – effectiveness of 14
– – heat stability 18
– – light stability 19
– – photo stabilization 15
– – processing stability 18
– – testing 18
– – thermal stabilization 15
– surface improvers
– – acrylics 38
– – fluorocarbon compounds 38
– – melt fracture 37
– – organic salts 38
– technical opportunities 43
– technical threats 44
– turnover of 14
agricultural waste 2
Akzo process 86
Alicyclobacillus sp. 151
alkali catalyzed methanolysis 107
aminolysis, polyurethanes 229
ammonolysis, of nylon 6 6, 103
antibacterial additives 29
antiblocking additives 39, 40
antifogging agents 41

Recycling of Polymers: Methods, Characterization and Applications, First Edition. Edited by Raju Francis.

antioxidants 15, 45
– consumption of 14
– function of 21
– secondary antioxidants 22
antistatic agents 26
aramid fiber 218
asbestos containing materials (ACM) 175
asbestos fibers 174
Australian and New Zealand Environment and Conservation Council (ANZECC) 116

b

Banbury mixers 38
BASF feedstock recycling process 88
batch reactors 65
benzyltrimethylammonium bromide 103
biocomposites, with biopolymer matrix 243
biological process, rubber recycling 148
bioplastics 122
bis(2-hydroxyethyl) terephthalamide (BHETA) 131
bis(2-hydroxyethyl) terephthalate (BHET) 131
black pigment 27
black plastics 116, 121
blending system 12
blocking 39
blowing agent
– expanded polystyrene 40
– mechanism of 41
bottle recycling, PET 236
bulk molding compounds, mechanical properties 219

c

cadmium 27
caprolactam 105
carbon blacks 35
carbon fibers 35, 192
– chemical recycling 195
– mechanical recycling 192
– pyrolysis process 193
– thermal recycling 189
carbon nanotubes 35
catalytic pyrolysis 105
catalyting cracking, polypropylene 71
cellulose nanocrystals (CNC) 165
Ceriporiopsis subvermispora 152
chemical depolymerization 60
chemical process, rubber recycling 143
chemical recycling 4, 59
– fibers 190, 195
– HDPE 238
– of nylon 101–102
– of polycarbonate, *see* polycarbonate
– of polyethylene, *see* polyethylene
– polymer blends and composite materials 215
– poly (methylmethacrylate)
– – chemical/feedstock recycling 92
– – dissolution/reprecipitation 91
– – thermal pyrolysis 91
– of polypropylene
– – catalyting cracking 71
– – co-pyrolysis 70
– – pyrolysis 69
– of polystyrene
– – catalytic degradation 77
– – cracking mechanism 79
– – dissolution technique 77
– – municipal solid waste 76
– – thermo-chemical recycling 79
– polyurethanes 227
– of polyvinyl chloride (PVC), *see* polyvinyl chloride (PVC)
– PLA 124
– plastics 119, 121
chlorinated fluorocarbons 48
chlorinated polyethylene(CPE) 20
civil engineering, rubber recycling 153
clinoptilolites 129
cobalt aluminates 27
colouring agents
– advancement in 47
– dyes 27
– pigments 27
commercial waste 2
Commonwealth Environmental Protection Agency (CEPA) 116
compatibilizers 13, 19
composites 8
condensation polymerisation 101
conical spouted bed reactor (CSBR) 65
co-polymeric system 48
cotton fibers 167
Council for Textile Recycling 169
coupling agents 29

d

Design for Environment (DFE) 76
desulfurisation of rubber 151
devulcanization method 145, 146
dibenzylidene sorbitol 33
dicumyl peroxide (DCP) 78
differential scanning calorimetry (DSC) 18
diphenyl disulphide (DD) 145
dissolution 77
– poly(methyl methacrylate) 91

dough molding compound (DMC) 214
dyes 27
dynamic thermal stability time (DTST) 17, 18

e
energy recovery 6, 60
energy production, rubber recycling 154
environmental scanning electron microscope (ESEM)
– of *T ferrooxidans* 149
– *Sphingomonas* sp. 150
epoxy resins 211
– cured 216
– formation 211
ethoxylated sorbitan ester 42
ethylbenzene 78
ethylene glycol, glycolysis in 108, 109
ethylene-propylene rubber (EPR) 20
expanded polystyrene (EPS) 40, 76, 77, 241
external lubricants 28
extruders 38
extrusion process 128

f
fatty acid amides 39
feedstock recycling 59, 92
– plastics 116, 120
– sustainability benefits 117
fiber reinforced plastics (FRPs) 7
fiber reinforced polymer composites (FRPCs) 163
fibers
– aramid 218
– carbon fiber-geethy 192
– chemical recycling 190, 195
– classification of 164
– fluidized bed process 189
– mechanical recycling 188, 192
– natural 164
– pyrolysis process 193
– synthetic 176
– thermal recycling 189
fillers 13, 25
– advancement in 48
– carbon blacks and carbon fibres 35
– carbon nanotubes 35
– inorganic/organic 33, 34
– nanoclays 35
fixed bed reactor 66
flame retardants (FRs) 45
– additive types 27
– advancement in 46
– reactive types 28
fluid catalytic cracking (FCC) 66, 69, 70, 82, 127
fluidized bed process, fibers 189
fluidized bed reactors 65
fluidized–bed recycling process 216
fluorocarbon compounds 38
FRPCs, *see* fiber reinforced polymer composites (FRPCs)

g
gasification 60
– hermolysis 64
– polyurethanes 230
general rubber goods (GRG) sector 141
glass fiber (GF) 187, 243
– chemical recycling 190
– fluidized bed process 189
– mechanical properties of 25
– mechanical recycling 188
glass fibre reinforced plastics (GFRP) 188
glass reinforced phenolic (GWP) 214
glass-mat reinforced thermoplastics (GMT) 181
glycolysis 228
– of PET 131
grinding 112
– machines description 213
– with water jet 243

h
halogen systems 28
HDPE, *see* high-density polyethylene (HDPE)
health problems, in children 252
heat distortion temperature (HDT) modifier 23
heat stabilisers 45
heat stability 18
heterogeneous polymers 12
hexamethylene diamine (HMDA) 104
high-density polyethylene (HDPE) 19–21, 116, 121
– chemical recycling 238
– cotton fiber 168
– degradation of 128
– recycling 237
– thermal cracking of 128
hindered amine light stabilisers (HALS) 16
hindered amine thermal stabilizers (HATS) 49
homogeneous polymeric materials 12
hydrate systems 28
hydrocracking, thermolysis 64
hydrocyclones 112

hydrogenation 60
– polyurethanes 230
– thermolysis 64
hydrolysis
– of nylon 4 6, 103
– of nylon 6 102
– of nylon 6 6, 103
– polyurethanes 229
hydroxybenzophenone 15
hydroxyphenylbenzotriazole 15
hygrothermal ageing, effect of 124

i
impact modifiers 13, 23, 45
incineration 115
– for energy recovery 216, 231
– polymer 252
inorganic brighteners 36
inorganic nucleating agents 32
internal lubricants 28
ionic liquids 106
iron oxides 27

k
kenaf 165

l
Level sensors 12
life-cycle analysis (LCA) study 236
life-cycle assessment (LCA) 117, 134
light stability 19
light stabilizers 45
– consumption of 14
low-density polyethylene (LDPE) 126–127
lubricants 28

m
mechanical recycling 4
– fibers 188, 192
– PLA 122, 124
– plastics 116, 117, 119, 120, 121
– polymer blends and composite materials 213
mechanochemical technique, rubber recycling 145
melt extrusion technique 37
melt flow index (MFI) 16, 17, 126, 127
melt fracture 37
methanolysis
– alkali catalyzed methanolysis 107
– ionic liquids 106
methylene diphenyl diissocyanate (MDI) 224
microcrystalline cellulose (MCC)
– kenaf 165
– PVC film 168
microwave pyrolysis 217
microwave technique, rubber recycling 146
mineral fillers 13
mixed plastic recycling processes 86
mixed plastics 8
modifiers 25
monomer methyl methacrylate (MMA) 91
municipal solid waste (MSW) 4, 76, 116, 119, 185
municipal waste 2

n
nanoclays 35
nanofibrillated cotton 170
natural fibers 7, 164
– asbestos 174
– cotton 167
– kenaf 165
natural rubber (NR) 141, 142
– chemical oxidation reaction 143
– chemical structure 142
– drawbacks 142
– world production/consumption 142
nitrous oxide, tyre recycling 144
N-methylmorpholine *N*-oxide (NMMO) 169
noise reduction coefficient (NRC) process 86
NTK process 87
nucleating agents
– coupled nucleating agents 33
– defining 32
– phosphates 33
– sorbitol compounds 33
nylon 177
– chemical recycling 101–102
– thermal recycling 102
– nylon 4, 6, hydrolysis of 103
– nylon 6
– – depolymerization 105
– – hydrolysis of 102
– nylon 6, 6
– – ammonolysis of 103
– – depolymerization of 104
– – hydrolysis of 103
– – recovery of monomers 104

o
oil absorbent, rubber recycling 154
olefin metathesis, degradation 144
optical brighteners
– chemical structure of 37
– enhancing stability of 37
– examples of 36

– inorganic brighteners 36
– organic brighteners 36
– selection of 36
organic antiblocking agents 39
organic brighteners 36
organic peroxides 45
organic phosphates 33
organic salts 38
oriented polystyrene (OPS) 76

p

persistent organic pollutants (POPs) 55
phenol, thermal cracking of HDPE 128
phosphates 33
phosphites 15
photo stabilization
– quenchers 15
– radical traps 16
– UV absorbers 15
pigments 27
plasticizers 45, 47, 28
plastics 2, 115
– additives in 6
– applications 134
– chemical recycling 119, 121
– classification 115, 120
– composites 8
– consumption of 14
– durable 117
– feedstock recycling 116, 120
– mechanical recycling 116, 119–121
– mixed plastics 8
– MSW 4
– PET 129
– PLA 122
– pollution 236
– polyethylene 126
– polymer in 3
– polypropylene 128
– polystyrene 129
– prime grades 3
– production of 3, 115
– PVC 125
– recycling of 3, 134, 235, 248, 251
– reuse 120
– thermoplastics 3
– thermosets 3
– waste handling 61
plastic solid waste (PSW) 244, 251, 252
plastic waste management scenario 117
poly(L-lactic acid) (PLLA) 124
polyamide (PA) 37, 101
polycarbonate (PC), chemical recycling 105, 106
– alkali catalyzed methanolysis 107
– glycolysis in ethylene glycol 108
– hydrolysis 108
– ionic liquids 106
– methanolysis 106–107
– polycarbonate wastes 106
– pyrolysis 109
– recycling problems 109
poly(methyl methacrylate) (PMMA), chemical recycling
– chemical/feedstock recycling 92
– dissolution/reprecipitation 91
– thermal pyrolysis 91
polyester 182
polyether polyols 224, 225
polyethylene (PE) 126
polyethylene, chemical recycling
– batch reactors 65
– degradative extrusion 64
– fixed bed reactor 66
– fluidized bed reactors 65
– gasification 64
– hydrogenation 64
– pyrolysis 63
– thermolysis 63–64
polyethylene high density (PEHD) 237
polyethylene terephthalate (PET) 129, 182
– bottle-to-bottle recycling 133
– chain extending reaction 184
– glycolysis of 131
– mechanical properties 25, 26
– pyrolysis 187
– recycling 236
polyhydroxyalkanoates (PHA) 82
poly lactic acid (PLA) 122
– chemical recycling 124
– degradation of 124
– mechanical properties 123
– mechanical recycling 122, 124
– zero viscosity of 123
polymer
– classes of 209
– recycling of 57, 251
polymer blends and composite materials 209
– application of recyclates 218
– chemical recycling 215
– mechanical recycling 213
– thermal recycling 216
polymeric materials, recyclability of 117
polymeric phenolic phosphites 15
polymer incineration systems 252
polymerization techniques 209
polymer recycling 11

polyolefins
– catalytic degradation 66
– thermal cracking 66
polypropylene(PP) 128, 166, 167, 240
– chemical recycling
– – acid strength and textural properties 68
– – catalyting cracking 71
– – co-pyrolysis 70
– – heterogeneous catalysis 68
– – pyrolysis 69
– – thermal cracking 68
– mechanical properties 25, 26
– moulding-grade 16
– stabilization 16, 17
– tensile strength of 20, 21
polystyrene (PS) 129, 231
– catalytic degradation 77
– cracking mechanism 79
– dissolution technique 77
– municipal solid waste 76
– thermal degradation 81
– thermo-chemical recycling 79
– thermo-oxidative degradation 81
polyurethanes (PUR) 223
– aminolysis 229
– chemical recycling 227
– energy recovery by incineration 231
– foam applications 225
– gasification 230
– glycolysis 228
– hydrogenation 230
– hydrolysis 229
– linkages in 232
– mechanical recycling 226
– particle size distribution 227
– properties of 223
– pyrolysis 230
– thermo chemical 230
polyvinyl chloride (PVC) 17, 253, 125
– chemical recycling 85
– – BASF feedstock recycling process 88
– – chlorine atoms of 83
– – combustion of 84
– – copyrolysis of 85
– – corrosion of process equipment 85
– – dehydrochlorination 84
– – mechanochemical reaction 85
– – mixed plastic recycling processes 86
– – mixed PVC wastes world initiatives 88
– – neutralization of HCl 83
– – nucleophilic substitution 86
– – oxidative degradation of 87
– – steam pressure 83
– – thermal cracking 84
– – Veba Combi cracking process 90
– mechanical properties 25, 26
– recycling 18
– stabilization 16, 239
– tensile strength of 20, 21
– thermal stability of 17
post-consumer plastic recycling stage 11
primary antioxidants 22
primary recycling 4
processing stability 18
protective charring systems 28
PSW, *see* plastic solid waste (PSW)
Pyrococcus furiosus 148
pyrolysis 102, 217
– catalytic pyrolysis 105
– co-pyrolysis 70
– fibers 193
– of polycarbonate 109
– polypropylene 69
– polyurethanes 230
– thermal pyrolysis 106
– thermolysis 63
pyrolysis reactor 118

q

quaternary recycling 60
quenchers 15
quinone 123

r

radical traps 16
recycling 2
– abrasion 112
– advantages 110
– automated techniques 111
– challenges in 11
– chemical recycling 4, 59
– cleaning stage 6
– codes 5
– collection stage 6
– density sorting methods 111
– energy recovery 6
– equipments for 12
– feedstock recycling 59
– fillers and modifiers 12
– grinding 112
– heavy medium separation 111
– hydrocyclones 112
– mechanical recycling 4, 58
– of plastics 3
– of polymers 57
– primary recycling 4, 58
– restabilization 12

– rubber 7
– size reduction stage 6
reinforcements 13
reinforcing agents
– carbon blacks and carbon fibres 35
– carbon nanotubes 35
– inorganic/organic 33, 34
– nanoclays 35
reprecipitation, poly(methyl methacrylate) 91
Resinicium bicolor 149
response surface methodology (RSM) 173
rubber recycling 7
– applications 152
– biological process 148
– chemical process 143
– civil engineering applications 153
– energy production 154
– mechanochemical technique 145
– MW technique 146
– oil absorbent 154
– sound insulation materials 152
– twin screw extruder 148
– ultrasonic technique 146
– zinc fertilizer 155

s
SEM
– cotton fiber 169
– Sisal fiber 173
– supercritical fluids 198
sheet molding compounds (SMC) 190, 213
– manufacturing process 212, 213
silicoaluminophosphate (SAPO) 67
single screw extruder 122, 129
solid state polycondensation (SSP) 133
solid state post-condensation (SSP) reaction 133
sorbitol compounds 33
sound insulation materials, for rubber recycling 152
spectroscopes methods 127
Sphingomonas sp. 150
– ESEM 150
stabiliser technology 46
stabilizing agents
– combinations for applications of polymers 19
– heat stability 18
– light stability 19
– photo stabilization 15
– processing stability 18
– thermal stabilization 15
structure reaction injection molding (SRIM) recycling 227
styrene-butadiene rubber (SBR) 146, 151
styrene-ethylene-propylene block copolymer (SEP) 20
supercritical fluids (SCFs)
– carbon fibers 196
– glass fibers 190
surface improvers
– acrylics 38
– fluorocarbon compounds 38
– melt fracture 37
– organic salts 38
synthetic fibers 176
– glass fiber-geethy 187
– nylons 177
– polyester 182
synthetic rubbers (SR) 141, 142
– examples 142
– world production/consumption 142

t
TDF, *see* tyre derived fuel (TDF)
telechelic polymers 144, 145
tensile stress 24
tertiary recycling 59
thermal cracking method 121
thermal pyrolysis 106
thermal recycling
– fibers 189
– nylon 102
– polymer blends and composite materials 216
thermal stabilization 15
thermo chemical, PUR 230
thermogravimetric analysis (TGA) 72
thermolysis
– gasification 64
– hydrogenation 64
– pyrolysis 63
thermo-oxidative degradation 55
thermoplastics 3, 115, 120
– biocomposites 116
– examples 115
thermosets 3
– chemical recycling 215
– examples 115
– matrix 116
– melting point of 115
Thiobacillus ferrooxidans (*T ferrooxidans*) 149–150
tire sector 141
titanium pigments 36
toluene 91

toluene Diisocyante (TDI) 224
1,3,5-triphenylhexane (TPH) 77
twin screw extruder 121
– rubber recycling 148
tyre derived fuel (TDF) 244
tyre recycling 144

u
ultrasonic technique, rubber recycling 146
ultrasound-assisted extraction (UAE) 173
UV absorbers 15

v
Veba Combi cracking process 90
vinylidene fluoride homopolymers 38
volumetric devices 12

w
waste
– agricultural 2
– commercial 2
– municipal 2
– sources of 2
waste electrical and electronic equipment (WEEE) 76, 250
waste-management strategy 249
water jet, grinding with 243
white pigments 27

x
X-ray photoelectron spectroscopy (XPS) 151
xylene 91

z
zero order recycling technique 60
zinc fertilizer 155
zinc salts 38